BIG DATA AND SOCIAL SCIENCE

A Practical Guide to Methods and Tools

Chapman & Hall/CRC

Statistics in the Social and Behavioral Sciences Series

Aims and scope

Large and complex datasets are becoming prevalent in the social and behavioral sciences and statistical methods are crucial for the analysis and interpretation of such data. This series aims to capture new developments in statistical methodology with particular relevance to applications in the social and behavioral sciences. It seeks to promote appropriate use of statistical, econometric and psychometric methods in these applied sciences by publishing a broad range of reference works, textbooks and handbooks.

The scope of the series is wide, including applications of statistical methodology in sociology, psychology, economics, education, marketing research, political science, criminology, public policy, demography, survey methodology and official statistics. The titles included in the series are designed to appeal to applied statisticians, as well as students, researchers and practitioners from the above disciplines. The inclusion of real examples and case studies is therefore essential.

Published Titles

Analyzing Spatial Models of Choice and Judgment with R
David A. Armstrong II, Ryan Bakker, Royce Carroll, Christopher Hare, Keith T. Poole, and Howard Rosenthal

Analysis of Multivariate Social Science Data, Second Edition
David J. Bartholomew, Fiona Steele, Irini Moustaki, and Jane I. Galbraith

Latent Markov Models for Longitudinal Data
Francesco Bartolucci, Alessio Farcomeni, and Fulvia Pennoni

Statistical Test Theory for the Behavioral Sciences
Dato N. M. de Gruijter and Leo J. Th. van der Kamp

Multivariable Modeling and Multivariate Analysis for the Behavioral Sciences
Brian S. Everitt

Multilevel Modeling Using R
W. Holmes Finch, Jocelyn E. Bolin, and Ken Kelley

Big Data and Social Science: A Practical Guide to Methods and Tools
Ian Foster, Rayid Ghani, Ron S. Jarmin, Frauke Kreuter, and Julia Lane

Ordered Regression Models: Parallel, Partial, and Non-Parallel Alternatives
Andrew S. Fullerton and Jun Xu

Bayesian Methods: A Social and Behavioral Sciences Approach, Third Edition
Jeff Gill

Multiple Correspondence Analysis and Related Methods
Michael Greenacre and Jorg Blasius

Applied Survey Data Analysis
Steven G. Heeringa, Brady T. West, and Patricia A. Berglund

Informative Hypotheses: Theory and Practice for Behavioral and Social Scientists
Herbert Hoijtink

Generalized Structured Component Analysis: A Component-Based Approach to Structural Equation Modeling
Heungsun Hwang and Yoshio Takane

Bayesian Psychometric Modeling
Roy Levy and Robert J. Mislevy

Statistical Studies of Income, Poverty and Inequality in Europe: Computing and Graphics in R Using EU-SILC
Nicholas T. Longford

Foundations of Factor Analysis, Second Edition
Stanley A. Mulaik

Linear Causal Modeling with Structural Equations
Stanley A. Mulaik

Age–Period–Cohort Models: Approaches and Analyses with Aggregate Data
Robert M. O'Brien

Handbook of International Large-Scale Assessment: Background, Technical Issues, and Methods of Data Analysis
Leslie Rutkowski, Matthias von Davier, and David Rutkowski

Generalized Linear Models for Categorical and Continuous Limited Dependent Variables
Michael Smithson and Edgar C. Merkle

Incomplete Categorical Data Design: Non-Randomized Response Techniques for Sensitive Questions in Surveys
Guo-Liang Tian and Man-Lai Tang

Handbook of Item Response Theory, Volume 1: Models
Wim J. van der Linden

Handbook of Item Response Theory, Volume 2: Statistical Tools
Wim J. van der Linden

Handbook of Item Response Theory, Volume 3: Applications
Wim J. van der Linden

Computerized Multistage Testing: Theory and Applications
Duanli Yan, Alina A. von Davier, and Charles Lewis

Chapman & Hall/CRC
Statistics in the Social and Behavioral Sciences Series

BIG DATA AND SOCIAL SCIENCE

A Practical Guide to Methods and Tools

Edited by

Ian Foster
University of Chicago
Argonne National Laboratory

Rayid Ghani
University of Chicago

Ron S. Jarmin
U.S. Census Bureau

Frauke Kreuter
University of Maryland
University of Manheim
Institute for Employment Research

Julia Lane
New York University
American Institutes for Research

CRC Press is an imprint of the
Taylor & Francis Group, an informa business
A CHAPMAN & HALL BOOK

CRC Press
Taylor & Francis Group
6000 Broken Sound Parkway NW, Suite 300
Boca Raton, FL 33487-2742

Printed on acid-free paper
Version Date: 20160414

International Standard Book Number-13: 978-1-4987-5140-7 (Hardback)

Library of Congress Cataloging-in-Publication Data

Names: Foster, Ian, 1959- editor.
Title: Big data and social science : a practical guide to methods and tools / edited by Ian Foster, University of Chicago, Illinois, USA, Rayid Ghani, University of Chicago, Illinois, USA, Ron S. Jarmin, U.S. Census Bureau, USA, Frauke Kreuter, University of Maryland, USA, Julia Lane, New York University, USA.
Description: Boca Raton, FL : CRC Press, [2017] | Series: Chapman & Hall/CRC statistics in the social and behavioral sciences series | Includes bibliographical references and index.
Identifiers: LCCN 2016010317 | ISBN 9781498751407 (alk. paper)
Subjects: LCSH: Social sciences--Data processing. | Social sciences--Statistical methods. | Data mining. | Big data.
Classification: LCC H61.3 .B55 2017 | DDC 300.285/6312--dc23
LC record available at https://lccn.loc.gov/2016010317

Visit the Taylor & Francis Web site at
http://www.taylorandfrancis.com

and the CRC Press Web site at
http://www.crcpress.com

Printed in Canada

Contents

Preface

The class on which this book is based was created in response to a very real challenge: how to introduce new ideas and methodologies about economic and social measurement into a workplace focused on producing high-quality statistics. We are deeply grateful for the inspiration and support of Census Bureau Director John Thompson and Deputy Director Nancy Potok in designing and implementing the class content and structure.

As with any book, there are many people to be thanked. We are grateful to Christina Jones, Ahmad Emad, Josh Tokle from the American Institutes for Research, and Jonathan Morgan from Michigan State University, who, together with Alan Marco and Julie Caruso from the US Patent and Trademark Office, Theresa Leslie from the Census Bureau, Brigitte Raumann from the University of Chicago, and Lisa Jaso from Summit Consulting, actually made the class happen.

We are also grateful to the students of three "Big Data for Federal Statistics" classes in which we piloted this material, and to the instructors and speakers beyond those who contributed as authors to this edited volume—Dan Black, Nick Collier, Ophir Frieder, Lee Giles, Bob Goerge, Laure Haak, Madian Khabsa, Jonathan Ozik, Ben Shneiderman, and Abe Usher. The book would not exist without them.

We thank Trent Buskirk, Davon Clarke, Chase Coleman, Stephanie Eckman, Matt Gee, Laurel Haak, Jen Helsby, Madian Khabsa, Ulrich Kohler, Charlotte Oslund, Rod Little, Arnaud Sahuguet, Tim Savage, Severin Thaler, and Joe Walsh for their helpful comments on drafts of this material.

We also owe a great debt to the copyeditor, Richard Leigh; the project editor, Charlotte Byrnes; and the publisher, Rob Calver, for their hard work and dedication.

Editors

Ian Foster is a Professor of Computer Science at the University of Chicago and a Senior Scientist and Distinguished Fellow at Argonne National Laboratory.

Ian has a long record of research contributions in high-performance computing, distributed systems, and data-driven discovery. He has also led US and international projects that have produced widely used software systems and scientific computing infrastructures. He has published hundreds of scientific papers and six books on these and other topics. Ian is an elected fellow of the American Association for the Advancement of Science, the Association for Computing Machinery, and the British Computer Society. His awards include the British Computer Society's Lovelace Medal and the IEEE Tsutomu Kanai award.

Rayid Ghani is the Director of the Center for Data Science and Public Policy and a Senior Fellow at the Harris School of Public Policy and the Computation Institute at the University of Chicago. Rayid is a reformed computer scientist and wannabe social scientist, but mostly just wants to increase the use of data-driven approaches in solving large public policy and social challenges. He is also passionate about teaching practical data science and started the Eric and Wendy Schmidt Data Science for Social Good Fellowship at the University of Chicago that trains computer scientists, statisticians, and social scientists from around the world to work on data science problems with social impact.

Before joining the University of Chicago, Rayid was the Chief Scientist of the Obama 2012 Election Campaign, where he focused on data, analytics, and technology to target and influence voters, donors, and volunteers. Previously, he was a Research Scientist and led the Machine Learning group at Accenture Labs. Rayid did his graduate work in machine learning at Carnegie Mellon University and is actively involved in organizing data science related conferences and workshops. In his ample free time, Rayid works with non-profits and government agencies to help them with their data, analytics, and digital efforts and strategy.

Ron S. Jarmin is the Assistant Director for Research and Methodology at the US Census Bureau. He formerly was the Bureau's Chief Economist and Chief of the Center for Economic Studies and a Research Economist. He holds a PhD in economics from the University of Oregon and has published papers in the areas of industrial organization, business dynamics, entrepreneurship, technology and firm performance, urban economics, data access, and statistical disclosure avoidance. He oversees a broad research program in statistics, survey methodology, and economics to improve economic and social measurement within the federal statistical system.

Frauke Kreuter is a Professor in the Joint Program in Survey Methodology at the University of Maryland, Professor of Methods and Statistics at the University of Mannheim, and head of the statistical methods group at the German Institute for Employment Research in Nuremberg. Previously she held positions in the Department of Statistics at the University of California Los Angeles (UCLA), and the Department of Statistics at the Ludwig-Maximillian's University of Munich. Frauke serves on several advisory boards for National Statistical Institutes around the world and within the Federal Statistical System in the United States. She recently served as the co-chair of the Big Data Task Force of the American Association for Public Opinion Research. She is a Gertrude Cox Award winner, recognizing statisticians in early- to mid-career who have made significant breakthroughs in statistical practice, and an elected fellow of the American Statistical Association. Her textbooks on *Data Analysis Using Stata* and *Practical Tools for Designing and Weighting Survey Samples* are used at universities worldwide, including Harvard University, Johns Hopkins University, Massachusetts Institute of Technology, Princeton University, and the University College London. Her Massive Open Online Course in Questionnaire Design attracted over 70,000 learners within the first year. Recently Frauke launched the international long-distance professional education program sponsored by the German Federal Ministry of Education and Research in Survey and Data Science.

Julia Lane is a Professor at the New York University Wagner Graduate School of Public Service and at the NYU Center for Urban Science and Progress, and she is a NYU Provostial Fellow for Innovation Analytics.

Julia has led many initiatives, including co-founding the UMETRICS and STAR METRICS programs at the National Science Foundation. She conceptualized and established a data enclave at NORC/University of Chicago. She also co-founded the creation and permanent establishment of the Longitudinal Employer-Household Dynamics Program at the US Census Bureau and the Linked Employer Employee Database at Statistics New Zealand. Julia has published over 70 articles in leading journals, including *Nature* and *Science*, and authored or edited ten books. She is an elected fellow of the American Association for the Advancement of Science and a fellow of the American Statistical Association.

Contributors

Stefan Bender
Deutsche Bundesbank
Frankfurt, Germany

Paul P. Biemer
RTI International
Raleigh, NC, USA
University of North Carolina
Chapel Hill, NC, USA

Jordan Boyd-Graber
University of Colorado
Boulder, CO, USA

Ahmad Emad
American Institutes for Research
Washington, DC, USA

Pascal Heus
Metadata Technology North America
Knoxville, TN, USA

Christina Jones
American Institutes for Research
Washington, DC, USA

Evgeny Klochikhin
American Institutes for Research
Washington, DC, USA

Jonathan Scott Morgan
Michigan State University
East Lansing, MI, USA

Cameron Neylon
Curtin University
Perth, Australia

Jason Owen-Smith
University of Michigan
Ann Arbor, MI, USA

Catherine Plaisant
University of Maryland
College Park, MD, USA

Malte Schierholz
University of Mannheim
Mannheim, Germany

Claudio Silva
New York University
New York, NY, USA

Joshua Tokle
Amazon
Seattle, WA, USA

Huy Vo
City University of New York
New York, NY, USA

M. Adil Yalçın
University of Maryland
College Park, MD, USA

Chapter 1

Introduction

This section provides a brief overview of the goals and structure of the book.

1.1 Why this book?

The world has changed for empirical social scientists. The new types of "big data" have generated an entire new research field—that of data science. That world is dominated by computer scientists who have generated new ways of creating and collecting data, developed new analytical and statistical techniques, and provided new ways of visualizing and presenting information. These new sources of data and techniques have the potential to transform the way applied social science is done.

Research has certainly changed. Researchers draw on data that are "found" rather than "made" by federal agencies; those publishing in leading academic journals are much less likely today to draw on preprocessed survey data (Figure 1.1).

The way in which data are used has also changed for both government agencies and businesses. Chief data officers are becoming as common in federal and state governments as chief economists were decades ago, and in cities like New York and Chicago, mayoral offices of data analytics have the ability to provide rapid answers to important policy questions [233]. But since federal, state, and local agencies lack the capacity to do such analysis themselves [8], they must make these data available either to consultants or to the research community. Businesses are also learning that making effective use of their data assets can have an impact on their bottom line [56].

And the jobs have changed. The new job title of "data scientist" is highlighted in job advertisements on CareerBuilder.com and Burning-glass.com—in the same category as statisticians, economists, and other quantitative social scientists if starting salaries are useful indicators.

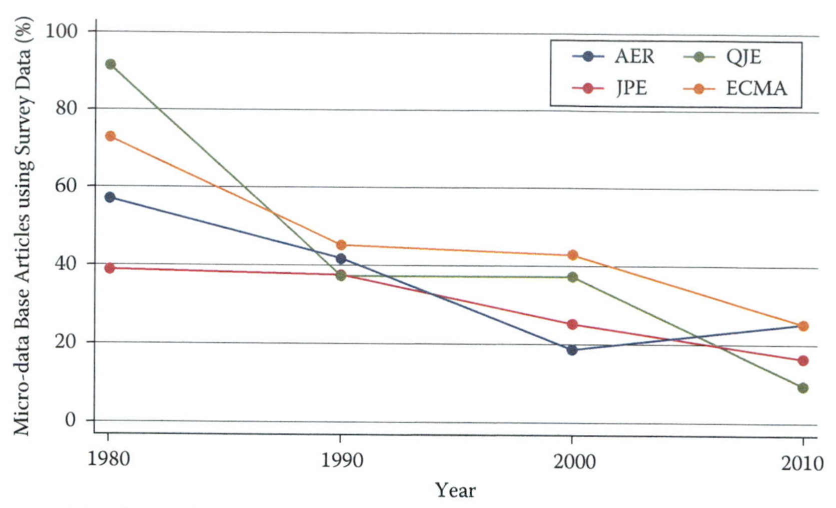

Figure 1.1. Use of pre-existing survey data in publications in leading journals, 1980–2010 [74]

The goal of this book is to provide social scientists with an understanding of the key elements of this new science, its value, and the opportunities for doing better work. The goal is also to identify the many ways in which the analytical toolkits possessed by social scientists can be brought to bear to enhance the generalizability of the work done by computer scientists.

We take a pragmatic approach, drawing on our experience of working with data. Most social scientists set out to solve a real-world social or economic problem: they frame the problem, identify the data, do the analysis, and then draw inferences. At all points, of course, the social scientist needs to consider the ethical ramifications of their work, particularly respecting privacy and confidentiality. The book follows the same structure. We chose a particular problem—the link between research investments and innovation—because that is a major social science policy issue, and one in which social scientists have been addressing using big data techniques. While the example is specific and intended to show how abstract concepts apply in practice, the approach is completely generalizable. The web scraping, linkage, classification, and text analysis methods on display here are canonical in nature. The inference and privacy and confidentiality issues are no different than in any other study involving human subjects, and the communication of results through visualization is similarly generalizable.

1.2 Defining big data and its value

There are almost as many definitions of big data as there are new types of data. One approach is to define big data as anything too big to fit onto your computer. Another approach is to define it as data with high volume, high velocity, and great variety. We choose the description adopted by the American Association of Public Opinion Research: "The term 'Big Data' is an imprecise description of a rich and complicated set of characteristics, practices, techniques, ethical issues, and outcomes all associated with data" [188].

▶ This topic is discussed in more detail in Chapter 5.

The value of the new types of data for social science is quite substantial. Personal data has been hailed as the "new oil" of the twenty-first century, and the benefits to policy, society, and public opinion research are undeniable [139]. Policymakers have found that detailed data on human beings can be used to reduce crime, improve health delivery, and manage cities better [205]. The scope is broad indeed: one of this book's editors has used such data to not only help win political campaigns but also show its potential for public policy. Society can gain as well—recent work shows data-driven businesses were 5% more productive and 6% more profitable than their competitors [56]. In short, the vision is that social science researchers can potentially, by using data with high velocity, variety, and volume, increase the scope of their data collection efforts while at the same time reducing costs and respondent burden, increasing timeliness, and increasing precision [265].

Example: New data enable new analyses

Spotshotter data, which have fairly detailed information for each gunfire incident, such as the precise timestamp and the nearest address, as well as the type of shot, can be used to improve crime data [63]; Twitter data can be used to improve predictions around job loss, job gain, and job postings [17]; and eBay postings can be used to estimate demand elasticities [104].

But most interestingly, the new data can change the way we think about measuring and making inferences about behavior. For example, it enables the capture of information on the subject's entire environment—thus, for example, the effect of fast food caloric labeling in health interventions [105]; the productivity of a cashier if he is within eyesight of a highly productive cashier but not otherwise [252]. So it offers the potential to understand the effects of complex environmental inputs on human behavior. In addition, big data, by its very nature, enables us to study the tails of a distribution in a way that is not possible with small data. Much of interest in human behavior is driven by the tails of the distribution—health care costs by small numbers of ill people [356], economic activity and employment by a small number of firms [93, 109]—and is impossible to study with the small sample sizes available to researchers.

Instead we are still faced with the same challenges and responsibilities as we were before in the survey and small data collection environment. Indeed, social scientists have a great deal to offer to a (data) world that is currently looking to computer scientists to provide answers. Two major areas to which social scientists can contribute, based on decades of experience and work with end users, are inference and attention to data quality.

1.3 Social science, inference, and big data

The goal of empirical social science is to make inferences about a population from available data. That requirement exists regardless of the data source—and is a guiding principle for this book. For probability-based survey data, methodology has been developed to overcome problems in the data generating process. A guiding principle for survey methodologists is the total survey error framework, and statistical methods for weighting, calibration, and other forms of adjustment are commonly used to mitigate errors in the survey process. Likewise for "broken" experimental data, techniques like propensity score adjustment and principal stratification are widely used to fix flaws in the data generating process. Two books provide frameworks for survey quality [35, 143].

▶ This topic is discussed in more detail in Chapter 10.

Across the social sciences, including economics, public policy, sociology, management, (parts of) psychology and the like, we can identify three categories of analysis with three different inferential goals: description, causation, and prediction.

Description The job of many social scientists is to provide descriptive statements about the population of interest. These could be univariate, bivariate, or even multivariate statements. Chapter 6 on machine learning will cover methods that go beyond simple descriptive statistics, known as *unsupervised learning* methods.

Descriptive statistics are usually created based on census data or sample surveys to generate some summary statistics like a mean, median, or a graphical distribution to describe the population of interest. In the case of a census, the work ends right there. With sample surveys the point estimates come with measures of uncertainties (standard errors). The estimation of standard errors has been worked out for most descriptive statistics and most common survey designs, even complex ones that include multiple layers of sampling and disproportional selection probabilities [154, 385].

Example: Descriptive statistics

The US Bureau of Labor Statistics surveys about 60,000 households a month and from that survey is able to describe national employment and unemployment levels. For example, in November 2015, total nonfarm payroll employment increased by 211,000 in November, and the unemployment rate was unchanged at 5.0%. Job gains occurred in construction, professional and technical services, and health care. Mining and information lost jobs [57].

Proper inference, even for purely descriptive purposes, from a sample to the population rests usually on knowing that everyone from the target population had the chance to be included in the survey, and knowing the selection probability for each element in the population. The latter does not necessarily need to be known prior to sampling, but eventually a probability is assigned for each case. Getting the selection probabilities right is particularly important when reporting totals [243]. Unfortunately in practice, samples that start out as probability samples can suffer from a high rate of nonresponse. Because the survey designer cannot completely control which units respond, the set of units that ultimately respond cannot be considered to be a probability sample [257]. Nevertheless, starting with a probability sample provides some degree of comfort that a sample will have limited coverage errors (nonzero probability of being in the sample), and there are methods for dealing with a variety of missing data problems [240].

Causation In many cases, social scientists wish to test hypotheses, often originating in theory, about relationships between phenomena of interest. Ideally such tests stem from data that allow causal inference: typically randomized experiments or strong nonexperimental study designs. When examining the effect of X on Y, knowing how cases were selected into the sample or data set is much less important in the estimation of causal effects than for descriptive studies, for example, population means. What is important is that all elements of the inferential population have a chance of being selected for the treatment [179]. In the debate about probability and nonprobability surveys, this distinction is often overlooked. Medical researchers have operated with unknown study selection mechanisms for years: for example, randomized trials that enroll only selected samples.

Example: New data and causal inference

One of the major risks with using big data without thinking about the data source is the misallocation of resources. Overreliance on, say, Twitter data in targeting resources after hurricanes can lead to the misallocation of resources towards young, Internet-savvy people with cell phones, and away from elderly or impoverished neighborhoods [340]. Of course, all data collection approaches have had similar risks. Bad survey methodology led the *Literary Digest* to incorrectly call the 1936 election [353]. Inadequate understanding of coverage, incentive and quality issues, together with the lack of a comparison group, has hampered the use of administrative records—famously in the case of using administrative records on crime to make inference about the role of death penalty policy in crime reduction [95].

Of course, in practice it is difficult to ensure that results are generalizable, and there is always a concern that the treatment effect on the treated is different than the treatment effect in the full population of interest [365]. Having unknown study selection probabilities makes it even more difficult to estimate population causal effects, but substantial progress is being made [99, 261]. As long as we are able to model the selection process, there is no reason not to do causal inference from so-called nonprobability data.

Prediction Forecasting or prediction tasks are a little less common among applied social science researchers as a whole, but are certainly an important element for users of official statistics—in particular, in the context of social and economic indicators—as generally for decision-makers in government and business. Here, similar to the causal inference setting, it is of utmost importance that we do know the process that generated the data, and we can rule out any unknown or unobserved systematic selection mechanism.

Example: Learning from the flu

"Five years ago [in 2009], a team of researchers from Google announced a remarkable achievement in one of the world's top scientific journals, *Nature*. Without needing the results of a single medical check-up, they were nevertheless able to track the spread of influenza across the US. What's more, they could do it more quickly than the Centers for Disease Control and Prevention (CDC). Google's tracking had only a day's delay, compared with the week or more it took for the CDC to assemble a picture based on reports from doctors' surgeries. Google was faster because it was tracking the outbreak by finding a correlation between what people searched for online and whether they had flu symptoms. . . .

"Four years after the original *Nature* paper was published, *Nature News* had sad tidings to convey: the latest flu outbreak had claimed an unexpected victim: Google Flu Trends. After reliably providing a swift and accurate account of flu outbreaks for several winters, the theory-free, data-rich model had lost its nose for where flu was going. Google's model pointed to a severe outbreak but when the slow-and-steady data from the CDC arrived, they showed that Google's estimates of the spread of flu-like illnesses were overstated by almost a factor of two.

"The problem was that Google did not know—could not begin to know—what linked the search terms with the spread of flu. Google's engineers weren't trying to figure out what caused what. They were merely finding statistical patterns in the data. They cared about correlation rather than causation" [155].

1.4 Social science, data quality, and big data

Most data in the real world are noisy, inconsistent, and suffers from missing values, regardless of its source. Even if data collection is cheap, the costs of creating high-quality data from the source—cleaning, curating, standardizing, and integrating—are substantial.

▶ This topic is discussed in more detail in Chapter 3.

Data quality can be characterized in multiple ways [76]:

- Accuracy: How accurate are the attribute values in the data?
- Completeness: Is the data complete?
- Consistency: How consistent are the values in and between the database(s)?
- Timeliness: How timely is the data?
- Accessibility: Are all variables available for analysis?

Social scientists have decades of experience in transforming messy, noisy, and unstructured data into a well-defined, clearly structured, and quality-tested data set. Preprocessing is a complex and time-consuming process because it is "hands-on"—it requires judgment and cannot be effectively automated. A typical workflow comprises multiple steps from data definition to parsing and ends with filtering. It is difficult to overstate the value of preprocessing for any data analysis, but this is particularly true in big data. Data need to be parsed, standardized, deduplicated, and normalized.

Parsing is a fundamental step taken regardless of the data source, and refers to the decomposition of a complex variable into components. For example, a freeform address field like "1234 E 56th St" might be broken down into a street number "1234" and a street name "E 56th St." The street name could be broken down further to extract the cardinal direction "E" and the designation "St." Another example would be a combined full name field that takes the form of a comma-separated last name, first name, and middle initial as in "Miller, David A." Splitting these identifiers into components permits the creation of more refined variables that can be used in the matching step.

In the simplest case, the distinct parts of a character field are delimited. In the name field example, it would be easy to create the separate fields "Miller" and "David A" by splitting the original field at the comma. In more complex cases, special code will have to be written to parse the field. Typical steps in a parsing procedure include:

1. Splitting fields into tokens (words) on the basis of delimiters,
2. Standardizing tokens by lookup tables and substitution by a standard form,
3. Categorizing tokens,
4. Identifying a pattern of anchors, tokens, and delimiters,
5. Calling subroutines according to the identified pattern, therein mapping of tokens to the predefined components.

Standardization refers to the process of simplifying data by replacing variant representations of the same underlying observation by a default value in order to improve the accuracy of field comparisons. For example, "First Street" and "1st St" are two ways of writing the same street name, but a simple string comparison of these values will return a poor result. By standardizing fields—and using the same standardization rules across files!—the number of true matches that are wrongly classified as nonmatches (i.e., the number of false nonmatches) can be reduced.

Some common examples of standardization are:

- Standardization of different spellings of frequently occurring words: for example, replacing common abbreviations in street names (Ave, St, etc.) or titles (Ms, Dr, etc.) with a common form. These kinds of rules are highly country- and language-specific.

- General standardization, including converting character fields to all uppercase and removing punctuation and digits.

Deduplication consists of removing redundant records from a single list, that is, multiple records from the same list that refer to the same underlying entity. After deduplication, each record in the first list will have at most one true match in the second list and vice versa. This simplifies the record linkage process and is necessary if the goal of record linkage is to find the best set of one-to-one links (as opposed to a list of all possible links). One can deduplicate a list by applying record linkage techniques described in this chapter to link a file to itself.

Normalization is the process of ensuring that the fields that are being compared across files are as similar as possible in the sense that they could have been generated by the same process. At minimum, the same standardization rules should be applied to both files. For additional examples, consider a salary field in a survey. There are number different ways that salary could be recorded: it might be truncated as a privacy-preserving measure or rounded to the nearest thousand, and missing values could be imputed with the mean or with zero. During normalization we take note of exactly how fields are recorded.

1.5 New tools for new data

The new data sources that we have discussed frequently require working at scales for which the social scientist's familiar tools are not designed. Fortunately, the wider research and data analytics community has developed a wide variety of often more scalable and flexible tools—tools that we will introduce within this book.

Relational database management systems (DBMSs) are used throughout business as well as the sciences to organize, process,

► This topic is discussed in more detail in Chapter 4.

and search large collections of structured data. NoSQL DBMSs are used for data that is extremely large and/or unstructured, such as collections of web pages, social media data (e.g., Twitter messages), and clinical notes. Extensions to these systems and also specialized single-purpose DBMSs provide support for data types that are not easily handled in statistical packages such as geospatial data, networks, and graphs.

Open source programming systems such as Python (used extensively throughout this book) and R provide high-quality implementations of numerous data analysis and visualization methods, from regression to statistics, text analysis, network analysis, and much more. Finally, parallel computing systems such as Hadoop and Spark can be used to harness parallel computer clusters for extremely large data sets and computationally intensive analyses.

These various components may not always work together as smoothly as do integrated packages such as SAS, SPSS, and Stata, but they allow researchers to take on problems of great scale and complexity. Furthermore, they are developing at a tremendous rate as the result of work by thousands of people worldwide. For these reasons, the modern social scientist needs to be familiar with their characteristics and capabilities.

1.6 The book's "use case"

This book is about the uses of big data in social science. Our focus is on working through the use of data as a social scientist normally approaches research. That involves thinking through how to use such data to address a question from beginning to end, and thereby learning about the associated tools—rather than simply engaging in coding exercises and then thinking about how to apply them to a potpourri of social science examples.

There are many examples of the use of big data in social science research, but relatively few that feature all the different aspects that are covered in this book. As a result, the chapters in the book draw heavily on a use case based on one of the first large-scale big data social science data infrastructures. This infrastructure, based on UMETRICS* data housed at the University of Michigan's Institute for Research on Innovation and Science (IRIS) and enhanced with data from the US Census Bureau, provides a new quantitative analysis and understanding of science policy based on large-scale computational analysis of new types of data.

★ UMETRICS: Universities Measuring the Impact of Research on Innovation and Science [228]

► iris.isr.umich.edu

The infrastructure was developed in response to a call from the President's Science Advisor (Jack Marburger) for a *science of science policy* [250]. He wanted a scientific response to the questions that he was asked about the impact of investments in science.

Example: The Science of Science Policy

Marburger wrote [250]: "How much should a nation spend on science? What kind of science? How much from private versus public sectors? Does demand for funding by potential science performers imply a shortage of funding or a surfeit of performers? These and related science policy questions tend to be asked and answered today in a highly visible advocacy context that makes assumptions that are deserving of closer scrutiny. A new 'science of science policy' is emerging, and it may offer more compelling guidance for policy decisions and for more credible advocacy. . . .

"Relating R&D to innovation in any but a general way is a tall order, but not a hopeless one. We need econometric models that encompass enough variables in a sufficient number of countries to produce reasonable simulations of the effect of specific policy choices. This need won't be satisfied by a few grants or workshops, but demands the attention of a specialist scholarly community. As more economists and social scientists turn to these issues, the effectiveness of science policy will grow, and of science advocacy too."

Responding to this policy imperative is a tall order, because it involves using all the social science and computer science tools available to researchers. The new digital technologies can be used to capture the links between the inputs into research, the way in which those inputs are organized, and the subsequent outputs [396, 415]. The social science questions that are addressable with this data infrastructure include the effect of research training on the placement and earnings of doctoral recipients, how university trained scientists and engineers affect the productivity of the firms they work for, and the return on investments in research. Figure 1.2 provides an abstract representation of the empirical approach that is needed: data about grants, the people who are funded on grants, and the subsequent scientific and economic activities.

First, data must be captured on what is funded, and since the data are in text format, computational linguistics tools must be applied (Chapter 7). Second, data must be captured on who is funded, and how they interact in teams, so network tools and analysis must be used (Chapter 8). Third, information about the type of results must be gleaned from the web and other sources (Chapter 2).

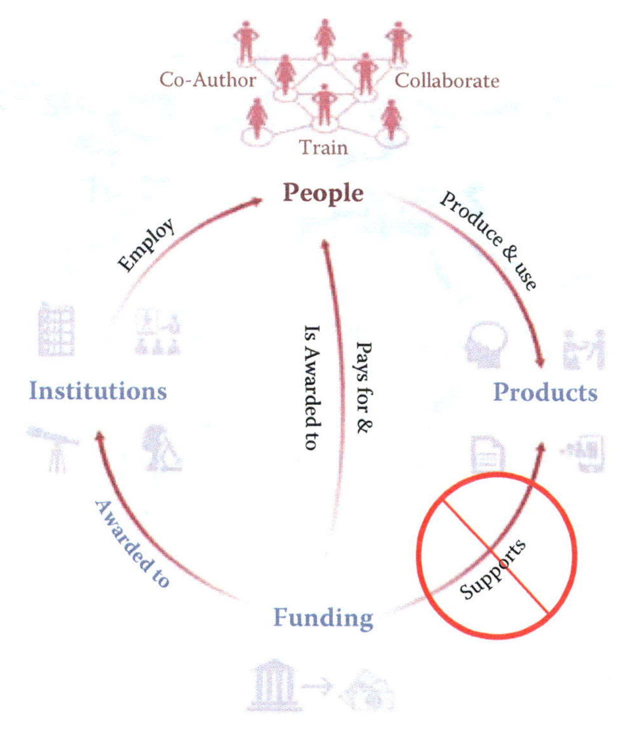

Figure 1.2. A visualization of the complex links between what and who is funded, and the results; tracing the direct link between funding and results is misleading and wrong

Finally, the disparate complex data sets need to be stored in databases (Chapter 4), integrated (Chapter 3), analyzed (Chapter 6), and used to make inferences (Chapter 10).

The use case serves as the thread that ties many of the ideas together. Rather than asking the reader to learn how to code "hello world," we build on data that have been put together to answer a real-world question, and provide explicit examples based on that data. We then provide examples that show how the approach generalizes.

For example, the text analysis chapter (Chapter 7) shows how to use natural language processing to describe *what* research is being done, using proposal and award text to identify the research topics in a portfolio [110, 368]. But then it also shows how the approach can be used to address a problem that is not just limited to science policy—the conversion of massive amounts of knowledge that is stored in text to usable information.

Similarly, the network analysis chapter (Chapter 8) gives specific examples using the UMETRICS data and shows how such data can be used to create new units of analysis—the networks of researchers who do science, and the networks of vendors who supply research inputs. It also shows how networks can be used to study a wide variety of other social science questions.

In another example, we use APIs* provided by publishers to describe the results generated by research funding in terms of publications and other measures of scientific impact, but also provide code that can be repurposed for many similar APIs.

★ Application Programming Interfaces

And, of course, since all these new types of data are provided in a variety of different formats, some of which are quite large (or voluminous), and with a variety of different timestamps (or velocity), we discuss how to store the data in different types of data formats.

1.7 The structure of the book

We organize the book in three parts, based around the way social scientists approach doing research. The first set of chapters addresses the new ways to capture, curate, and store data. The second set of chapters describes what tools are available to process and classify data. The last set deals with analysis and the appropriate handling of data on individuals and organizations.

1.7.1 Part I: Capture and curation

The four chapters in Part I (see Figure 1.3) tell you how to capture and manage data.

Chapter 2 describes how to extract information from social media about the transmission of knowledge. The particular application will be to develop links to authors' articles on Twitter using PLOS articles and to pull information about authors and articles from web sources by using an API. You will learn how to retrieve link data from bookmarking services, citations from Crossref, links from Facebook, and information from news coverage. In keeping with the social science grounding that is a core feature of the book, the chapter discusses what data can be captured from online sources, what is potentially reliable, and how to manage data quality issues.

Big data differs from survey data in that we must typically combine data from multiple sources to get a complete picture of the activities of interest. Although computer scientists may sometimes

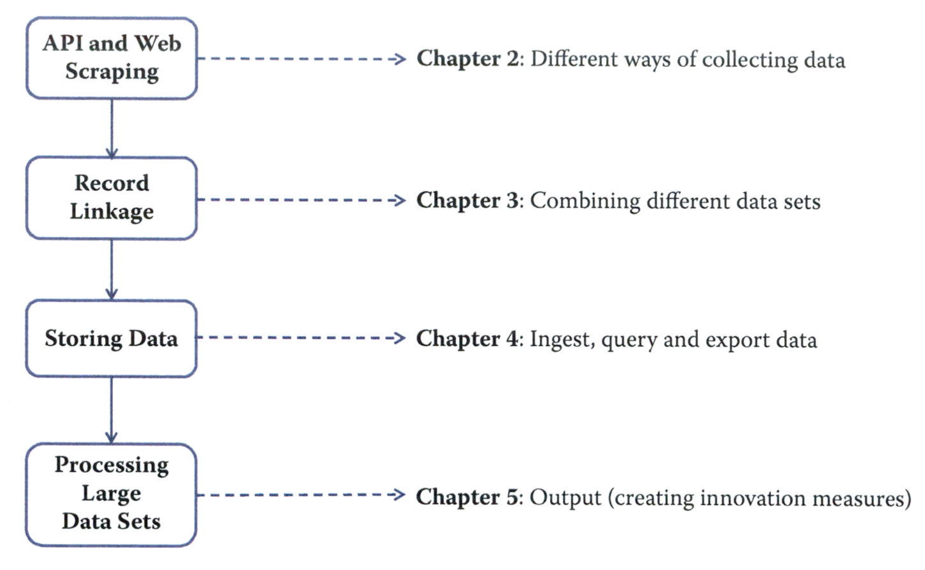

Figure 1.3. The four chapters of Part I focus on data *capture* and *curation*

simply "mash" data sets together, social scientists are rightfully concerned about issues of missing links, duplicative links, and erroneous links. Chapter 3 provides an overview of traditional rule-based and probabilistic approaches to data linkage, as well as the important contributions of machine learning to the linkage problem.

Once data have been collected and linked into different files, it is necessary to store and organize it. Social scientists are used to working with one analytical file, often in statistical software tools such as SAS or Stata. Chapter 4, which may be the most important chapter in the book, describes different approaches to storing data in ways that permit rapid and reliable exploration and analysis.

Big data is sometimes defined as data that are too big to fit onto the analyst's computer. Chapter 5 provides an overview of clever programming techniques that facilitate the use of data (often using parallel computing). While the focus is on one of the most widely used big data programming paradigms and its most popular implementation, Apache Hadoop, the goal of the chapter is to provide a conceptual framework to the key challenges that the approach is designed to address.

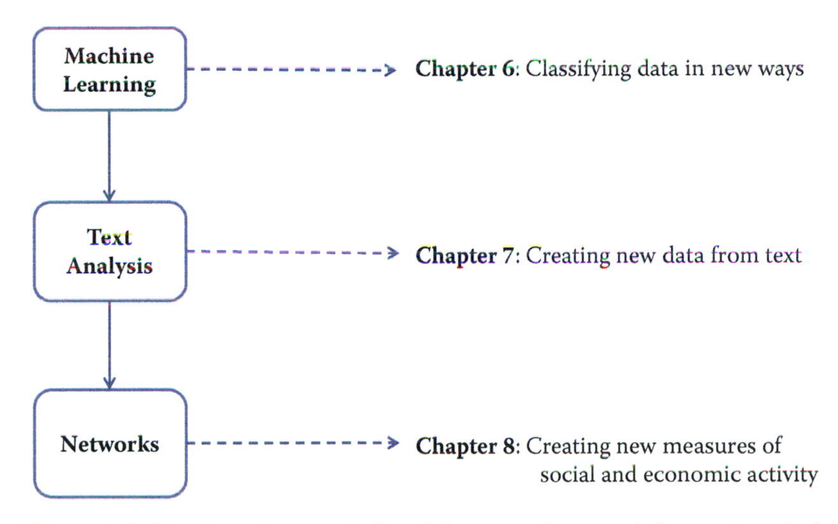

Figure 1.4. The four chapters in Part II focus on data *modeling* and *analysis*

1.7.2 Part II: Modeling and analysis

The three chapters in Part II (see Figure 1.4) introduce three of the most important tools that can be used by social scientists to do new and exciting research: machine learning, text analysis, and social network analysis.

Chapter 6 introduces machine learning methods. It shows the power of machine learning in a variety of different contexts, particularly focusing on clustering and classification. You will get an overview of basic approaches and how those approaches are applied. The chapter builds from a conceptual framework and then shows you how the different concepts are translated into code. There is a particular focus on random forests and support vector machine (SVM) approaches.

Chapter 7 describes how social scientists can make use of one of the most exciting advances in big data—text analysis. Vast amounts of data that are stored in documents can now be analyzed and searched so that different types of information can be retrieved. Documents (and the underlying activities of the entities that generated the documents) can be categorized into topics or fields as well as summarized. In addition, machine translation can be used to compare documents in different languages.

Social scientists are typically interested in describing the activities of individuals and organizations (such as households and firms) in a variety of economic and social contexts. The frames within which data are collected have typically been generated from tax or other programmatic sources. The new types of data permit new units of analysis—particularly network analysis—largely enabled by advances in mathematical graph theory. Thus, Chapter 8 describes how social scientists can use network theory to generate measurable representations of patterns of relationships connecting entities. As the author points out, the value of the new framework is not only in constructing different right-hand-side variables but also in studying an entirely new unit of analysis that lies somewhere between the largely atomistic actors that occupy the markets of neo-classical theory and the tightly managed hierarchies that are the traditional object of inquiry of sociologists and organizational theorists.

1.7.3 Part III: Inference and ethics

The four chapters in Part III (see Figure 1.5) cover three advanced topics relating to data inference and ethics—information visualization, errors and inference, and privacy and confidentiality—and introduce the workbooks that provide access to the practical exercises associated with the text.

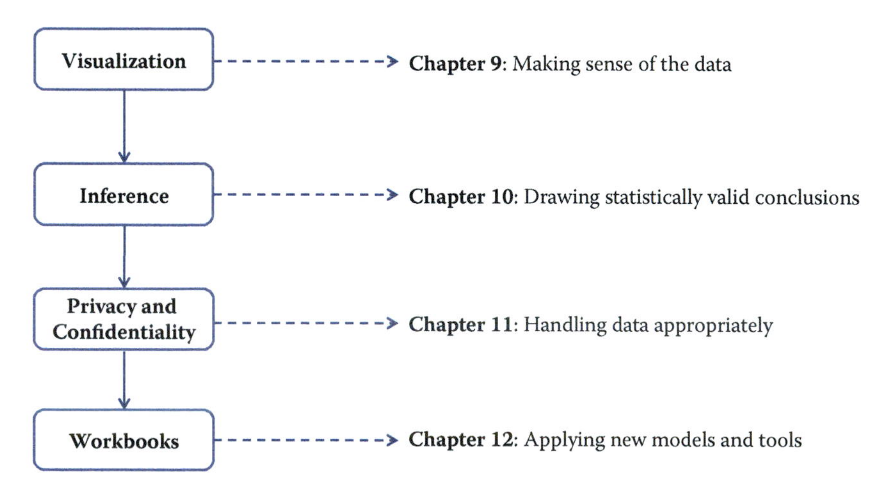

Figure 1.5. The four chapters in Part III focus on *inference* and *ethics*

Chapter 9 introduces information visualization methods and describes how you can use those methods to explore data and communicate results so that data can be turned into interpretable, actionable information. There are many ways of presenting statistical information that convey content in a rigorous manner. The goal of this chapter is to explore different approaches and examine the information content and analytical validity of the different approaches. It provides an overview of effective visualizations.

Chapter 10 deals with inference and the errors associated with big data. Social scientists know only too well the cost associated with bad data—we highlighted the classic *Literary Digest* example in the introduction to this chapter, as well as the more recent Google Flu Trends. Although the consequences are well understood, the new types of data are so large and complex that their properties often cannot be studied in traditional ways. In addition, the data generating function is such that the data are often selective, incomplete, and erroneous. Without proper data hygiene, errors can quickly compound. This chapter provides a systematic way to think about the error framework in a big data setting.

Chapter 11 addresses the issue that sits at the core of any study of human beings—privacy and confidentiality. In a new field, like the one covered in this book, it is critical that many researchers have access to the data so that work can be replicated and built on—that there be a scientific basis to data science. Yet the rules that social scientists have traditionally used for survey data, namely anonymity and informed consent, no longer apply when the data are collected in the wild. This concluding chapter identifies the issues that must be addressed for responsible and ethical research to take place.

Finally, Chapter 12 provides an overview of the practical work that accompanies each chapter—the workbooks that are designed, using Jupyter notebooks, to enable students and interested practitioners to apply the new techniques and approaches in selected chapters. We hope you have a lot of fun with them.

▶ See jupyter.org.

1.8 Resources

For more information on the science of science policy, see Husbands et al.'s book for a full discussion of many issues [175] and the online resources at the eponymous website [352].

This book is above all a *practical* introduction to the methods and tools that the social scientist can use to make sense of big data, and thus programming resources are also important. We make extensive use of the Python programming language and the MySQL database management system in both the book and its supporting workbooks. We recommend that any social scientist who aspires to work with large data sets become proficient in the use of these two systems, and also one more, GitHub. All three, fortunately, are quite accessible and are supported by excellent online resources. Time spent mastering them will be repaid many times over in more productive research.

▶ Read this! http://bit.ly/1VgytVV

For Python, Alex Bell's *Python for Economists* (available online [31]) provides a wonderful 30-page introduction to the use of Python in the social sciences, complete with XKCD cartoons. Economists Tom Sargent and John Stachurski provide a very useful set of lectures and examples at http://quant-econ.net/. For more detail, we recommend Charles Severance's *Python for Informatics: Exploring Information* [338], which not only covers basic Python but also provides material relevant to web data (the subject of Chapter 2) and MySQL (the subject of Chapter 4). This book is also freely available online and is supported by excellent online lectures and exercises.

For MySQL, Chapter 4 provides introductory material and pointers to additional resources, so we will not say more here.

We also recommend that you master GitHub. A version control system is a tool for keeping track of changes that have been made to a document over time. GitHub is a hosting service for projects that use the Git version control system. As Strasser explains [363], Git/GitHub makes it straightforward for researchers to create digital lab notebooks that record the data files, programs, papers, and other resources associated with a project, with automatic tracking of the changes that are made to those resources over time. GitHub also makes it easy for collaborators to work together on a project, whether a program or a paper: changes made by each contributor are recorded and can easily be reconciled. For example, we used GitHub to create this book, with authors and editors checking in changes and comments at different times and from many time zones. We also use GitHub to provide access to the supporting workbooks. Ram [314] provides a nice description of how Git/GitHub can be used to promote reproducibility and transparency in research.

One more resource that is outside the scope of this book but that you may well want to master is the cloud [21,236]. It used to be that when your data and computations became too large to analyze on your laptop, you were out of luck unless your employer (or a friend) had a larger computer. With the emergence of cloud storage and computing services from the likes of Amazon Web Services, Google, and Microsoft, powerful computers are available to anyone with a credit card. We and many others have had positive experiences using such systems for the analysis of urban [64], environmental [107], and genomic [32] data analysis and modeling, for example. Such systems may well represent the future of research computing.

Part I

Capture and Curation

Chapter 2

Working with Web Data and APIs

Cameron Neylon

This chapter will show you how to extract information from social media about the transmission of knowledge. The particular application will be to develop links to authors' articles on Twitter using PLOS articles and to pull information using an API. You will get link data from bookmarking services, citations from Crossref, links from Facebook, and information from news coverage. The examples that will be used are from Twitter. In keeping with the social science grounding that is a core feature of the book, it will discuss what can be captured, what is potentially reliable, and how to manage data quality issues.

2.1 Introduction

A tremendous lure of the Internet is the availability of vast amounts of data on businesses, people, and their activity on social media. But how can we capture the information and make use of it as we might make use of more traditional data sources? In this chapter, we begin by describing how web data can be collected, using the use case of UMETRICS and research output as a readily available example, and then discuss how to think about the scope, coverage, and integration issues associated with its collection.

Often a big data exploration starts with information on people or on a group of people. The web can be a rich source of additional information. It can also act as pointers to new sources of information, allowing a pivot from one perspective to another, from one kind of query to another. Often this is exploratory. You have an existing core set of data and are looking to augment it. But equally this exploration can open up whole new avenues. Sometimes the data are completely unstructured, existing as web pages spread across a site, and sometimes they are provided in a machine-readable form. The challenge is in having a sufficiently diverse toolkit to bring all of this information together.

Using the example of data on researchers and research outputs, we will explore obtaining information directly from web pages (*web scraping*) as well as explore the uses of APIs—web services that allow an interaction with, and retrieval of, structured data. You will see how the crucial pieces of integration often lie in making connections between disparate data sets and how in turn making those connections requires careful quality control. The emphasis throughout this chapter is on the importance of focusing on the purpose for which the data will be used as a guide for data collection. While much of this is specific to data about research and researchers, the ideas are generalizable to wider issues of data and public policy.

2.2 Scraping information from the web

With the range of information available on the web, our first question is how to access it. The simplest approach is often to manually go directly to the web and look for data files or other information. For instance, on the NSF website [268] it is possible to obtain data dumps of all grant information. Sometimes data are available only on web pages or we only want a subset of this information. In this case web scraping is often a viable approach.

Web scraping involves using a program to download and process web pages directly. This can be highly effective, particularly where tables of information are made available online. It is also useful in cases where it is desirable to make a series of very similar queries. In each case we need to look at the website, identify how to get the information we want, and then process it. Many websites deliberately make this difficult to prevent easy access to their underlying data.

2.2.1 Obtaining data from the HHMI website

Let us suppose we are interested in obtaining information on those investigators that are funded by the Howard Hughes Medical Institute (HHMI). HHMI has a website that includes a search function for funded researchers, including the ability to filter by field, state, and role. But there does not appear to be a downloadable data set of this information. However, we can automate the process with code to create a data set that you might compare with other data.

This process involves first understanding how to construct a URL that will do the search we want. This is most easily done by playing with search functionality and investigating the URL structures that are returned. Note that in many cases websites are not helpful here. However, with HHMI if we do a general search and play with the structure of the URL, we can see some of the elements of the URL that we can think of as a query. As we want to see *all* investigators, we do not need to limit the search, and so with some fiddling we come up with a URL like the following. (We have broken the one-line URL into three lines for ease of presentation.)

```
http://www.hhmi.org/scientists/browse?
   kw=&sort_by=field_scientist_last_name&
   sort_order=ASC&items_per_page=20&page=0
```

The `requests` module, available natively in Jupyter Python notebooks, is a useful set of tools for handling interactions with websites. It lets us construct the request that we just presented in terms of a base URL and query terms, as follows:

```
>> BASE_URL = "http://www.hhmi.org/scientists/browse"
>> query = {
            "kw" : "",
            "sort_by" : "field_scientist_last_name",
            "sort_order" : "ASC",
            "items_per_page" : 20,
            "page" : None
           }
```

With our request constructed we can then make the call to the web page to get a response.

```
>> import requests
>> response = requests.get(BASE_URL, params=query)
```

The first thing to do when building a script that hits a web page is to make sure that your call was successful. This can be checked by looking at the response code that the web server sent—and, obviously, by checking the actual HTML that was returned. A `200` code means success and that everything should be OK. Other codes may mean that the URL was constructed wrongly or that there was a server error.

```
>> response.status_code
200
```

With the page successfully returned, we now need to process the text it contains into the data we want. This is not a trivial exercise. It is possible to search through and find things, but there

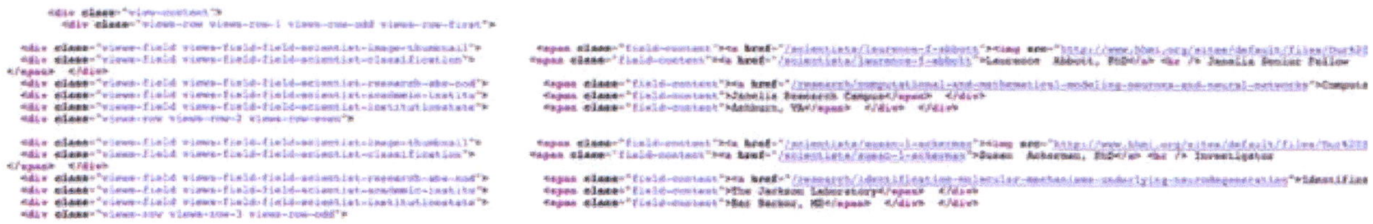

Figure 2.1. Source HTML from the portion of an HHMI results page containing information on HHMI investigators; note that the webscraping results in badly formatted html which is difficult to read.

★ Python features many useful libraries; BeautifulSoup is particularly helpful for webscraping.

are a range of tools that can help with processing HTML and XML data. Among these one of the most popular is a module called BeautifulSoup* [319], which provides a number of useful functions for this kind of processing. The module documentation provides more details.

We need to check the details of the page source to find where the information we are looking for is kept (see, for example, Figure 2.1). Here, all the details on HHMI investigators can be found in a <div> element with the class attribute view-content. This structure is not something that can be determined in advance. It requires knowledge of the structure of the page itself. Nested inside this <div> element are another series of divs, each of which corresponds to one investigator. These have the class attribute view-rows. Again, there is nothing obvious about finding these, it requires a close examination of the page HTML itself for any specific case you happen to be looking at.

We first process the page using the BeautifulSoup module (into the variable soup) and then find the div element that holds the information on investigators (investigator_list). As this element is unique on the page (I checked using my web browser), we can use the find method. We then process that div (using find_all) to create an iterator object that contains each of the page segments detailing a single investigator (investigators).

```
>> from bs4 import BeautifulSoup
>> soup = BeautifulSoup(response.text, "html5lib")
>> investigator_list = soup.find('div', class_ = "view-content")
>> investigators = investigator_list.find_all("div", class_ = "
    views-row")
```

As we specified in our query parameters that we wanted 20 results per page, we should check whether our list of page sections has the right length.

```
>> len(investigators)
20
```

```
# Given a request response object, parse for HHMI investigators
def scrape(page_response):
   # Obtain response HTML and the correct <div> from the page
   soup = BeautifulSoup(response.text, "html5lib")
   inv_list = soup.find('div', class_ = "view-content")

   # Create a list of all the investigators on the page
   investigators = inv_list.find_all("div", class_ = "views-row")

   data = [] # Make the data object to store scraping results

   # Scrape needed elements from investigator list
   for investigator in investigators:
       inv = {} # Create a dictionary to store results

       # Name and role are in same HTML element; this code
       # separates them into two data elements
       name_role_tag = investigator.find("div",
           class_ = "views-field-field-scientist-classification")
       strings = name_role_tag.stripped_strings
       for string,a in zip(strings, ["name", "role"]):
           inv[a] = string

       # Extract other elements from text of specific divs or from
       # class attributes of tags in the page (e.g., URLs)
       research_tag = investigator.find("div",
          class_ = "views-field-field-scientist-research-abs-nod")
       inv["research"] = research_tag.text.lstrip()
       inv["research_url"] = "http://hhmi.org"
          + research_tag.find("a").get("href")
       institution_tag = investigator.find("div",
          class_ = "views-field-field-scientist-academic-institu")
       inv["institute"] = institution_tag.text.lstrip()
       town_state_tag = investigator.find("div",
           class_ = "views-field-field-scientist-institutionstate"
    )
       inv["town"], inv["state"] = town_state_tag.text.split(",")
       inv["town"] = inv.get("town").lstrip()
       inv["state"] = inv.get("state").lstrip()

       thumbnail_tag = investigator.find("div",
          class_ = "views-field-field-scientist-image-thumbnail")
       inv["thumbnail_url"] = thumbnail_tag.find("img")["src"]
       inv["url"] = "http://hhmi.org"
          + thumbnail_tag.find("a").get("href")

       # Add the new data to the list
       data.append(inv)
   return data
```

Listing 2.1. Python code to parse for HHMI investigators

Finally, we need to process each of these segments to obtain the data we are looking for. This is the actual "scraping" of the page to get the information we want. Again, this involves looking closely at the HTML itself, identifying where the information is held, what tags can be used to find it, and often doing some postprocessing to clean it up (removing spaces, splitting different elements up).

Listing 2.1 provides a function to handle all of this. The function accepts the response object from the requests module as its input, processes the page text to soup, and then finds the `investigator_list` as above and processes it into an actual list of the investigators. For each investigator it then processes the HTML to find and clean up the information required, converting it to a dictionary and adding it to our growing list of data.

Let us check what the first two elements of our data set now look like. You can see two dictionaries, one relating to Laurence Abbott, who is a senior fellow at the HHMI Janelia Farm Campus, and one for Susan Ackerman, an HHMI investigator based at the Jackson Laboratory in Bar Harbor, Maine. Note that we have also obtained URLs that give more details on the researcher and their research program (`research_url` and `url` keys in the dictionary) that could provide a useful input to textual analysis or topic modeling (see Chapter 7).

```
>> data = scrape(response)
>> data[0:2]
[{'institute': u'Janelia Research Campus ',
  'name': u'Laurence Abbott, PhD',
  'research': u'Computational and Mathematical Modeling of Neurons
     and Neural... ',
  'research_url': u'http://hhmi.org/research/computational-and-
    mathematical-modeling-neurons-and-neural-networks',
  'role': u'Janelia Senior Fellow',
  'state': u'VA ',
  'thumbnail_url': u'http://www.hhmi.org/sites/default/files/Our
    %20Scientists/Janelia/Abbott-112x112.jpg',
  'town': u'Ashburn',
  'url': u'http://hhmi.org/scientists/laurence-f-abbott'},
 {'institute': u'The Jackson Laboratory ',
  'name': u'Susan Ackerman, PhD',
  'research': u'Identification of the Molecular Mechanisms
    Underlying... ',
  'research_url': u'http://hhmi.org/research/identification-
    molecular-mechanisms-underlying-neurodegeneration',
  'role': u'Investigator',
  'state': u'ME ',
  'thumbnail_url':
u'http://www.hhmi.org/sites/default/files/Our%20Scientists/
    Investigators/Ackerman-112x112.jpg',
```

```
  'town': u'Bar Harbor',
  'url': u'http://hhmi.org/scientists/susan-l-ackerman'}]
```

So now we know we can process a page from a website to generate usefully structured data. However, this was only the first page of results. We need to do this for each page of results if we want to capture all the HHMI investigators. We could just look at the number of pages that our search returned manually, but to make this more general we can actually scrape the page to find that piece of information and use that to calculate how many pages we need to work through.

The number of results is found in a `div` with the class "view-headers" as a piece of free text ("Showing 1-20 of 493 results"). We need to grab the text, split it up (I do so based on spaces), find the right number (the one that is before the word "results") and convert that to an integer. Then we can divide by the number of items we requested per page (20 in our case) to find how many pages we need to work through. A quick mental calculation confirms that if page 0 had results 1-20, page 24 would give results 481-493.

```
>> # Check total number of investigators returned
>> view_header = soup.find("div", class_ = "view-header")
>> words = view_header.text.split(" ")
>> count_index = words.index("results.") - 1
>> count = int(words[count_index])

>> # Calculate number of pages, given count & items_per_page
>> num_pages = count/query.get("items_per_page")
>> num_pages
24
```

Then it is a simple matter of putting the function we constructed earlier into a loop to work through the correct number of pages. As we start to hit the website repeatedly, we need to consider whether we are being polite. Most websites have a file in the root directory called robots.txt that contains guidance on using programs to interact with the website. In the case of http://hhmi.org the file states first that we are allowed (or, more properly, not forbidden) to query http://www.hhmi.org/scientists/ programmatically. Thus, you can pull down all of the more detailed biographical or research information, if you so desire. The file also states that there is a requested "Crawl-delay" of 10. This means that if you are making repeated queries (as we will be in getting the 24 pages), you should wait for 10 seconds between each query. This request is easily accommodated by adding a timed delay between each page request.

```
>> for page_num in range(num_pages):
>> # We already have page zero and we need to go to 24:
>> # range(24) is [0,1,...,23]
>>     query["items_per_page"] = page_num + 1
>>     page = requests.get(BASE_URL, params=query)
>> # We use extend to add list for each page to existing list
>>     data.extend(scrape(page))
>> print "Retrieved and scraped page number:", query.get("
   items_per_page")
>> time.sleep(10) # robots.txt at hhmi.org specifies a crawl delay
     of 10 seconds
Retrieved and scraped page number: 1
Retrieved and scraped page number: 2
...
Retrieved and scraped page number: 24
```

Finally we can check that we have the right number of results after our scraping. This should correspond to the 493 records that the website reports.

```
>> len(data)
493
```

2.2.2 Limits of scraping

While scraping websites is often necessary, is can be a fragile and messy way of working. It is problematic for a number of reasons: for example, many websites are designed in ways that make scraping difficult or impossible, and other sites explicitly prohibit this kind of scripted analysis. (Both reasons apply in the case of the NSF and Grants.gov websites, which is why we use the HHMI website in our example.)

In many cases a better choice is to process a data dump from an organization. For example, the NSF and Wellcome Trust both provide data sets for each year that include structured data on all their awarded grants. In practice, integrating data is a continual challenge of figuring out what is the easiest way to proceed, what is allowed, and what is practical and useful. The selection of data will often be driven by pragmatic rather than theoretical concerns.

Increasingly, however, good practice is emerging in which organizations provide APIs to enable scripted and programmatic access to the data they hold. These tools are much easier and generally more effective to work with. They are the focus of much of the rest of this chapter.

2.3 New data in the research enterprise

The new forms of data we are discussing in this chapter are largely available because so many human activities—in this case, discussion, reading, and bookmarking—are happening online. All sorts of data are generated as a side effect of these activities. Some of that data is public (social media conversations), some private (IP addresses requesting specific pages), and some intrinsic to the service (the identity of a user who bookmarks an article). What exactly are these new forms of data? There are broadly two new directions that data availability is moving in. The first is information on new forms of research output, data sets, software, and in some cases physical resources. There is an interest across the research community in expanding the set of research outputs that are made available and, to drive this, significant efforts are being made to ensure that these nontraditional outputs are seen as legitimate outputs. In particular there has been a substantial policy emphasis on data sharing and, coupled with this, efforts to standardize practice around data citation. This is applying a well-established measure (citation) to a new form of research output.

The second new direction, which is more developed, takes the alternate route, providing *new forms of information on existing types of output*, specifically research articles. The move online of research activities, including discovery, reading, writing, and bookmarking, means that many of these activities leave a digital trace. Often these traces are public or semi-public and can be collected and tracked. This certainly raises privacy issues that have not been comprehensively addressed but also provides a rich source of data on who is doing what with research articles.

There are a wide range of potential data sources, so it is useful to categorize them. Figure 2.2 shows one possible categorization, in which data sources are grouped based on the level of engagement and the stage of use. It starts from the left with "views," measures of online views and article downloads, followed by "saves" where readers actively collect articles into a library of their own, through online *discussion* forums such as blogs, social media and new commentary, formal scholarly *recommendations*, and, finally, formal *citations*.

These categories are a useful way to understand the classes of information available and to start digging into the sources they can be obtained from. For each category we will look at the *kind* of usage that the indicator is a proxy for, which *users* are captured by

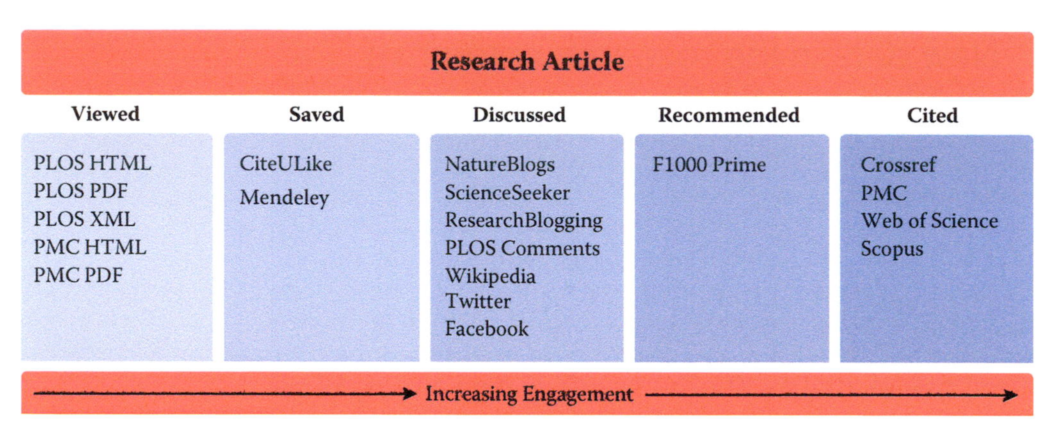

Figure 2.2. Classes of online activity related to research journal articles. Reproduced from Lin and Fenner [237], under a Creative Commons Attribution v 3.0 license

the indicator, the *limitations* that the indicator has as a measure, and the *sources* of data. We start with the familiar case of formal literature citations to provide context.

Example: Citations

Most quantitative analyses of research have focused on citations from research articles to other research articles. Many familiar measures—such as Impact Factors, Scimago Journal Rank, or Eigenfactor—are actually measures of journal rather than article performance. However, information on citations at the article level is increasingly the basis for much bibliometric analysis.

- Kind of usage
    - Citing a scholarly work is a signal from a researcher that a specific work has relevance to, or has influenced, the work they are describing.
    - It implies significant engagement and is a measure that carries some weight.
- Users
    - Researchers, which means usage by a specific group for a fairly small range of purposes.
    - With high-quality data, there are some geographical, career, and disciplinary demographic details.
- Limitations
    - The citations are slow to accumulate, as they must pass through a peer-review process.

  - It is seldom clear from raw data why a paper is being cited.
  - It provides a limited view of usage, as it only reflects reuse in research, not application in the community.

- Sources

  - Public sources of citation data include PubMed Central and Europe PubMed Central, which mine publicly available full text to find citations.
  - Proprietary sources of citation data include Thomson Reuters' Web of Knowledge and Elsevier's Scopus.
  - Some publishers make citation data collected by Crossref available.

Example: Page views and downloads

A major new source of data online is the number of times articles are viewed. Page views and downloads can be defined in different ways and can be reached via a range of paths. Page views are an immediate measure of usage. Viewing a paper may involve less engagement than citation or bookmarking, but it can capture interactions with a much wider range of users.

The possibility of drawing demographic information from downloads has significant potential for the future in providing detailed information on who is reading an article, which may be valuable for determining, for example, whether research is reaching a target audience.

- Kind of usage

  - It counts the number of people who have clicked on an article page or downloaded an article.

- Users

  - Page views and downloads report on use by those who have access to articles. For publicly accessible articles this could be anyone; for subscription articles it is likely to be researchers.

- Limitations

  - Page views are calculated in different ways and are not directly comparable across publishers. Standards are being developed but are not yet widely applied.
  - Counts of page views cannot easily distinguish between short-term visitors and those who engage more deeply with an article.
  - There are complications if an article appears in multiple places, for example at the journal website and a repository.

- Sources
  - Some publishers and many data repositories make page view data available in some form. Publishers with public data include PLOS, Nature Publishing Group, Ubiquity Press, Co-Action Press, and Frontiers.
  - Data repositories, including Figshare and Dryad, provide page view and download information.
  - PubMed Central makes page views of articles hosted on that site available to depositing publishers. PLOS and a few other publishers make this available.

Example: Analyzing bookmarks

Tools for collecting and curating personal collections of literature, or web content, are now available online. They make it easy to make copies and build up indexes of articles. Bookmarking services can choose to provide information on the number of people who have bookmarked a paper.

Two important services targeted at researchers are Mendeley and CiteULike. Mendeley has the larger user base and provides richer statistics. Data include the number of users that who bookmarked a paper, groups that have collected a paper, and in some cases demographics of users, which can include discipline, career stage, and geography.

Bookmarks accumulate rapidly after publication and provide evidence of scholarly interest. They correlate quite well with the eventual number of citations. There are also public bookmarking services that provide a view onto wider interest in research articles.

- Kind of usage
  - Bookmarking is a purposeful act. It may reveal more interest than a page view, but less than a citation.
  - Its uses are different from those captured by citations.
  - The bookmarks may include a variety of documents, such as papers for background reading, introductory material, position or policy papers, or statements of community positions.
- Users
  - Academic-focused services provide information on use by researchers.
  - Each service has a different user profile in, for instance, sciences or social sciences.

    - All services have a geographical bias towards North America and Europe.
    - There is some demographic information, for instance, on countries where users are bookmarking the most.

- Limitations
    - There is bias in coverage of services; for instance, Mendeley has good coverage of biomedical literature.
    - It can only report on activities of signed-up users.
    - It is not usually possible to determine why a bookmark has been created.

- Sources
    - Mendeley and CiteULike both have public APIs that provide data that are freely available for reuse.
    - Most consumer bookmarking services provide some form of API, but this often has restrictions or limitations.

Example: Discussions on social media

Social media are one of the most valuable new services producing information about research usage. A growing number of researchers, policymakers, and technologists are on these services discussing research.

There are three major features of social media as a tool. First, among a large set of conversations, it is possible to discover a discussion about a specific paper. Second, Twitter makes it possible to identify groups discussing research and to learn whether they were potential targets of the research. Third, it is possible to reconstruct discussions to understand what paths research takes to users.

In the future it will be possible to identify target audiences and to ask whether they are being reached and how modified distribution might maximize that reach. This could be a powerful tool, particularly for research with social relevance.

Twitter provides the most useful data because discussions and the identity of those involved are public. Connections between users and the things they say are often available, making it possible to identify communities discussing work. However, the 140-character limit on Twitter messages ("tweets") does not support extended critiques. Facebook has much less publicly available information—but being more private, it can be a site for frank discussion of research.

- Kind of usage
    - Those discussing research are showing interest potentially greater than page views.

- Often users are simply passing on a link or recommending an article.
- It is possible to navigate to tweets and determine the level and nature of interest.
- Conversations range from highly technical to trivial, so numbers should be treated with caution.
- Highly tweeted or Facebooked papers also tend to have significant bookmarking and citation.
- Professional discussions can be swamped when a piece of research captures public interest.

- Users
  - The user bases and data sources for Twitter and Facebook are global and public.
  - There are strong geographical biases.
  - A rising proportion of researchers use Twitter and Facebook for professional activities.
  - Many journalists, policymakers, public servants, civil society groups, and others use social media.

- Limitations
  - Frequent lack of explicit links to papers is a serious limitation.
  - Use of links is biased towards researchers and against groups not directly engaged in research.
  - There are demographic issues and reinforcement effects—retweeting leads to more retweeting in preference to other research—so analysis of numbers of tweets or likes is not always useful.

Example: Recommendations

A somewhat separate form of usage is direct expert recommendations. The best-known case of this is the F1000 service on which experts offer recommendations with reviews of specific research articles. Other services such as collections or personal recommendation services may be relevant here as well.

- Kind of usage
  - Recommendations from specific experts show that particular outputs are worth looking at in detail, are important, or have some other value.
  - Presumably, recommendations are a result of in-depth reading and a high level of engagement.

- Users
  - Recommendations are from a selected population of experts depending on the service in question.
  - In some cases this might be an algorithmic recommendation service.
- Limitations
  - Recommendations are limited to the interests of the selected population of experts.
  - The recommendation system may be biased in terms of the interests of recommenders (e.g., towards—or away from—new theories vs. developing methodology) as well as their disciplines.
  - Recommendations are slow to build up.

2.4 A functional view

The descriptive view of data types and sources is a good place to start, but it is subject to change. Sources of data come and go, and even the classes of data types may expand and contract in the medium to long term. We also need a more functional perspective to help us understand how these sources of data relate to activities in the broader research enterprise.

Consider Figure 1.2 in Chapter 1. The research enterprise has been framed as being made up of people who are generating outputs. The data that we consider in this chapter relate to connections between outputs, such as citations between research articles and tweets referring to articles. These connections are themselves created by people, as shown in Figure 2.3. The people in turn may be classed as belonging to certain categories or communities. What is interesting, and expands on the simplified picture of Figure 1.2, is that many of these people are not professional researchers. Indeed, in some cases they may not be people at all but automated systems of some kind. This means we need to expand the set of actors we are considering. As described above, we are also expanding the range of outputs (or objects) that we are considering as well.

In the simple model of Figure 2.3, there are three categories of things (nodes on the graph): objects, people, and the communities they belong to. Then there are the relationships between these elements (connections between nodes). Any given data source may provide information on different parts of this graph, and the information available is rarely complete or comprehensive. Data from

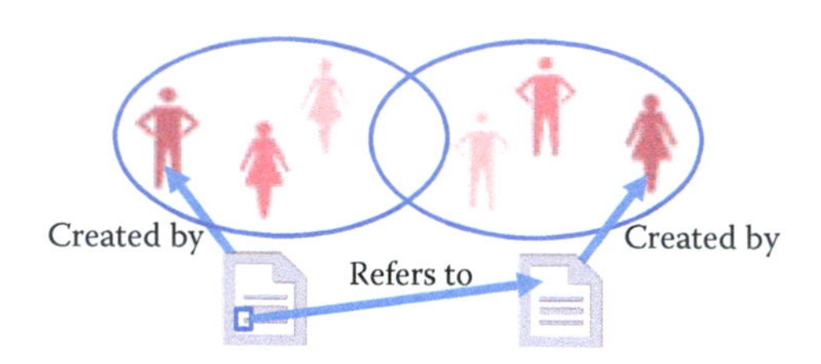

Figure 2.3. A simplified model of online interactions between research outputs and the objects that refer to them

▶ See Section 3.2.

different sources can also be difficult to integrate. As with any data integration, combining sources relies on being able to confidently identify those nodes that are common between data sources. Therefore identifying unique objects and people is critical to making progress.

These data are not necessarily public but many services choose to make some data available. An important characteristic of these data sources is that they are completely in the gift of the service provider. Data availability, its presentation, and upstream analysis can change without notice. Data are sometimes provided as a dump but is also frequently provided through an API.

An API is simply a tool that allows a program to interface with a service. APIs can take many different forms and be of varying quality and usefulness. In this section we will focus on one common type of API and examples of important publicly available APIs relevant to research communications. We will also cover combining APIs and the benefits and challenges of bringing multiple data sources together.

2.4.1 Relevant APIs and resources

There is a wide range of other sources of information that can be used in combination with the APIs featured above to develop an overview of research outputs and of where and how they are being used. There are also other tools that can allow deeper analysis of the outputs themselves. Table 2.1 gives a partial list of key data sources and APIs that are relevant to the analysis of research outputs.

2.4.2 RESTful APIs, returned data, and Python wrappers

The APIs we will focus on here are all examples of RESTful services. REST stands for Representational State Transfer [121, 402], but for

Table 2.1. Popular sources of data relevant to the analysis of research outputs

Source	Description	API	Free
	Bibliographic Data		
PubMed	An online index that combines bibliographic data from Medline and PubMed Central. PubMed Central and Europe PubMed Central also provide information.	Y	Y
Web of Science	The bibliographic database provided by Thomson Reuters. The ISI Citation Index is also available.	Y	N
Scopus	The bibliographic database provided by Elsevier. It also provides citation information.	Y	N
Crossref	Provides a range of bibliographic metadata and information obtained from members registering DOIs.	Y	Y
Google Scholar	Provides a search index for scholarly objects and aggregates citation information.	N	Y
Microsoft Academic Search	Provides a search index for scholarly objects and aggregates citation information. Not as complete as Google Scholar, but has an API.	Y	Y
	Social Media		
Altmetric.com	A provider of aggregated data on social media and mainstream media attention of research outputs. Most comprehensive source of information across different social media and mainstream media conversations.	Y	N
Twitter	Provides an API that allows a user to search for recent tweets and obtain some information on specific accounts.	Y	Y
Facebook	The Facebook API gives information on the number of pages, likes, and posts associated with specific web pages.	Y	Y
	Author Profiles		
ORCID	Unique identifiers for research authors. Profiles include information on publication lists, grants, and affiliations.	Y	Y
LinkedIn	CV-based profiles, projects, and publications.	Y	*
	Funder Information		
Gateway to Research	A database of funding decisions and related outputs from Research Councils UK.	Y	Y
NIH Reporter	Online search for information on National Institutes of Health grants. Does not provide an API but a downloadable data set is available.	N	Y
NSF Award Search	Online search for information on NSF grants. Does not provide an API but downloadable data sets by year are available.	N	Y

* The data are restricted: sometimes fee based, other times not.

our purposes it is most easily understood as a means of transferring data using web protocols. Other forms of API require additional tools or systems to work with, but RESTful APIs work directly over the web. This has the advantage that a human user can also with relative ease play with the API to understand how it works. Indeed, some websites work simply by formatting the results of API calls.

As an example let us look at the Crossref API. This provides a range of information associated with Digital Object Identifiers (DOIs) registered with Crossref. DOIs uniquely identify an object, and Crossref DOIs refer to research objects, primarily (but not entirely) research articles. If you use a web browser to navigate to http://api.crossref.org/works/10.1093/nar/gni170, you should receive back a webpage that looks something like the following. (We have laid it out nicely to make it more readable.)


```
{ "status" : "ok",
  "message-type" : "work",
  "message-version" : "1.0.0",
  "message" :
   { "subtitle": [],
     "subject" : ["Genetics"],
     "issued" : { "date-parts" : [[2005,10,24]] },
     "score" : 1.0,
     "prefix" : "http://id.crossref.org/prefix/10.1093",
     "author" : [ "affiliation" : [],
                  "family" : "Whiteford",
                  "given" : "N."}],
     "container-title" : ["Nucleic Acids Research"],
     "reference-count" : 0,
     "page" : "e171-e171",
     "deposited" : {"date-parts" : [[2013,8,8]],
                    "timestamp" : 1375920000000},
     "issue" : "19",
     "title" :
       ["An analysis of the feasibility of short read sequencing"]
    ,
     "type" : "journal-article",
     "DOI" : "10.1093/nar/gni170",
     "ISSN" : ["0305-1048","1362-4962"],
     "URL" : "http://dx.doi.org/10.1093/nar/gni170",
     "source" : "Crossref",
     "publisher" : "Oxford University Press (OUP)",
     "indexed" : {"date-parts" : [[2015,6,8]],
                  "timestamp" : 1433777291246},
     "volume" : "33",
     "member" : "http://id.crossref.org/member/286"
   }
}
```


This is a package of JavaScript Object Notation (JSON)* data returned in response to a query. The query is contained entirely in the URL, which can be broken up into pieces: the root URL (http://api.crossref.org) and a data "query," in this case made up of a "field" (`works`) and an identifier (the DOI `10.1093/nar/gni170`). The Crossref API provides information about the article identified with this specific DOI.

★ JSON is an open standard way of storing and exchanging data.

2.5 Programming against an API

Programming against an API involves constructing HTTP requests and parsing the data that are returned. Here we use the Crossref API to illustrate how this is done. Crossref is the provider of DOIs used by many publishers to uniquely identify scholarly works. Crossref is not the only organization to provide DOIs. The scholarly communication space DataCite is another important provider. The documentation is available at the Crossref website [394].

Once again the `requests` Python library provides a series of convenience functions that make it easier to make HTTP calls and to process returned JSON. Our first step is to import the module and set a base URL variable.

```
>> import requests
>> BASE_URL = "http://api.crossref.org/"
```

A simple example is to obtain metadata for an article associated with a specific DOI. This is a straightforward call to the Crossref API, similar to what we saw earlier.

```
>> doi = "10.1093/nar/gni170"
>> query = "works/"
>> url = BASE_URL + query + doi
>> response = requests.get(url)
>> url
http://api.crossref.org/works/10.1093/nar/gni170
>> response.status_code
200
```

The `response` object that the `requests` library has created has a range of useful information, including the URL called and the response code from the web server (in this case 200, which means everything is OK). We need the JSON body from the response object (which is currently text from the perspective of our script) converted to a Python dictionary. The `requests` module provides a convenient function for performing this conversion, as the following code shows. (All strings in the output are in Unicode, hence the `u'` notation.)

```
>> response_dict = response.json()
>> response_dict
{ u'message' :
  { u'DOI' : u'10.1093/nar/gni170',
    u'ISSN' : [ u'0305-1048', u'1362-4962' ],
    u'URL' : u'http://dx.doi.org/10.1093/nar/gni170',
    u'author' : [ {u'affiliation' : [],
                   u'family' : u'Whiteford',
```

```
                        u'given' : u'N.'} ],
    u'container-title' : [ u'Nucleic Acids Research' ],
    u'deposited' : { u'date-parts' : [[2013, 8, 8]],
                        u'timestamp' : 1375920000000 },
    u'indexed' : { u'date-parts' : [[2015, 6, 8]],
                        u'timestamp' : 1433777291246 },
    u'issue' : u'19',
    u'issued' : { u'date-parts' : [[2005, 10, 24]] },
    u'member' : u'http://id.crossref.org/member/286',
    u'page' : u'e171-e171',
    u'prefix' : u'http://id.crossref.org/prefix/10.1093',
    u'publisher' : u'Oxford University Press (OUP)',
    u'reference-count' : 0,
    u'score' : 1.0,
    u'source' : u'Crossref',
    u'subject' : [u'Genetics'],
    u'subtitle' : [],
    u'title' : [u'An analysis of the feasibility of short read
    sequencing'],
    u'type' : u'journal-article',
    u'volume' : u'33'
  },
  u'message-type' : u'work',
  u'message-version' : u'1.0.0',
  u'status' : u'ok'
}
```


This data object can now be processed in whatever way the user wishes, using standard manipulation techniques.

The Crossref API can, of course, do much more than simply look up article metadata. It is also valuable as a search resource and for cross-referencing information by journal, funder, publisher, and other criteria. More details can be found at the Crossref website.

2.6 Using the ORCID API via a wrapper

ORCID, which stands for "Open Research and Contributor Identifier" (see orcid.org; see also [145]), is a service that provides unique identifiers for researchers. Researchers can claim an ORCID profile and populate it with references to their research works, funding and affiliations. ORCID provides an API for interacting with this information. For many APIs there is a convenient Python wrapper that can be used. The ORCID-Python wrapper works with the ORCID v1.2 API to make various API calls straightforward. This wrapper only works with the public ORCID API and can therefore only access publicly available data.

Using the API and wrapper together provides a convenient means of getting this information. For instance, given an ORCID, it is straightforward to get profile information. Here we get a list of publications associated with my ORCID and look at the the first item on the list.

```
>> import orcid
>> cn = orcid.get("0000-0002-0068-716X")
>> cn
<Author Cameron Neylon, ORCID 0000-0002-0068-716X>
>> cn.publications[0]
<Publication "Principles for Open Scholarly Infrastructures-v1">
```

The wrapper has created Python objects that make it easier to work with and manipulate the data. It is common to take the return from an API and create objects that behave as would be expected in Python. For instance, the `publications` object is a list populated with publications (which are also Python-like objects). Each publication in the list has its own attributes, which can then be examined individually. In this case the external IDs attribute is a list of further objects that include a DOI for the article and the ISSN of the journal the article was published in.

```
>> len(cn.publications)
70
>> cn.publications[12].external_ids
[<ExternalID DOI:10.1371/journal.pbio.1001677>, <ExternalID ISSN
    :1545-7885>]
```

As a simple example of data processing, we can iterate over the list of publications to identify those for which a DOI has been provided. In this case we can see that of the 70 publications listed in this ORCID profile (at the time of testing), 66 have DOIs.

```
>> exids = []
>> for pub in cn.publications:
       if pub.external_ids:
       exids = exids + pub.external_ids
>> DOIs = [exid.id for exid in exids if exid.type == "DOI"]
>> len(DOIs)
66
```

Wrappers generally make operating with an API simpler and cleaner by abstracting away the details of making HTTP requests. Achieving the same by directly interacting with the ORCID API would require constructing the appropriate URLs and parsing the returned data into a usable form. Where a wrapper is available it is generally much easier to use. However, wrappers may not be actively developed and may lag the development of the API. Where possible, use a wrapper that is directly supported or recommended by the API provider.

2.7 Quality, scope, and management

The examples in the previous section are just a small dip into the surface of the data available, but we already can see a number of issues that are starting to surface. A great deal of care needs to be taken when using these data, and a researcher will need to apply subject matter knowledge as well as broader data management expertise. Some of the core issues are as follows:

▶ See Chapter 10.

Integration In the examples given above with Crossref and ORCID, we used a known identifier (a DOI or an ORCID). Integrating data from Crossref to supplement the information from an ORCID profile is possible, but it depends on the linking of identifiers. Note that for the profile data we obtained, only 66 or the 70 items had DOIs. Data integration across multiple data sources that reference DOIs is straightforward for those objects that have DOIs, and messy or impossible for those that do not. In general, integration is possible, but it depends on a means of cross-referencing between data sets. Unique identifiers that are common to both are extremely powerful but only exist in certain cases (see also Chapter 3).

Coverage Without a population frame, it is difficult to know whether the information that can be captured is comprehensive. For example, "the research literature" is at best a vague concept. A variety of indexes, some openly available (PubMed, Crossref), some proprietary (Scopus, Web of Knowledge, many others), cover different partially overlapping segments of this corpus of work. Each index has differing criteria for inclusion and differing commitments to completeness. Sampling of "the literature" is therefore impossible, and the choice of index used for any study can make a substantial difference to the conclusions.

Completeness Alongside the question of coverage (how broad is a data source?), with web data and opt-in services we also need to probe the completeness of a data set. In the example above, 66 of 70 objects have a DOI *registered*. This does not mean that those four other objects do not have a DOI, just that there are none included in the ORCID record. Similarly, ORCID profiles only exist for a subset of researchers at this stage. Completeness feeds into integration challenges. While many researchers have a Twitter profile and many have an ORCID profile, only a small subset of ORCID profiles provide a link to a Twitter profile. See below for a worked example.

Scope In survey data sets, the scope is defined by the question being asked. This is not the case with much of these new data. For example, the challenges listed above for research articles, traditionally considered the bedrock of research outputs, at least in the natural sciences, are much greater for other forms of research outputs. Increasingly, the data generated from research projects, software, materials, and tools, as well as reports and presentations, are being shared by researchers in a variety of settings. Some of these are formal mechanisms for publication, such as large disciplinary databases, books, and software repositories, and some are highly informal. Any study of (a subset of) these outputs has as its first challenge the question of how to limit the corpus to be studied.

Source and validity The challenges described above relate to the identification and counting of outputs. As we start to address questions of how these outputs are being used, the issues are compounded. To illustrate some of the difficulties that can arise, we examine the number of citations that have been reported for a single sample article on a biochemical methodology [68]. This article has been available for eight years and has accumulated a reasonable number of citations for such an article over that time.

However, the exact number of citations identified varies radically, depending on the data source. Scopus finds 40, while Web of Science finds only 38. A Google Scholar search performed on the same date identified 59. These differences relate to the size of the corpus from which inward citations are being counted. Web of Science has the smallest database, with Scopus being larger and Google Scholar substantially larger again. Thus the size of the index not only affects output counting, it can also have a substantial effect on any analysis that uses that corpus. Alongside the size of the corpus, the means of analysis can also have an effect. For the same article, PubMed Central reports 10 citations but Europe PubMed Central reports 18, despite using a similar corpus. The distinction lies in differences in the methodology used to mine the corpus for citations.

Identifying the underlying latent variable These issues multiply as we move into newer forms of data. These sparse and incomplete sources of data require different treatment than more traditional structured and comprehensive forms of data. They are more useful as a way of identifying activities than of quantifying or comparing them. Nevertheless, they can provide new insight into the processes of knowledge dissemination and community building that are occurring online.

2.8 Integrating data from multiple sources

We often must work across multiple data sources to gather the information needed to answer a research question. A common pattern is to search in one location to create a list of identifiers and then use those identifiers to query another API. In the ORCID example above, we created a list of DOIs from a single ORCID profile. We could use those DOIs to obtain further information from the Crossref API and other sources. This models a common path for analysis of research outputs: identifying a corpus and then seeking information on its performance.

In this example, we will build on the ORCID and Crossref examples to collect a set of work identifiers from an ORCID profile and use a range of APIs to identify additional metadata as well as information on the performance of those articles. In addition to the ORCID API, we will use the PLOS Lagotto API. Lagotto is the software that was built to support the Article Level Metrics program at PLOS, the open access publisher, and its API provides information on various metrics of PLOS articles. A range of other publishers and service providers, including Crossref, also provide an instance of this API, meaning the same tools can be used to collect information on articles from a range of sources.

2.8.1 The Lagotto API

The module `pyalm` is a wrapper for the Lagotto API, which is served from a range of hosts. We will work with two instances in particular: one run by PLOS, and the Crossref DOI Event Tracker (DET, recently renamed Crossref Event Data) pilot service. We first need to provide the details of the URLs for these instances to our wrapper. Then we can obtain some information for a single DOI to see what the returned data look like.

```
>> import pyalm
>> pyalm.config.APIS = {'plos' : {'url' :
>>         'http://alm.plos.org/api/v5/articles'},
>>         'det' : {'url' :
```

```
>>          'http://det.labs.crossref.org/api/v5/articles'}
>>          }
>> det_alm_test = pyalm.get_alm('10.1371/journal.pbio.1001677',
>>          info='detail', instance='det')
det_alm_test
{ 'articles' : [<ArticleALM Expert Failure: Re-evaluating Research
     Assessment, DOI 10.1371/journal.pbio.1001677>],
  'meta' : {u'error' : None, u'page' : 1,
           u'total' : 1, u'total_pages' : 1}
}
```

The library returns a Python dictionary containing two elements. The articles key contains the actual data and the meta key includes general information on the results of the interaction with the API. In this case the library has returned one page of results containing one object (because we only asked about one DOI). If we want to collect a lot of data, this information helps in the process of paging through results. It is common for APIs to impose some limit on the number of results returned, so as to ensure performance. By default the Lagotto API has a limit of 50 results.

The articles key holds a list of ArticleALM objects as its value. Each ArticleALM object has a set of internal attributes that contain information on each of the metrics that the Lagotto instance collects. These are derived from various data providers and are called *sources*. Each can be accessed by name from a dictionary called "sources." The `iterkeys()` function provides an iterator that lets us loop over the set of keys in a dictionary. Within the source object there is a range of information that we will dig into.

```
>> article = det_alm_test.get('articles')[0]
>> article.title
u'Expert Failure: Re-evaluating Research Assessment'
>> for source in article.sources.iterkeys():
>>    print source, article.sources[source].metrics.total
reddit 0
datacite 0
pmceuropedata 0
wikipedia 1
pmceurope 0
citeulike 0
pubmed 0
facebook 0
wordpress 0
pmc 0
mendeley 0
crossref 0
```

The DET service only has a record of citations to this article from Wikipedia. As we will see below, the PLOS service returns more results. This is because some of the sources are not yet being queried by DET.

Because this is a PLOS paper we can also query the PLOS Lagotto instance for the same article.

```
>> plos_alm_test = pyalm.get_alm('10.1371/journal.pbio.1001677',
     info='detail', instance='plos')
>> article_plos = plos_alm_test.get('articles')[0]
>> article_plos.title
u'Expert Failure: Re-evaluating Research Assessment'
>> for source in article_plos.sources.iterkeys():
>>     print source, article_plos.sources[source].metrics.total
datacite 0
twitter 130
pmc 610
articlecoveragecurated 0
pmceurope 1
pmceuropedata 0
researchblogging 0
scienceseeker 0
copernicus 0
f1000 0
wikipedia 1
citeulike 0
wordpress 2
openedition 0
reddit 0
nature 0
relativemetric 125479
figshare 0
facebook 1
mendeley 14
crossref 3
plos_comments 2
articlecoverage 0
counter 12551
scopus 2
pubmed 1
orcid 3
```

The PLOS instance provides a greater range of information but also seems to be giving larger numbers than the DET instance in many cases. For those sources that are provided by both API instances, we can compare the results returned.

```
>> for source in article.sources.iterkeys():
>>     print source, article.sources[source].metrics.total,
>>     article_plos.sources[source].metrics.total
reddit 0 0
datacite 0 0
pmceuropedata 0 0
wikipedia 1 1
pmceurope 0 1
citeulike 0 0
pubmed 0 1
```

```
facebook 0 1
wordpress 0 2
pmc 0 610
mendeley 0 14
crossref 0 3
```

The PLOS Lagotto instance is collecting more information and has a wider range of information sources. Comparing the results from the PLOS and DET instances illustrates the issues of coverage and completeness discussed previously. The data may be sparse for a variety of reasons, and it is important to have a clear idea of the strengths and weaknesses of a particular data source or aggregator. In this case the DET instance is returning information for some sources for which it is does not yet have data.

We can dig deeper into the events themselves that the metrics.total count aggregates. The API wrapper collects these into an event object within the source object. These contain the JSON returned from the API in most cases. For instance, the Crossref source is a list of JSON objects containing information on an article that cites our article of interest. The first citation event in the list is a citation from the *Journal of the Association for Information Science and Technology* by Du et al.

```
>> article_plos.sources['crossref'].events[0]
{u'event' :
   {u'article_title' : u'The effects of research level and article
     type on the differences between citation metrics and F1000
    recommendations',
    u'contributors' :
      {u'contributor' :
         [ { u'contributor_role' : u'author',
             u'first_author' : u'true',
             u'given_name' : u'Jian',
             u'sequence' : u'first',
             u'surname' : u'Du' },
           { u'contributor_role' : u'author',
             u'first_author' : u'false',
             u'given_name' : u'Xiaoli',
             u'sequence' : u'additional',
             u'surname' : u'Tang'},
           { u'contributor_role' : u'author',
             u'first_author' : u'false',
             u'given_name' : u'Yishan',
             u'sequence' : u'additional',
             u'surname' : u'Wu'} ]
         },
     u'doi' : u'10.1002/asi.23548',
     u'first_page' : u'n/a',
     u'fl_count' : u'0',
     u'issn' : u'23301635',
```

```
      u'journal_abbreviation' : u'J Assn Inf Sci Tec',
      u'journal_title' : u'Journal of the Association for
     Information Science and Technology',
      u'publication_type' : u'full_text',
      u'year' : u'2015'
    },
   u'event_csl' : {
     u'author' :
            [ { u'family' : u'Du', u'given' : u'Jian'},
              {u'family' : u'Tang', u'given' : u'Xiaoli'},
              {u'family' : u'Wu', u'given' : u'Yishan'} ],
     u'container-title' : u'Journal of the Association for
     Information Science and Technology',
     u'issued' : {u'date-parts' : [[2015]]},
     u'title' : u'The Effects Of Research Level And Article Type
     On The Differences Between Citation Metrics And F1000
     Recommendations',
     u'type' : u'article-journal',
     u'url' : u'http://doi.org/10.1002/asi.23548'
    },
  u'event_url' : u'http://doi.org/10.1002/asi.23548'
}
```

Another source in the PLOS data is Twitter. In the case of the Twitter events (individual tweets), this provides the text of the tweet, user IDs, user names, URL of the tweet, and the date. We can see from the length of the events list that there are at least 130 tweets that link to this article.

```
>> len(article_plos.sources['twitter'].events)
130
```

Again, noting the issues of coverage, scope, and completeness, it is important to consider the limitations of these data. This is a lower bound as it represents search results returned by searching the Twitter API for the DOI or URL of the article. Other tweets that discuss the article may not include a link, and the Twitter search API also has limitations that can lead to incomplete results. The number must therefore be seen as both incomplete and a lower bound.

We can look more closely at data on the first tweet on the list. Bear in mind that the order of the list is not necessarily special. This is not the first tweet about this article chronologically.

```
>> article_plos.sources['twitter'].events[0]
{ u'event' : {u'created_at': u'2013-10-08T21:12:28Z',
  u'id' : u'387686960585641984',
  u'text' : u'We have identified the Higgs boson; it is surely not
      beyond our reach to make research assessment useful http://t
     .co/Odcm8dVRSU#PLOSBiology',
```

```
  u'user' : u'catmacOA',
  u'user_name' : u'Catriona MacCallum',
  u'user_profile_image' :
  u'http://a0.twimg.com/profile_images/1779875975/
    CM_photo_reduced_normal.jpg'},
  u'event_time' : u'2013-10-08T21:12:28Z',
  u'event_url' : u'http://twitter.com/catmacOA/status
    /387686960585641984'
}
```

We could use the Twitter API to understand more about this person. For instance, we could look at their Twitter followers and whom they follow, or analyze the text of their tweets for topic modeling. Much work on social media interactions is done with this kind of data, using forms of network and text analysis described elsewhere in this book.

▶ See Chapters 7 and 8.

A different approach is to integrate these data with information from another source. We might be interested, for instance, in whether the author of this tweet is a researcher, or whether they have authored research papers. One thing we could do is search the ORCID API to see if there are any ORCID profiles that link to this Twitter handle.

```
>> twitter_search = orcid.search("catmacOA")
>> for result in twitter_search:
>>     print unicode(result)
>>     print result.researcher_urls}
<Author Catriona MacCallum, ORCID 0000-0001-9623-2225>
[<Website twitter [http://twitter.com/catmacOA]>]
```

So the person with this Twitter handle seems to have an ORCID profile. That means we can also use ORCID to gather more information on their outputs. Perhaps they have authored work which is relevant to our article?

```
>> cm = orcid.get("0000-0001-9623-2225")
>> for pub in cm.publications[0:5]:
>>     print pub.title
The future is open: opportunities for publishers and institutions
Open Science and Reporting Animal Studies: Who's Accountable?
Expert Failure: Re-evaluating Research Assessment
Why ONE Is More Than 5
Reporting Animal Studies: Good Science and a Duty of Care
```

From this analysis we can show that this tweet is actually from one of my co-authors of the article.

To make this process easier we write the convenience function shown in Listing 2.2 to go from a Twitter user handle to try and find an ORCID for that person.

```
# Take a twitter handle or user name and return an ORCID
def twitter2orcid(twitter_handle,
                  resp = 'orcid', search_depth = 10):
    search = orcid.search(twitter_handle)
    s = [r for r in search]
    orc = None
    i = 0
    while i < search_depth and orc == None and i < len(s):
        arr = [('twitter.com' in website.url)
               for website in s[i].researcher_urls]
        if True in arr:
            index = arr.index(True)
            url = s[i].researcher_urls[index].url
            if url.lower().endswith(twitter_handle.lower()):
                orc = s[i].orcid
                return orc
        i+=1
     return None
```

Listing 2.2. Python code to find ORCID for Twitter handle

Let us do a quick test of the function.

```
>> twitter2orcid('catmacOA')
u'0000-0001-9623-2225'
```

2.8.2 Working with a corpus

In this case we will continue as previously to collect a set of works from a single ORCID profile. This collection could just as easily be a date range, or subject search at a range of other APIs. The target is to obtain a set of identifiers (in this case DOIs) that can be used to precisely query other data sources. This is a general pattern that reflects the issues of scope and source discussed above. The choice of how to construct a corpus to analyze will strongly affect the results and the conclusions that can be drawn.

```
>> # As previously, collect DOIs available from an ORCID profile
>> cn = orcid.get("0000-0002-0068-716X")
>> exids = []
>> for pub in cn.publications:
>>     if pub.external_ids:
>>         exids = exids + pub.external_ids
>> DOIs = [exid.id for exid in exids if exid.type == "DOI"]
>> len(DOIs)
66
```

We have recovered 66 DOIs from the ORCID profile. Note that we have not obtained an identifier for every work, as not all have DOIs. This result illustrates an important point about data integration. In practice it is generally not worth the effort of attempting to integrate data on objects unless they have a unique identifier or key that can be used in multiple data sources, hence the focus on DOIs and ORCIDs in these examples. Even in our search of the ORCID API for profiles that are associated with a Twitter account, we used the Twitter handle as a unique ID to search on.

While it is possible to work with author names or the titles of works directly, disambiguating such names and titles is substantially more difficult than working with unique identifiers. Other chapters (in particular, Chapter 3) deal with issues of data cleaning and disambiguation. Much work has been done on this basis, but increasingly you will see that the first step in any analysis is simply to discard objects without a unique ID that can be used across data sources.

We can obtain data for these from the DET API. As is common with many APIs, there is a limit to how many queries can be simultaneously run, in this case 50, so we divide our query into batches.

```
>> batches = [DOIs[0:50], DOIs[51:-1]]
>> det_alms = []
>> for batch in batches:
>>     alms_response = pyalm.get_alm(batch, info="detail",
    instance="det")
>>     det_alms.extend(alms_response.get('articles'))
>> len(det_alms)
24
```

The DET API only provides information on a subset of Crossref DOIs. The process that Crossref has followed to populate its database has focused on more recently published articles, so only 24 responses are received in this case for the 66 DOIs we queried on. A good exercise would be to look at which of the DOIs are found and which are not. Let us see how much interesting data is available in the subset of DOIs for which we have data.

```
>> for r in [d for d in det_alms if d.sources['wikipedia'].metrics
    .total != 0]:
>>     print r.title
>>     print '      ', r.sources['pmceurope'].metrics.total, '
    pmceurope citations'
>>     print '      ', r.sources['wikipedia'].metrics.total, '
    wikipedia citations'
Architecting the Future of Research Communication: Building the
    Models and Analytics for an Open Access Future
```

```
      1 pmceurope citations
      1 wikipedia citations
Expert Failure: Re-evaluating Research Assessment
      0 pmceurope citations
      1 wikipedia citations
LabTrove: A Lightweight, Web Based, Laboratory "Blog" as a Route
    towards a Marked Up Record of Work in a Bioscience Research
    Laboratory
      0 pmceurope citations
      1 wikipedia citations
The lipidome and proteome of oil bodies from Helianthus annuus (
    common sunflower)
      2 pmceurope citations
      1 wikipedia citations
```

As discussed above, this shows that the DET instance, while it provides information on a greater number of DOIs, has less complete data on each DOI at this stage. Only four of the 24 responses have Wikipedia references. You can change the code to look at the full set of 24, which shows only sparse data. The PLOS Lagotto instance provides more data but only on PLOS articles. However, it does provide data on all PLOS articles, going back earlier than the set returned by the DET instance. We can collect the set of articles from the profile published by PLOS.

```
>> plos_dois = []
>> for doi in DOIs:
>>     # Quick and dirty, should check Crossref API for publisher
>>     if doi.startswith('10.1371'):
>>         plos_dois.append(doi)
>> len(plos_dois)
7

>> plos_alms = pyalm.get_alm(plos_dois, info='detail', instance='
    plos').get('articles')
>> for article in plos_alms:
>>     print article.title
>>     print '      ', article.sources['crossref'].metrics.total, '
    Crossref citations'
>>     print '      ', article.sources['twitter'].metrics.total, '
    tweets'
Architecting the Future of Research Communication: Building the
    Models and Analytics for an Open Access Future
      2 Crossref citations
      48 tweets
Expert Failure: Re-evaluating Research Assessment
      3 Crossref citations
      130 tweets
LabTrove: A Lightweight, Web Based, Laboratory "Blog" as a Route
    towards a Marked Up Record of Work in a Bioscience Research
    Laboratory
      6 Crossref citations
```

```
      1 tweets
More Than Just Access: Delivering on a Network-Enabled Literature
      4 Crossref citations
      95 tweets
Article-Level Metrics and the Evolution of Scientific Impact
      24 Crossref citations
      5 tweets
Optimal Probe Length Varies for Targets with High Sequence
    Variation: Implications for Probe
Library Design for Resequencing Highly Variable Genes
      2 Crossref citations
      1 tweets
Covalent Attachment of Proteins to Solid Supports and Surfaces via
     Sortase-Mediated Ligation
      40 Crossref citations
      0 tweets
```

From the previous examples we know that we can obtain information on citing articles and tweets associated with these 66 articles. From that initial corpus we now have a collection of up to 86 related articles (cited and citing), a few hundred tweets that refer to (some of) those articles, and perhaps 500 people if we include authors of both articles and tweets. Note how for each of these links our query is limited, so we have a subset of all the related objects and agents. At this stage we probably have duplicate articles (one article might cite multiple in our set of seven) and duplicate people (authors in common between articles and authors who are also tweeting).

These data could be used for network analysis, to build up a new corpus of articles (by following the citation links), or to analyze the links between authors and those tweeting about the articles. We do not pursue an in-depth analysis here, but will gather the relevant objects, deduplicate them as far as possible, and count how many we have in preparation for future analysis.

```
>> # Collect all citing DOIs & author names from citing articles
>> citing_dois = []
>> citing_authors = []
>> for article in plos_alms:
>>     for cite in article.sources['crossref'].events:
>>         citing_dois.append(cite['event']['doi'])
>>         # Use 'extend' because the element is a list
>>         citing_authors.extend(cite['event_csl']['author'])
>> print '\nBefore de-deduplication:'
>> print '   ', len(citing_dois), 'DOIs'
>> print '   ', len(citing_authors), 'citing authors'
>>
>> # Easiest way to deduplicate is to convert to a Python set
>> citing_dois = set(citing_dois)
>> citing_authors = set([author['given'] + author['family'] for
```

```
    author in citing_authors])
>> print '\nAfter de-deduplication:'
>> print '  ', len(citing_dois), 'DOIs'
>> print '  ', len(citing_authors), 'citing authors'

Before de-deduplication:
  81 DOIs
  346 citing authors

After de-deduplication:
  78 DOIs
  278 citing authors
```

```
>> # Collect all tweets, usernames; check for ORCIDs
>> tweet_urls = set()
>> twitter_handles = set()
>> for article in plos_alms:
>>     for tweet in article.sources['twitter'].events:
>>         tweet_urls.add(tweet['event_url'])
>>         twitter_handles.add(tweet['event']['user'])
>> # No need to explicitly deduplicate as we created sets directly
>> print len(tweet_urls), 'tweets'
>> print len(twitter_handles), 'Twitter users'

280 tweets
210 Twitter users
```

It could be interesting to look at which Twitter users interact most with the articles associated with this ORCID profile. To do that we would need to create not a set but a list, and then count the number of duplicates in the list. The code could be easily modified to do this. Another useful exercise would be to search ORCID for profiles corresponding to citing authors. The best way to do this would be to obtain ORCIDs associated with each of the citing articles. However, because ORCID data are sparse and incomplete, there are two limitations here. First, the author may not have an ORCID. Second, the article may not be explicitly linked to another article. Try searching ORCID for the DOIs associated with each of the citing articles.

In this case we will look to see how many of the Twitter handles discussing these articles are associated with an ORCID profile we can discover. This in turn could lead to more profiles and more cycles of analysis to build up a network of researchers interacting through citation and on Twitter. Note that we have inserted a delay between calls. This is because we are making a larger number of API calls (one for each Twitter handle). It is considered polite to keep the pace at which calls are made to an API to a reasonable level. The ORCID API does not post suggested limits at the moment, but delaying for a second between calls is reasonable.

```
>> tweet_orcids = []
>> for handle in twitter_handles:
>>     orc = twitter2orcid(handle)
>>     if orc:
>>         tweet_orcids.append(orc)
>>     time.sleep(1) # wait one second between each call to the
    ORCID API
>> print len(tweet_orcids)
12
```

In this case we have identified 12 ORCID profiles that we can link positively to tweets about this set of articles. This is a substantial underestimate of the likely number of ORCIDs associated with these tweets. However, relatively few ORCIDs have Twitter accounts registered as part of the profile. To gain a broader picture a search and matching strategy would need to be applied. Nevertheless, for these 12 we can look more closely into the profiles.

The first step is to obtain the actual profile information for each of the 12 ORCIDs that we have found. Note that at the moment what we have is the ORCIDs themselves, not the retrieved profiles.

```
>> orcs = []
>> for id in tweet_orcids:
>>     orcs.append(orcid.get(id))
```

With the profiles retrieved we can then take a look at who they are, and check that we do in fact have sensible Twitter handles associated with them. We could use this to build up the network of related authors and Twitter users for further analysis.

```
>> for orc in orcs:
>>     i = [('twitter.com' in website.url) for website in orc.
    researcher_urls].index(True)
>>     twitter_url = orc.researcher_urls[i].url
>>     print orc.given_name, orc.family_name, orc.orcid,
    twitter_url
Catriona MacCallum 0000-0001-9623-2225 http://twitter.com/catmacOA
John Dupuis 0000-0002-6066-690X https://twitter.com/dupuisj
Johannes Velterop 0000-0002-4836-6568 https://twitter.com/
    Villavelius
Stuart Lawson 0000-0002-1972-8953 https://twitter.com/Lawsonstu
Nelson Piedra 0000-0003-1067-8707 http://www.twitter.com/nopiedra
Iryna Kuchma 0000-0002-2064-3439 https://twitter.com/irynakuchma
Frank Huysmans 0000-0002-3468-9032 https://twitter.com/fhuysmans
Salvatore Salvi VICIDOMINI 0000-0001-5086-7401 https://twitter.com
    /SalViVicidomini
William Gunn 0000-0002-3555-2054 http://twitter.com/mrgunn
Stephen Curry 0000-0002-0552-8870 https://twitter.com/
    Stephen_Curry
Cameron Neylon 0000-0002-0068-716X http://twitter.com/
```

```
    cameronneylon
Graham Steel 0000-0003-4681-8011 https://twitter.com/McDawg
```

2.9 Working with the graph of relationships

▶ See Chapter 8.

In the above examples we started with the profile of an individual, used this to create a corpus of works, which in turn led us to other citing works (and their authors) and commentary about those works on Twitter (and the people who wrote those comments). Along the way we built up a graph of relationships between objects and people. In this section we will look at this model of the data and how it reveals limitations and strengths of these forms of data and what can be done with them.

2.9.1 Citation links between articles

A citation in a research article (or a policy document or working paper) defines a relationship between that citing article and the cited article. The exact form of the relationship is generally poorly defined, at least at the level of large-scale data sets. A citation might be referring to previous work, indicating the source of data, or supporting (or refuting) an idea. While efforts have been made to codify citation types, they have thus far gained little traction.

In our example we used a particular data source (Crossref) for information about citations. As previously discussed, this will give different results than other sources (such as Thomson Reuters, Scopus, or Google Scholar) because other sources look at citations from a different set of articles and collect them in a different way. The completeness of the data will always be limited. We could use the data to clearly connect the citing articles and their authors because author information is generally available in bibliographic metadata. However, we would have run into problems if we had only had names. ORCIDs can provide a way to uniquely identify authors and ensure that our graph of relationships is clean.

A *citation* is a reference from an object of one type to an object of the same type. We also sought to link social media activity with specific articles. Rather than a link between objects that are the same (articles) we started to connect different kinds of objects together. We are also expanding the scope of the communities (i.e., people) that might be involved. While we focused on the question of which Twitter handles were connected with researchers, we could

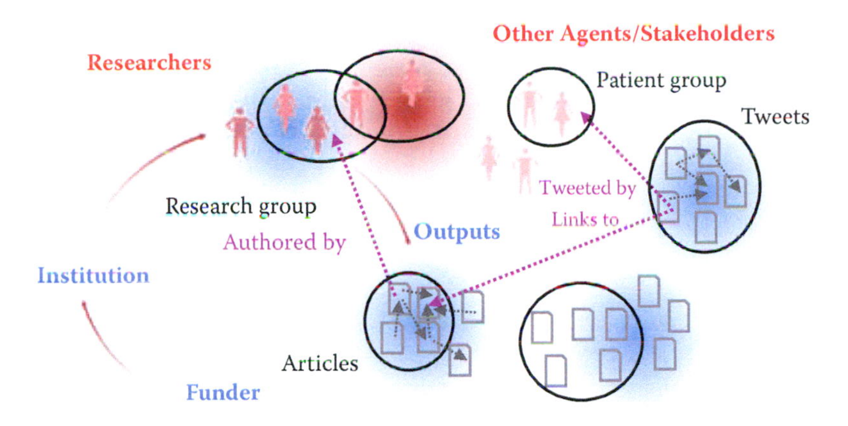

Figure 2.4. A functional view of proxies and relationships

just as easily have focused on trying to discover which comments came from people who are *not* researchers.

We used the Lagotto API at PLOS to obtain this information. The PLOS API in turn depends on the Twitter Search API. A tweet that refers explicitly to a research article, perhaps via a Crossref DOI, can be discovered, and a range of services do these kinds of checks. These services generally rely either on Twitter Search or, more generally, on a search of "the firehose," a dump of all Twitter data that are available for purchase. The distinction is important because Twitter Search does not provide either a complete or a consistent set of results. In addition, there will be many references to research articles that do not contain a unique identifier, or even a link. These are more challenging to discover. As with citations, the completeness of any data set will always be limited.

However, the set of all tweets is a more defined set of objects than the set of all "articles." Twitter is a specific social media service with a defined scope. "Articles" is a broad class of objects served by a very wide range of services. Twitter is clearly a subset of all discussions and is highly unlikely to be representative of "all discussions." Equally the set of all objects with a Crossref DOI, while defined, is unlikely to be representative of all articles.

Expanding on Figure 2.3, we show in Figure 2.4 agents and actors (people) and outputs. We place both agents and outputs into categories that may be more or less well defined. In practice our analysis is limited to those objects that we discover by using some "selector" (circles in this diagram), which may or may not have a close correspondence with the "real" categories (shown with graded shapes). Our aim is to identify, aggregate, and in some cases count the relationships between and within categories of objects; for instance, citations are relationships between formal research outputs. A tweet may have a relationship ("links to") with a specific formally published research output. Both tweets and formal research outputs relate to specific agents ("authors") of the content.

2.9.2 Categories, sources, and connections

We can see in this example a distinction between categories of objects of interest (articles, discussions, people) and sources of information on subsets of those categories (Crossref, Twitter, ORCID). Any analysis will depend on one or more data sources, and in turn be limited by the coverage of those data sources. The selectors used to generate data sets from these sources will have their own limitations.

Similar to a query on a structured data set, the selector itself may introduce bias. The crucial difference between filtering on a comprehensive (or at least representative) data set and the data sources we are discussing here is that these data sources are by their very nature incomplete. Survey data may include biases introduced in the way that the survey itself is structured or the sampling is designed, but the intent is to be comprehensive. Many of these new forms of data make no attempt to be comprehensive or avowedly avoid such an attempt.

Understanding this incompleteness is crucial to understanding the forms of inference that can be made from these data. Sampling is only possible within a given source or corpus, and this limits the conclusions that can be drawn to the scope of that corpus. It is frequently possible to advance a plausible argument to claim that such findings are more broadly applicable, but it is crucial to avoid assuming that this is the case. In particular, it is important to be clear about what data sources a finding applies to and where the boundary between the strongly evidenced finding and a claim about its generalization lies. Much of the literature on scholarly communications and research impact is poor on this point.

If this is an issue in identifying the objects of interest, it is even more serious when seeking to identify the relationships between them, which are, after all generally the thing of interest. In some cases there are reasonably good sources of data between objects of the same class (at least those available from the same data sources) such as citations between journal articles or links between tweets.

However, as illustrated in this chapter, detecting relationships between tweets and articles is much more challenging.

These issues can arise both due to the completeness of the data source itself (e.g., ORCID currently covers only a subset of researchers; therefore, the set of author–article relationships is limited) or due to the challenges of identification (e.g., in the Twitter case above) or due to technical limitations at source (the difference between the Twitter search API and the firehose). In addition, because the source data and the data services are both highly dynamic and new, there is often a mismatch. Many services tracking Twitter data only started collecting data relatively recently. There is a range of primary and secondary data sources working to create more complete data sets. However, once again it is important to treat all of these data as sparse and limited as well as highly dynamic and changeable.

2.9.3 Data availability and completeness

With these caveats in hand and the categorization discussed above, we can develop a mapping of what data sources exist, what objects those data sources inform us about, the completeness of those data sources, and how well the relationships between the different data sources are tracked. Broadly speaking, data sources concern themselves with either agents (mostly people) or objects (articles, books, tweets, posts), while additionally providing additional data about the relationships of the agents or objects that they describe with other objects or agents.

The five broad types of data described above are often treated as ways of categorizing the data source. They are more properly thought of as relationships between objects, or between objects and agents. Thus, for example, citations are relationships between articles; the tweets that we are considering are actually relationships between specific Twitter posts and articles; and "views" are an event associating a reader (agent) with an article. The last case illustrates that often we do not have detailed information on the relationship but merely a count of them. Relationships between agents (such as co-authorship or group membership) can also be important.

With this framing in hand, we can examine which types of relationships we can obtain data on. We need to consider both the *quality* of data available and the *completeness* of the data availability. These metrics are necessarily subjective and any analysis will be a personal view of a particular snapshot in time. Nevertheless, some major trends are available.

We have growing and improving data on the relationships between a wide range of objects and agents and traditional scholarly outputs. Although it is sparse and incomplete in many places, *nontraditional* information on *traditional* outputs is becoming more available and increasingly rich. By contrast, references from traditional outputs to nontraditional outputs are weaker and data that allow us to understand the relationships between nontraditional outputs is very sparse.

In the context of the current volume, a major weakness is our inability to triangulate around people and communities. While it may be possible to collect a set of co-authors from a bibliographic data source and to identify a community of potential research users on Twitter or Facebook, it is extremely challenging to connect these different sets. If a community is discussing an article or book on social media, it is almost impossible to ascertain whether the authors (or, more generically, interested parties such as authors of cited works or funders) are engaged in that conversation.

2.9.4 The value of sparse dynamic data

Two clear messages arise from our analysis. These new forms of data are incomplete or sparse, both in quality and in coverage, and they change. A data source that is poor today may be much improved tomorrow. A query performed one minute may give different results the next. This can be both a strength and a weakness: data are up to the minute, giving a view of relationships as they form (and break), but it makes ensuring consistency within analyses and across analyses challenging. Compared to traditional surveys, these data sources cannot be relied on as either representative samples or to be stable.

A useful question to ask, therefore, is what kind of statements these data *can* support. Questions like this will be necessarily different from the questions that can be posed with high-quality survey data. More often they provide an existence proof that something has happened—but they cannot, conversely, show that it has not. They enable some forms of comparison and determination of the characteristics of activity in some cases.

Provide evidence that ... Because much of the data that we have is sparse, the absence of an indicator cannot reliably be taken to mean an absence of activity. For example, a lack of Mendeley bookmarks may not mean that a paper is not being saved by researchers, just that those who do save the article are not using Mendeley to do it. Similarly, a lack of tweets about an article does not mean the article is not being discussed. But we can use the data that do exist to show that some activity is occurring. Here are some examples:

- Provide evidence that relevant communities are aware of a specific paper. I identified the fact that a paper by Jewkes et al. [191] was mentioned by crisis centers, sexual health organizations, and discrimination support groups in South Africa when I was looking for University of Cape Town papers that had South African Twitter activity using Altmetric.com.

- Provide evidence that a relatively under-cited paper is having a research impact. There is a certain kind of research article, often a method description or a position paper, that is influential without being (apparently) heavily cited. For instance, the *PLoS One* article by Shen et al. [341] has a respectable 14,000 views and 116 Mendeley bookmarks, but a relatively (for the number of views) small number of WoS citations (19) compared to, say, another article, by Leahy et al. [231] and also in *PLoS One*, that is similar in age and number of views but has many more citations.

- Provide evidence of public interest in some topic. Many articles at the top of lists ordered by views or social media mentions are of ephemeral (or prurient) interest—the usual trilogy of sex, drugs, and rock and roll. However, if we dig a little deeper, a wide range of articles surface, often not highly cited but clearly of wider interest. For example, an article on Y-chromosome distribution in Afghanistan [146] has high page views and Facebook activity among papers with a Harvard affiliation but is not about sex, drugs, nor rock and roll. Unfortunately, because this is Facebook data we cannot see *who* is talking about it, which limits our ability to say *which* groups are talking about it, which could be quite interesting.

Compare ... Comparisons using social media or download statistics need real care. As noted above, the data are sparse so it is important that comparisons are fair. Also, comparisons need to be on the basis of something that the data can actually tell you: for example, "which article is discussed more by this online community," *not* "which article is discussed more."

- Compare the extent to which these articles are discussed by this online patient group, or possibly specific online communities in general. Here the online communities might be a proxy for a broader community, or there might be a specific interest in knowing whether the dissemination strategy reaches this community. It is clear that in the longer term social media will be a substantial pathway for research to reach a wide range of audiences, and understanding which communities are discussing what research will help us to optimize the communication.

- Compare the readership of these articles in these countries. One thing that most data sources are weak on at the moment is demographics, but in principle the data are there. Are these articles that deal with diseases of specific areas actually being viewed by readers in those areas? If not, why not? Do they have Internet access, could lay summaries improve dissemination, are they going to secondary online sources instead?

- Compare the communities discussing these articles online. Is most conversation driven by science communicators or by researchers? Are policymakers, or those who influence them, involved? What about practitioner communities? These comparisons require care, and simple counting rarely provides useful information. But understanding which people within which networks are driving conversations can give insight into who is aware of the work and whether it is reaching target audiences.

What flavor is it? Priem et al. [310] provide a thoughtful analysis of the PLOS Article Level Metrics data set. They used principal component analysis to define different "flavors of impact" based on the way different combinations of signals seemed to point to different kinds of interest. Many of the above use cases are variants on this theme—what kind of article is this? Is it a policy piece, of public interest? Is it of interest to a niche research community or does it have wider public implications? Is it being used in education or in health practice? And to what extent are these different kinds of use independent from each other?

It is important to realize that these kinds of data are proxies of things that we do not truly understand. They are signals of the flow of information down paths that we have not mapped. To me this is the most exciting possibility and one we are only just starting to explore. What can these signals tell us about the underlying pathways down which information flows? How do different combinations of signals tell us about who is using that information now, and how they might be applying it in the future? Correlation analysis cannot answer these questions, but more sophisticated approaches might. And with that information in hand we could truly *design* scholarly communication systems to maximize their reach, value, and efficiency.

2.10 Bringing it together: Tracking pathways to impact

Collecting data on research outputs and their performance clearly has significant promise. However, there are a series of substantial challenges in how best to use these data. First, as we have seen, it is sparse and patchy. Absence of evidence cannot be taken as evidence of absence. But, perhaps more importantly, it is unclear in many cases what these various proxies actually *mean*. Of course this is also true of more familiar indicators like citations.

Finally, there is a challenge in how to effectively analyze these data. The sparse nature of the data is a substantial problem in itself, but in addition there are a number of significantly confounding effects. The biggest of these is time. The process of moving research outputs and their use online is still proceeding, and the uptake and penetration of online services and social media by researchers and other relevant communities has increased rapidly over the past few years and will continue to do so for some time.

These changes are occurring on a timescale of months, or even weeks, so any analysis must take into account how those changes may contribute to any observed signal. Much attention has focused on how different quantitative proxies correlate with each other. In essence this has continued the mistake that has already been made with citations. Focusing on proxies themselves implicitly makes the assumption that it is the proxy that matters, rather than the underlying process that is actually of interest. Citations are irrelevant; what matters is the influence that a piece of research has had. Citations are merely a proxy for a particular slice of influence, a (limited) indicator of the underlying process in which a research output is used by other researchers.

Of course, these are common challenges for many "big data" situations. The challenge lies in using large, but disparate and messy, data sets to provide insight while avoiding the false positives that will arise from any attempt to mine data blindly for correlations. Using the appropriate models and tools and careful validation of findings against other sources of data are the way forward.

2.10.1 Network analysis approaches

One approach is to use these data to dissect and analyze the (visible) network of relationships between agents and objects. This approach can be useful in defining how networks of collaborators change over time, who is in contact with whom, and how outputs are related to each other. This kind of analysis has been productive with citation graphs (see Eigenfactor for an example) as well as with small-scale analysis of grant programs (see, for instance, the Lattes analysis of the network grant program).

Network analysis techniques and visualization are covered in Chapter 8 (on networks) and clustering and categorization in Chapter 6 (on machine learning). Networks may be built up from any combination of outputs, actors/agents, and their relationships to each other. Analyses that may be particularly useful are those searching for highly connected (proxy for influential) actors or outputs, clustering to define categories that emerge from the data itself (as opposed to external categorization) and comparisons between networks, both between those built from specific nodes (people, outputs) and between networks that are built from data relating to different time frames.

Care is needed with such analyses to make sure that comparisons are valid. In particular, when doing analyses of different time frames, it is important to compare any change in the network characteristics that are due to general changes over time as opposed to specific changes. As noted above, this is particularly important with networks based on social media data, as any networks are likely to have increased in size and diversity over the past few years as more users interested in research have joined. It is important to distinguish in these cases between changes relating to a specific intervention or treatment and those that are environmental. As with any retrospective analysis, a good counterfactual sample is required.

2.10.2 Future prospects and new data sources

As the broader process of research moves online we are likely to have more and more information on what is being created, by whom, and when. As access to these objects increases, both through provision of open access to published work and through increased data sharing, it will become more and more feasible to mine the objects themselves to enrich the metadata. And finally, as the use of unique identifiers increases for both outputs and people, we will be able to cross-reference across data sources much more strongly.

Much of the data currently being collected is of poor quality or is inconsistently processed. Major efforts are underway to develop standards and protocols for initial processing, particularly for page view and usage data. Alongside efforts such as the Crossref DOI Event Tracker Service to provide central clearing houses for data, both consistency and completeness will continue to rise, making new and more comprehensive forms of analysis feasible.

Perhaps the most interesting prospect is new data that arise as more of the outputs and processes of research move online. As the availability of data outputs, software products, and even potentially the raw record of lab notebooks increases, we will have opportunities to query how (and how much) different reagents, techniques, tools, and instruments are being used. As the process of policy development and government becomes more transparent and better connected, it will be possible to watch in real time as research has its impact on the public sphere. And as health data moves online there will be opportunities to see how both chemical and behavioral interventions affect health outcomes in real time.

In the end all of this data will also be grist to the mill for further research. For the first time we will have the opportunity to treat the research enterprise as a system that is subject to optimization and engineering. Once again the challenges of what it is we are seeking to optimize for are questions that the data itself cannot answer, but in turn the data can better help us to have the debate about what matters.

2.11 Summary

The term *research impact* is difficult and politicized, and it is used differently in different areas. At its root it can be described as the change that a particular part of the research enterprise (e.g., research project, researcher, funding decision, or institute) makes in the world. In this sense, it maps well to standard approaches in the social sciences that seek to identify how an intervention has led to change.

The link between "impact" and the distribution of limited research resources makes its definition highly political. In fact, there are many forms of impact, and the different pathways to different kinds of change, further research, economic growth, improvement in health outcomes, greater engagement of citizenry, or environmental change may have little in common beyond our interest in how they can be traced to research outputs. Most public policy on research investment has avoided the difficult question of which impacts are most important.

In part this is due to the historical challenges of providing evidence for these impacts. We have only had good data on formal research outputs, primarily journal articles, and measures have focused on naïve metrics such as productivity or citations, or on qualitative peer review. Broader impacts have largely been evidenced through case studies, an expensive and nonscalable approach.

The move of research processes online is providing much richer and more diverse information on how research outputs are used and disseminated. We have the prospect of collecting much more information around the performance and usage of traditional research outputs as well as greater data on the growing diversity of nontraditional research outputs that are now being shared.

It is possible to gain quantitative information on the numbers of people looking at research, different groups talking about research (in different places), those citing research in different places, and recommendations and opinions on the value of work. These data are sparse and incomplete and its use needs to acknowledge these limitations, but it is nonetheless possible to gain new and valuable insights from analysis.

Much of this data is available from web services in the form of application programming interfaces. Well-designed APIs make it easy to search for, gather, and integrate data from multiple sources. A key aspect of successfully integrating data is the effective use and application of unique identifiers across data sets that allow straightforward cross-referencing. Key among the identifiers currently being used are ORCIDs to uniquely identify researchers and DOIs, from both Crossref and increasingly DataCite, to identify research outputs. With good cross-referencing it is possible to obtain rich data sets that can be used as inputs to many of the techniques described elsewhere in the book.

The analysis of this new data is a nascent field and the quality of work done so far has been limited. In my view there is a substantial opportunity to use these rich and diverse data sets to treat the underlying question of how research outputs flow from the academy to their sites of use. What are the underlying processes that lead to various impacts? This means treating these data sets as time domain signals that can be used to map and identify the underlying processes. This approach is appealing because it offers the promise of probing the actual process of knowledge diffusion while making fewer assumptions about what we think is happening.

2.12 Resources

We talked a great deal here about how to access publications and other resources via their DOIs. Paskin [297] provides a nice summary of the problems that DOIs solve and how they work.

ORCIDs are another key piece of this puzzle, as we have seen throughout this chapter. You might find some of the early articles describing the need for unique author IDs useful, such as Bourne et al. [46], as well as more recent descriptions [145]. More recent initiatives on expanding the scope of identifiers to materials and software have also been developed [24].

More general discussions of the challenges and opportunities of using metrics in research assessment may be found in recent reports such as the HEFCE Expert Group Report [405], and I have covered some of the broader issues elsewhere [274].

There are many good introductions to web scraping using BeautifulSoup and other libraries as well as API usage in general. Given the pace at which APIs and Python libraries change, the best and most up to date source of information is likely to be a web search.

In other settings, you may be concerned with assigning DOIs to data that you generate yourself, so that you and others can easily and reliably refer to and access that data in their own work. Here we face an embarrassment of riches, with many systems available that each meet different needs. Big data research communities such as climate science [404], high-energy physics [304], and astronomy [367] operate their own specialized infrastructures that you are unlikely to require. For small data sets, Figshare [122] and DataCite [89] are often used. The Globus publication service [71] permits an institution or community to build their own publication system.

2.13 Acknowledgements and copyright

Section 2.3 is adapted in part from Neylon et al. [275], copyright International Development Research Center, Canada, used here under a Creative Commons Attribution v 4.0 License.

Section 2.9.4 is adapted in part from Neylon [273], copyright PLOS, used here under a Creative Commons Attribution v 4.0 License.

Chapter 3

Record Linkage

Joshua Tokle and Stefan Bender

Big data differs from survey data in that it is typically necessary to combine data from multiple sources to get a complete picture of the activities of interest. Although computer scientists tend to simply "mash" data sets together, social scientists are rightfully concerned about issues of missing links, duplicative links, and erroneous links. This chapter provides an overview of traditional rule-based and probabilistic approaches, as well as the important contribution of machine learning to record linkage.

3.1 Motivation

Big data offers social scientists great opportunities to bring together many different types of data, from many different sources. Merging different data sets provides new ways of creating population frames that are generated from the digital traces of human activity rather than, say, tax records. These opportunities, however, create different kinds of challenges from those posed by survey data. Combining information from different sources about an individual, business, or geographic entity means that the social scientist must determine whether or not two entities on two different files are the same. This determination is not easy. In the UMETRICS data, if data are to be used to measure the impact of research grants, is David A. Miller from Stanford, CA, the same as David Andrew Miller from Fairhaven, NJ, in a list of inventors? Is Google the same as Alphabet if the productivity and growth of R&D-intensive firms is to be studied? Or, more generally, is individual A the same person as the one who appears on a list of terrorists that has been compiled? Does the product that a customer is searching for match the products that business B has for sale?

The consequences of poor record linkage decisions can be substantial. In the business arena, Christen reports that as much as 12% of business revenues are lost due to bad linkages [76]. In the security arena, failure to match travelers to a "known terrorist" list may result in those individuals entering the country, while overzealous matching could lead to numbers of innocent citizens being detained. In finance, incorrectly detecting a legitimate purchase as a fraudulent one annoys the customer, but failing to identify a thief will lead to credit card losses. Less dramatically, in the scientific arena when studying patenting behavior, if it is decided that two inventors are the same person, when in fact they are not, then records will be incorrectly grouped together and one researcher's productivity will be overstated. Conversely, if the records for one inventor are believed to correspond to multiple individuals, then that inventor's productivity will be understated.

This chapter discusses current approaches to joining multiple data sets together—commonly called *record linkage*. Other names associated with record linkage are entity disambiguation, entity resolution, co-reference resolution, statistical matching, and data fusion, meaning that records which are linked or co-referent can be thought of as corresponding to the same underlying entity. The number of names is reflective of a vast literature in social science, statistics, computer science, and information sciences. We draw heavily here on work by Winkler, Scheuren, and Christen, in particular [76, 77, 165]. To ground ideas, we use examples from a recent paper examining the effects of different algorithms on studies of patent productivity [387].

3.2 Introduction to record linkage

There are many reasons to link data sets. Linking to existing data sources to solve a measurement need instead of implementing a new survey results in cost savings (and almost certainly time savings as well) and reduced burden on potential survey respondents. For some research questions (e.g., a survey of the reasons for death of a longitudinal cohort of individuals) a new survey may not be possible. In the case of administrative data or other automatically generated data, the sample size is much greater than would be possible from a survey.

Record linkage can be used to compensate for data quality issues. If a large number of observations for a particular field are missing, it may be possible to link to another data source to fill in the missing values. For example, survey respondents might not want to share a sensitive datum like income. If the researcher has access to an official administrative list with income data, then those values can be used to supplement the survey [5].

Record linkage is often used to create new longitudinal data sets by linking the same entities over time [190]. More generally, linking separate data sources makes it possible to create a combined data set that is richer in coverage and measurement than any of the individual data sources [4].

Example: The Administrative Data Research Network

The UK's Administrative Data Research Network* (ADRN) is a major investment by the United Kingdom to "improve our knowledge and understanding of the society we live in . . . [and] provide a sound base for policymakers to decide how to tackle a range of complex social, economic and environmental issues" by linking administrative data from a variety of sources, such as health agencies, court records, and tax records in a confidential environment for approved researchers. The linkages are done by trusted third-party providers. [103]

★ "Administrative data" typically refers to data generated by the administration of a government program, as distinct from deliberate survey collection.

Linking is straightforward if each entity has a corresponding unique identifier that appears in the data sets to be linked. For example, two lists of US employees may both contain Social Security numbers. When a unique identifier exists in the data or can be created, no special techniques are necessary to join the data sets.

If there is no unique identifier available, then the task of identifying unique entities is challenging. One instead relies on fields that only partially identify the entity, like names, addresses, or dates of birth. The problem is further complicated by poor data quality and duplicate records, issues well attested in the record linkage literature [77] and sure to become more important in the context of big data. Data quality issues include input errors (typos, misspellings, truncation, extraneous letters, abbreviations, and missing values) as well as differences in the way variables are coded between the two data sets (age versus date of birth, for example). In addition to record linkage algorithms, we will discuss different data preprocessing steps that are necessary first steps for the best results in record linkage.

To find all possible links between two data sets it would be necessary to compare each record of the first data set with each record of the second data set. The computational complexity of this approach grows quadratically with the size of the data—an important consideration, especially in the big data context. To compensate for this complexity, the standard second step in record linkage, after preprocessing, is indexing or blocking, which creates subsets of similar records and reduces the total number of comparisons.

The outcome of the matching step is a set of predicted links—record pairs that are likely to correspond to the same entity. After these are produced, the final stage of the record linkage process is to evaluate the result and estimate the resulting error rates. Unlike other areas of application for predictive algorithms, ground truth or gold standard data sets are rarely available. The only way to create a reliable truth data set sometimes is through an expensive clerical review process that may not be viable for a given application. Instead, error rates must be estimated.

An input data set may contribute to the linked data in a variety of ways, such as increasing coverage, expanding understanding of the measurement or mismeasurement of underlying latent variables, or adding new variables to the combined data set. It is therefore important to develop a well-specified reason for linking the data sets, and to specify a loss function to proxy the cost of false negative matches versus false positive matches that can be used to guide match decisions. It is also important to understand the coverage of the different data sets being linked because differences in coverage may result in bias in the linked data. For example, consider the problem of linking Twitter data to a sample-based survey—elderly adults and very young children are unlikely to use Twitter and so the set of records in the linked data set will have a youth bias, even if the original sample was representative of the population. It is also essential to engage in critical thinking about what latent variables are being captured by the measures in the different data sets—an "occupational classification" in a survey data set may be very different from a "job title" in an administrative record or a "current position" in LinkedIn data.

▶ This topic is discussed in more detail in Chapter 10.

Example: Employment and earnings outcomes of doctoral recipients

A recent paper in *Science* matched UMETRICS data on doctoral recipients to Census data on earnings and employment outcomes. The authors note that some 20% of the doctoral recipients are not matched for several reasons: (i) the recipient does not have a job in the US, either for family reasons or because he/she goes back to his/her home country; (ii) he/she starts up a business rather than choosing employment; or (iii) it is not possible to uniquely match him/her to a Census Bureau record. They correctly note that there may be biases introduced in case (iii), because Asian names are more likely duplicated and harder to uniquely match [415]. Improving the linkage algorithm would increase the estimate of the effects of investments in research and the result would be more accurate.

Comparing the kinds of heterogeneous records associated with big data is a new challenge for social scientists, who have traditionally used a technique first developed in the 1960s to apply computers to the problem of medical record linkage. There is a reason why this approach has survived: it has been highly successful in linking survey data to administrative data, and efficient implementations of this algorithm can be applied at the big data scale. However, the approach is most effective when the two files being linked have a number of fields in common. In the new landscape of big data, there is a greater need to link files that have few fields in common but whose noncommon fields provide additional predictive power to determine which records should be linked. In some cases, when sufficient training data can be produced, more modern machine learning techniques may be applied.

The canonical record linkage workflow process is shown in Figure 3.1 for two data files, A and B. The goal is to identify all pairs of records in the two data sets that correspond to the same underlying individual. One approach is to compare all data units from file A

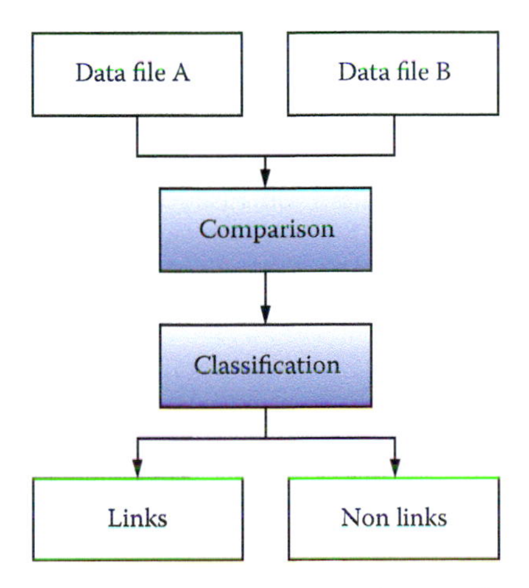

Figure 3.1. The preprocessing pipeline

with all units in file B and classify all of the comparison outcomes to decide whether or not the records match. In a perfect statistical world the comparison would end with a clear determination of links and nonlinks.

Alas, a perfect world does not exist, and there is likely to be noise in the variables that are common to both data sets and that will be the main identifiers for the record linkage. Although the original files A and B are the starting point, the identifiers must be preprocessed before they can be compared. Determining identifiers for the linkage and deciding on the associated cleaning steps are extremely important, as they result in a necessary reduction of the possible search space.

In the next section we begin our overview of the record linkage process with a discussion of the main steps in data preprocessing. This is followed by a section on approaches to record linkage that includes rule-based, probabilistic, and machine learning algorithms. Next we cover classification and evaluation of links, and we conclude with a discussion of data privacy in record linkage.

3.3 Preprocessing data for record linkage

As noted in the introductory chapter, all data work involves preprocessing, and data that need to be linked is no exception. Preprocessing refers to a workflow that transforms messy, noisy, and unstructured data into a well-defined, clearly structured, and quality-tested data set. Elsewhere in this book, we discuss general strategies for data preprocessing. In this section, we focus specifically on preprocessing steps relating to the choice of input fields for the record linkage algorithm. Preprocessing for any kind of a new data set is a complex and time-consuming process because it is "hands-on": it requires judgment and cannot be effectively automated. It may be tempting to minimize this demanding work under the assumption that the record linkage algorithm will account for issues in the data, but it is difficult to overstate the value of preprocessing for record linkage quality. As Winkler notes: "In situations of reasonably high-quality data, preprocessing can yield a greater improvement in matching efficiency than string comparators and 'optimized' parameters. In some situations, 90% of the improvement in matching efficiency may be due to preprocessing" [406].

> ► This topic (quality of data, preprocessing issues) is discussed in more detail in Section 1.4.

The first step in record linkage is to develop link keys, which are the record fields that will be used to predict link status. These can include common identifiers like first and last name. Survey and administrative data sets may include a number of clearly identifying variables like address, birth date, and sex. Other data sets, like transaction records or social media data, often will not include address or birth date but may still include other identifying fields like occupation, a list of interests, or connections on a social network. Consider this chapter's illustrative example of the US Patent and Trademark Office (USPTO) data [387]:

> USPTO maintains an online database of all patents issued in the United States. In addition to identifying information about the patent, the database contains each patent's list of inventors and assignees, the companies, organizations, individuals, or government agencies to which the patent is assigned. ... However, inventors and assignees in the USPTO database are not given unique identification numbers, making it difficult to track inventors and assignees across their patents or link their information to other data sources.

There are some basic precepts that are useful when considering identifying fields. The more different values a field can take, the less likely it is that two randomly chosen individuals in the population will agree on those values. Therefore, fields that exhibit a wider range of values are more powerful as link keys: names are much better link keys than sex or year of birth.

Example: Link keys in practice

"A Harvard professor has re-identified the names of more than 40 percent of a sample of anonymous participants in a high-profile DNA study, highlighting the dangers that ever greater amounts of personal data available in the Internet era could unravel personal secrets. ... Of the 1,130 volunteers Sweeney and her team reviewed, about 579 provided zip code, date of birth and gender, the three key pieces of information she needs to identify anonymous people combined with information from voter rolls or other public records. Of these, Sweeney succeeded in naming 241, or 42 percent of the total. The Personal Genome Project confirmed that 97 percent of the names matched those in its database if nicknames and first name variations were included" [369].

Complex link keys like addresses can be broken down into components so that the components can be compared independently of one another. This way, errors due to data quality can be further isolated. For example, assigning a single comparison value to the complex fields "1600 Pennsylvania" and "160 Pennsylvania Ave" is less informative than assigning separate comparison values to the street number and street name portions of those fields. A record linkage algorithm that uses the decomposed field can make more nuanced distinctions by assigning different weights to errors in each component.

Sometimes a data set can include different variants of a field, like legal first name and nickname. In these cases match rates can be improved by including all variants of the field in the record comparison. For example, if only the first list includes both variants, and the second list has a single "first name" field that could be either a legal first name or a nickname, then match rates can be improved by comparing both variants and then keeping the better of the two comparison outcomes. It is important to remember, however, that some record linkage algorithms expect field comparisons to be somewhat independent. In our example, using the outcome from both comparisons as separate inputs into the probabilistic model we describe below may result in a higher rate of false negatives. If a record has the same value in the legal name and nickname fields, and if that value happens to agree with the first name field in the second file, then the agreement is being double-counted. By the same token, if a person in the first list has a nickname that differs significantly from their legal first name, then a comparison of that record to the corresponding record will unfairly penalize the outcome because at least one of those name comparisons will show a low level of agreement.

Preprocessing serves two purposes in record linkage. First, to correct for issues in data quality that we described above. Second, to account for the different ways that the input files were generated, which may result in the same underlying data being recorded on different scales or according to different conventions.

Once preprocessing is finished, it is possible to start linking the records in the different data sets. In the next section we describe a technique to improve the efficiency of the matching step.

3.4 Indexing and blocking

There is a practical challenge to consider when comparing the records in two files. If both files are roughly the same size, say 100 records in the first and 100 records in the second file, then there are 10,000 possible comparisons, because the number of pairs is the product of the number of records in each file. More generally, if the number of records in each file is approximately n, then the total number of possible record comparisons is approximately n^2. Assuming that there are no duplicate records in the input files, the proportion of record comparisons that correspond to a link is only $1/n$. If we naively proceed with all n^2 possible comparisons, the linkage algorithm will spend the bulk of its time comparing records that are not matches. Thus it is possible to speed up record linkage significantly by skipping comparisons between record pairs that are not likely to be linked.

Indexing refers to techniques that determine which of the possible comparisons will be made in a record linkage application. The most used technique for indexing is blocking. In this approach you construct a "blocking key" for each record by concatenating fields or parts of fields. Two records with identical blocking keys are said to be in the same block, and only records in the same block are compared. This technique is effective because performing an exact comparison of two blocking keys is a relatively quick operation compared to a full record comparison, which may involve multiple applications of a fuzzy string comparator.

Example: Blocking in practice

Given two lists of individuals, one might construct the blocking key by concatenating the first letter of the last name and the postal code and then "blocking" on first character of last name and postal code. This reduces the total number of comparisons by only comparing those individuals in the two files who live in the same locality and whose last names begin with the same letter.

There are important considerations when choosing the blocking key. First, the choice of blocking key creates a potential bias in the linked data because true matches that do not share the same blocking key will not be found. In the example, the blocking strategy could fail to match records for individuals whose last name changed or who moved. Second, because blocking keys are compared exactly, there is an implicit assumption that the included fields will not have typos or other data entry errors. In practice, however, the blocking fields will exhibit typos. If those typos are not uniformly distributed over the population, then there is again the possibility of bias in the linked data set. One simple strategy for dealing with imperfect blocking keys is to implement multiple rounds of block-

▸ This topic is discussed in more detail in Chapter 10.

ing and matching. After the first set of matches is produced, a new blocking strategy is deployed to search for additional matches in the remaining record pairs.

Blocking based on exact field agreements is common in practice, but there are other approaches to indexing that attempt to be more error tolerant. For example, one may use clustering algorithms to identify sets of similar records. In this approach an index key, which is analogous to the blocking key above, is generated for both data sets and then the keys are combined into a single list. A distance function must be chosen and pairwise distances computed for all keys. The clustering algorithm is then applied to the combined list, and only record pairs that are assigned to the same cluster are compared. This is a theoretically appealing approach but it has the drawback that the similarity metric has to be computed for all pairs of records. Even so, computing the similarity measure for a pair of blocking keys is likely to be cheaper than computing the full record comparison, so there is still a gain in efficiency. Whang et al. [397] provide a nice review of indexing approaches.

In addition to reducing the computational burden of record linkage, indexing plays an important secondary role. Once implemented, the fraction of comparisons made that correspond to true links will be significantly higher. For some record linkage approaches that use an algorithm to find optimal parameters—like the probabilistic approach—having a larger ratio of matches to nonmatches will produce a better result.

3.5 Matching

The purpose of a record linkage algorithm is to examine pairs of records and make a prediction as to whether they correspond to the same underlying entity. (There are some sophisticated algorithms that examine sets of more than two records at a time [359], but pairwise comparison remains the standard approach.) At the core of every record linkage algorithm is a function that compares two records and outputs a "score" that quantifies the similarity between those records. Mathematically, the match score is a function of the output from individual field comparisons: agreement in the first name field, agreement in the last name field, etc. Field comparisons may be binary—indicating agreement or disagreement—or they may output a range of values indicating different levels of agreement. There are a variety of methods in the statistical and computer science literature that can be used to generate a match score, including nearest-neighbor matching, regression-based matching, and propensity score matching. The probabilistic approach to record linkage defines the match score in terms of a likelihood ratio [118].

Example: Matching in practice

Long strings, such as assignee and inventor names, are susceptible to typographical errors and name variations. For example, none of Sony Corporation, Sony Corporatoin and Sony Corp. will match using simple exact matching. Similarly, David vs. Dave would not match [387].

Comparing fields whose values are continuous is straightforward: often one can simply take the absolute difference as the comparison value. Comparing character fields in a rigorous way is more complicated. For this purpose, different mathematical definitions of the distance between two character fields have been defined. Edit distance, for example, is defined as the minimum number of edit operations—chosen from a set of allowed operations—needed to convert one string to another. When the set of allowed edit operations is single-character insertions, deletions, and substitutions, the corresponding edit distance is also known as the Levenshtein distance. When transposition of adjacent characters is allowed in addition to those operations, the corresponding edit distance is called the Levenshtein-Damerau distance.

Edit distance is appealing because of its intuitive definition, but it is not the most efficient string distance to compute. Another standard string distance known as Jaro-Winkler distance was developed with record linkage applications in mind and is faster to compute. This is an important consideration because in a typical record linkage application most of the algorithm run time will be spent performing field comparisons. The definition of Jaro-Winkler distance is less intuitive than edit distance, but it works as expected: words with more characters in common will have a higher Jaro-Winkler value than those with fewer characters in common. The output value is normalized to fall between 0 and 1. Because of its history in record linkage applications, there are some standard variants of Jaro-Winkler distance that may be implemented in record linkage software. Some variants boost the weight given to agreement in the first few characters of the strings being compared. Others decrease the score penalty for letter substitutions that arise from common typos.

Once the field comparisons are computed, they must be combined to produce a final prediction of match status. In the following sections we describe three types of record linkage algorithms: rule-based, probabilistic, and machine learning.

3.5.1 Rule-based approaches

A natural starting place is for a data expert to create a set of ad hoc rules that determine which pairs of records should be linked. In the classical record linkage setting where the two files have a number of identifying fields in common, this is not the optimal approach. However, if there are few fields in common but each file contains auxiliary fields that may inform a linkage decision, then an ad hoc approach may be appropriate.

Example: Linking in practice

Consider the problem of linking two lists of individuals where both lists contain first name, last name, and year of birth. Here is one possible linkage rule: link all pairs of records such that

- the Jaro–Winkler comparison of first names is greater than 0.9
- the Jaro–Winkler comparison of last names is greater than 0.9
- the first three digits of the year of birth are the same.

The result will depend on the rate of data errors in the year of birth field and typos in the name fields.

By *auxiliary field* we mean data fields that do not appear on both data sets, but which may nonetheless provide information about whether records should be linked. Consider a situation in which the first list includes an occupation field and the second list includes educational history. In that case one might create additional rules to eliminate matches where the education was deemed to be an unlikely fit for the occupation.

This method may be attractive if it produces a reasonable-looking set of links from intuitive rules, but there are several pitfalls. As the number of rules grows it becomes harder to understand the ways that the different rules interact to produce the final set of links. There is no notion of a threshold that can be increased or decreased depending on the tolerance for false positive and false negative errors. The rules themselves are not chosen to satisfy any kind of optimality, unlike the probabilistic and machine learning methods. Instead, they reflect the practitioner's domain knowledge about the data sets.

3.5.2 Probabilistic record linkage

In this section we describe the probabilistic approach to record linkage, also known as the Fellegi–Sunter algorithm [118]. This approach dominates in traditional record linkage applications and remains an effective and efficient way to solve the record linkage problem today.

In this section we give a somewhat formal definition of the statistical model underlying the algorithm. By understanding this model, one is better equipped to define link keys and record comparisons in an optimal way.

Example: Usefulness of probabilistic record linkage

In practice, it is typically the case that a researcher will want to combine two or more data sets containing records for the same individuals or units that possibly come from different sources. Unless the sources all contain the same unique identifiers, linkage will likely require matching on standardized text strings. Even standardized data are likely to contain small differences that preclude exact matching as in the matching example above. The Census Bureau's Longitudinal Business Database (LBD) links establishment records from administrative and survey sources. Exact numeric identifiers do most of the heavy lifting, but mergers, acquisitions, and other actions can break these linkages. Probabilistic record linkage on company names and/or addresses is used to fix these broken linkages that bias statistics on business dynamics [190].

Let A and B be two lists of individuals whom we wish to link. The product set $A \times B$ contains all possible pairs of records where the first element of the pair comes from A and the second element of the pair comes from B. A fraction of these pairs will be matches, meaning that both records in the pair represent the same underlying individual, but the vast majority of them will be nonmatches. In other words, $A \times B$ is the disjoint union of the set of matches M and the set of nonmatches U, a fact that we denote formally by $A \times B = M \cup U$.

Let γ be a vector-valued function on $A \times B$ such that, for $a \in A$ and $b \in B$, $\gamma(a, b)$ represents the outcome of a set of field comparisons between a and b. For example, if both A and B contain data on individuals' first names, last names, and cities of residence, then γ could be a vector of three binary values representing agreement in first name, last name, and city. In that case $\gamma(a, b) = (1, 1, 0)$ would mean that the records a and b agree on first name and last name, but disagree on city of residence.

For this model, the comparison outcomes in $\gamma(a, b)$ are not required to be binary, but they do have to be categorical: each component of $\gamma(a, b)$ should take only finitely many values. This means that a continuous comparison outcome—such as output from the Jaro–Winkler string comparator—has to be converted to an ordinal value representing levels of agreement. For example, one might create a three-level comparison, using one level for exact agreement, one level for approximate agreement defined as a Jaro–Winkler score greater than 0.85, and one level for nonagreement corresponding to a Jaro–Winkler score less than 0.85.

If a variable being used in the comparison has a significant number of missing values, it can help to create a comparison outcome level to indicate missingness. Consider two data sets that both have middle initial fields, and suppose that in one of the data sets the middle initial is filled in only about half of the time. When comparing records, the case where both middle initials are filled in but are not the same should be treated differently from the case where one of the middle initials is blank, because the first case provides more evidence that the records do not correspond to the same person. We handle this in the model by defining a three-level comparison for the middle initial, with levels to indicate "equal," "not equal," and "missing."

Probabilistic record linkage works by weighing the probability of seeing the result $\gamma(a, b)$ if (a, b) belongs to the set of matches M against the probability of seeing the result if (a, b) belongs to the set of nonmatches U. Conditional on M or U, the distribution of the individual comparisons defined by γ are assumed to be mutually independent. The parameters that define the marginal distributions of $\gamma|M$ are called *m-weights*, and similarly the marginal distributions of $\gamma|U$ are called *u-weights*.

In order to apply the Fellegi–Sunter method, it is necessary to choose values for these parameters, m-weights and u-weights. With labeled data—a pair of lists for which the match status is known—it is straightforward to solve for optimal values. Training data are not usually available, however, and the typical approach is to use expectation maximization to find optimal values.

We have noted that primary motivation for record linkage is to create a linked data set for analysis that will have a richer set of fields than either of the input data sets alone. A natural application is to perform a linear regression using a combination of variables from both files as predictors. With all record linkage approaches it is a challenge to understand how errors from the linkage process will manifest in the regression. Probabilistic record linkage has an advantage over rule-based and machine learning approaches in that there are theoretical results concerning coefficient bias and errors [221, 329]. More recently, Chipperfield and Chambers have developed an approach based on the bootstrap to account for record linkage errors when making inferences for cross-tabulated variables [75].

3.5.3 Machine learning approaches to linking

Computer scientists have contributed extensively in parallel literature focused on linking large data sets [76]. Their focus is on identifying potential links using approaches that are fast and scalable, and approaches are developed based on work in network algorithms and machine learning.

While simple blocking as described in Section 3.4 is standard in Fellegi–Sunter applications, computer scientists are likely to use the more sophisticated clustering approach to indexing. Indexing may also use network information to include, for example, records for individuals that have a similar place in a social graph. When linking lists of researchers, one might specify that comparisons should be made between records that share the same address, have patents in the same patent class, or have overlapping sets of coinventors. These approaches are known as semantic blocking, and the computational requirements are similar to standard blocking [76].

▶ This topic is discussed in more detail in Chapter 6.

In recent years machine learning approaches have been applied to record linkage following their success in other areas of prediction and classification. Computer scientists couch the analytical problem as one of entity resolution, even though the conceptual problem is identical. As Wick et al. [400] note:

> Entity resolution, the task of automatically determining which mentions refer to the same real-world entity, is a crucial aspect of knowledge base construction and management. However, performing entity resolution at large scales is challenging because (1) the inference algorithms must cope with unavoidable system scalability issues and (2) the search space grows exponentially in the number of mentions. Current conventional wisdom declares that performing coreference at these scales requires decom-

> posing the problem by first solving the simpler task of entity-linking (matching a set of mentions to a known set of KB entities), and then performing entity discovery as a post-processing step (to identify new entities not present in the KB). However, we argue that this traditional approach is harmful to both entity-linking and overall coreference accuracy. Therefore, we embrace the challenge of jointly modeling entity-linking and entity discovery as a single entity resolution problem.

Figure 3.2 provides a useful comparison between classical record linkage and learning-based approaches. In machine learning there is a predictive model and an algorithm for "learning" the optimal set of parameters to use in the predictive algorithm. The learning algorithm relies on a training data set. In record linkage, this would be a curated data set with true and false matches labeled as such. See [387] for an example and a discussion of how a training data set was created for the problem of disambiguating inventors in the USPTO database. Once optimal parameters are computed from the training data, the predictive model can be applied to unlabeled data to find new links. The quality of the training data set is critical; the model is only as good as the data it is trained on.

▶ See Chapter 6.

An example of a machine learning model that is popular for record linkage is the random forest model [50]. This is a classification model that fits a large number of classification trees to a labeled training data set. Each individual tree is trained on a bootstrap sample of all labeled cases using a random subset of predictor variables. After creating the classification trees, new cases are labeled by giving each tree a vote and keeping the label that receives the most votes. This highly randomized approach corrects for a problem with simple classification trees, which is that they may overfit to training data.

As shown in Figure 3.2, a major difference between probabilistic and machine learning approaches is the need for labeled training data to implement the latter approach. Usually training data are created through a painstaking process of clerical review. After an initial round of record linkage, a sample of record pairs that are not clearly matches or nonmatches is given to a research assistant who makes the final determination. In some cases it is possible to create training data by automated means. For example, when there is a subset of the complete data that contains strongly identifying fields. Suppose that both of the candidate lists contain name and date of birth fields and that in the first list the date of birth data are

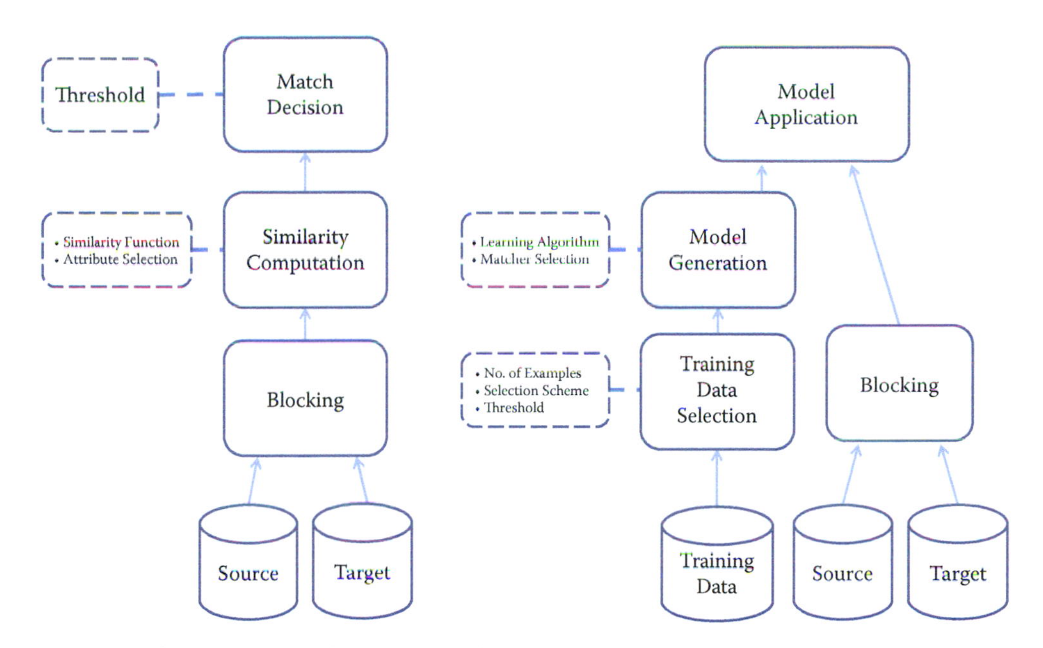

Figure 3.2. Probabilistic (left) vs. machine learning (right) approaches to linking. Source: Köpcke et al. [213]

complete, but in the second list only about 10% of records contain date of birth. For reasonably sized lists, name and date of birth together will be a nearly unique identifier. It is then possible to perform probabilistic record linkage on the subset of records with date of birth and be confident that the error rates would be small. If the subset of records with date of birth is representative of the complete data set, then the output from the probabilistic record linkage can be used as "truth" data.

Given a quality training data set, machine learning approaches may have advantages over probabilistic record linkage. Consider the random forest model. Random forests are more robust to correlated predictor variables, because only a random subset of predictors is included in any individual classification tree. The conditional independence assumption, to which we alluded in our discussion of the probabilistic model, can be dropped. An estimate of the generalization error can be computed in the form of "out-of-bag error." A measure of variable importance is computed that gives an idea of how powerful a particular field comparison is in terms of correctly predicting link status. Finally, unlike the Fellegi–Sunter model, predictor variables can be continuous.

The combination of being robust to correlated variables and providing a variable importance measure makes random forests a useful diagnostic tool for record linkage models. It is possible to refine the record linkage model iteratively, by first including many predictor variables, including variants of the same comparison, and then using the variable importance measure to narrow down the predictors to a parsimonious set.

There are many published studies on the effectiveness of random forests and other machine learning algorithms for record linkage. Christen and Ahmed et al. provide some pointers [77, 108].

3.5.4 Disambiguating networks

The problem of disambiguating entities in a network is closely related to record linkage: in both cases the goal is to consolidate multiple records corresponding to the same entity. Rather than finding the same entity in two data sets, however, the goal in network disambiguation is to consolidate duplicate records in a network data set. By network we mean that the data set contains not only typical record fields like names and addresses but also information about how entities relate to one another: entities may be coauthors, coinventors, or simply friends in a social network.

The record linkage techniques that we have described in this chapter can be applied to disambiguate a network. To do so, one must convert the network to a form that can be used as input into a record linkage algorithm. For example, when disambiguating a social network one might define a field comparison whose output gives the fraction of friends in common between two records. Ventura et al. demonstrated the relative effectiveness of the probabilistic method and machine learning approaches to disambiguating a database of inventors in the USPTO database [387]. Another approach is to apply clustering algorithms from the computer science literature to identify groups of records that are likely to refer to the same entity. Huang et al. [172] have developed a successful method based on an efficient computation of distance between individuals in the network. These distances are then fed into the DBSCAN clustering algorithm to identify unique entities.

3.6 Classification

Once the match score for a pair of records has been computed using the probabilistic or random forest method, a decision has to be made whether the pair should be linked. This requires classifying the pair as either a “true” or a “false” match. In most cases, a third classification is required—sending for manual review and classification.

3.6.1 Thresholds

In the probabilistic and random forest approaches, both of which output a "match score" value, a classification is made by establishing a threshold T such that all records with a match score greater than T are declared to be links. Because of the way these algorithms are defined, the match scores are not meaningful by themselves and the threshold used for one linkage application may not be appropriate for another application. Instead, the classification threshold must be established by reviewing the model output.

Typically one creates an output file that includes pairs of records that were compared along with the match score. The file is sorted by match score and the reviewer begins to scan the file from the highest match scores to the lowest. For the highest match scores the record pairs will agree on all fields and there is usually no question about the records being linked. However, as the scores decrease the reviewer will see more record pairs whose match status is unclear (or that are clearly nonmatches) mixed in with the clear matches. There are a number of ways to proceed, depending on the resources available and the goal of the project.

Rather than set a single threshold, the reviewer may set two thresholds $T_1 > T_2$. Record pairs with a match score greater than T_1 are marked as matches and removed from further consideration. The set of record pairs with a match score between T_1 and T_2 are believed to contain significant numbers of matches and nonmatches. These are sent to clerical review, meaning that research assistants will make a final determination of match status. The final set of links will include clear matches with a score greater than T_1 as well as the record pairs that pass clerical review. If the resources are available for this approach and the initial threshold T_1 is set sufficiently high, then the resulting data set will contain a minimal number of false positive links. The collection of record pairs with match scores between T_1 and T_2 is sometimes referred to as the clerical review region.

The clerical review region generally contains many more pairs than the set of clear matches, and it can be expensive and time-consuming to review each pair. Therefore, a second approach is to establish tentative threshold T and send only a sample of record pairs with scores in a neighborhood of T to clerical review. This results in data on the relative numbers of true matches and true nonmatches at different score levels, as well as the characteristics of record pairs that appear at a given level. Based on the review and the relative tolerance for false positive errors and false negative errors, a final threshold T' is set such that pairs with a score greater than T' are considered to be matches.

After viewing the results of the clerical review, it may be determined that the parameters to the record linkage algorithm could be improved to create a clearer delineation between matches and non-matches. For example, a research assistant may determine that many potential false positives appear near the tentative threshold because the current set of record linkage parameters is giving too much weight to agreement in first name. In this case the reviewer may decide to update the record linkage model to produce an improved set of match scores. The update may consist in an ad hoc adjustment of parameters, or the result of the clerical review may be used as training data and the parameter-fitting algorithm may be run again. An iterative approach like this is common when first linking two data sets because the clerical review process can improve one's understanding of the data sets involved.

Setting the threshold value higher will reduce the number of false positives (record pairs for which a link is incorrectly predicted) while increasing the number of false negatives (record pairs that should be linked but for which a link is not predicted). The proper tradeoff between false positive and false negative error rates will depend on the particular application and the associated loss function, but there are some general concerns to keep in mind. Both types of errors create bias, which can impact the generalizability of analyses conducted on the linked data set. Consider a simple regression on the linked data that includes fields from both data sets. If the threshold is too high, then the linked data will be biased toward records with no data entry errors or missing values, and whose fields did not change over time. This set of records may not be representative of the population as a whole. If a low threshold is used, then the set of linked records will contain more pairs that are not true links and the variables measured in those records are independent of each other. Including these records in a regression amounts to adding statistical noise to the data.

3.6.2 One-to-one links

In the probabilistic and machine learning approaches to record linkage that we have described, each record pair is compared and a link is predicted independently of all other record pairs. Because of the independence of comparisons, one record in the first file may be predicted to link to multiple records in the second file. Under the assumption that each input file has been deduplicated, at most one of these predictions can correspond to a true link. For many applications it is preferable to extract a set of "best" links with the property that each record in one file links to at most one record in the second file. A set of links with this property is said to be one-to-one.

One possible definition of "best" is a set of one-to-one links such that the sum of the match scores of all included links is maximal. This is an example of what is known as the *assignment problem* in combinatorial optimization. In the linear case above, where we care about the sum of match scores, the problem can be solved exactly using the Hungarian algorithm [216].

▶ This topic is discussed in more detail in Chapter 6.

3.7 Record linkage and data protection

In many social science applications data sets there is no need for data to include identifying fields like names and addresses. These fields may be left out intentionally out of concern for privacy, or they may simply be irrelevant to the research question. For record linkage, however, names and addresses are among the best possible identifiers. We describe two approaches to the problem of balancing needs for both effective record linkage and privacy.

▶ See Chapter 11.

The first approach is to establish a trusted third party or safe center. The concept of trusted third parties (TTPs) is well known in cryptography. In the case of record linkage, a third party takes a place between the data owners and the data users, and it is this third party that actually performs the linkage work. Both the data owners and data users trust the third party in the sense that it assumes responsibility for data protection (data owners) and data competence (data users) at the same time. No party other than the TTP learns about the private data of the other parties. After record linkage only the linked records are revealed, with no identifiers attached. The TTP ensures that the released linked data set cannot be relinked to any of the source data sets. Possible third parties are safe centers, which are operated by lawyers, or official trusted institutions like the US Census Bureau. Some countries like the UK and Germany are establishing new institutions specifically to act as TTPs for record linkage work.

The second approach is known as privacy-preserving record linkage. The goal of this approach is to find the same individual in separate data files without revealing the identity of the individual [80]. In privacy-preserving record linkage, cryptographic procedures are used to encrypt or hash identifiers before they are shared for record linkage. Many of these procedures require exact matching of the identifiers, however, and do not tolerate any errors in the original identifiers. This leads to information loss because it is not possible to account for typos or other small variations in hashed fields. To account for this, Schnell has developed a method to calculate string similarity of encrypted fields using bloom filters [330, 332].

In many countries these approaches are combined. For example, when the UK established the ADRN, the latter established the concept of trusted third parties. That third party is provided with data in which identifying fields have been hashed. This solves the challenge of trust between the different parties. Some authors argue that transparency of data use and informed consent will help to build trust. In the context of big data this is more challenging..

▶ This topic is discussed in more detail in Chapter 11.

3.8 Summary

Accurate record linkage is critical to creating high-quality data sets for analysis. However, outside of a few small centers for record linkage research, linking data sets historically relied on artisan approaches, particularly for parsing and cleaning data sets. As the creation and use of big data increases, so does the need for systematic record linkage. The history of record linkage is long by computer science standards, but new data challenges encourage the development of new approaches like machine learning methods, clustering algorithms, and privacy-preserving record linkage.

Record linkage stands on the boundary between statistics, information technology, and privacy. We are confident that there will continue to be exciting developments in this field in the years to come.

3.9 Resources

Out of many excellent resources on the subject, we note the following:

- We strongly recommend Christen's book [76].
- There is a wealth of information available on the ADRN website [103].
- Winkler has a series of high-quality survey articles [407].
- The German Record Linkage Center is a resource for research, software, and ongoing conference activities [331].

Chapter 4

Databases

Ian Foster and Pascal Heus

Once the data have been collected and linked into different files, it is necessary to store and organize them. Social scientists are used to working with one analytical file, often in SAS, Stata, SPSS, or R. This chapter, which may be the most important chapter in the book, describes different approaches to storing data in ways that permit rapid and reliable exploration and analysis.

4.1 Introduction

We turn now to the question of how to store, organize, and manage the data used in data-intensive social science. As the data with which you work grow in volume and diversity, effective data management becomes increasingly important if you are to avoid issues of scale and complexity from overwhelming your research processes. In particular, when you deal with data that get frequently updated, with changes made by different people, you will frequently want to use database management systems (DBMSs) instead of maintaining data in single files or within siloed statistical packages such as SAS, SPSS, Stata, and R. Indeed, we go so far as to say: if you take away *just one thing* from this book, it should be this: *Use a database!*

As we explain in this chapter, DBMSs provide an environment that greatly simplifies data management and manipulation. They require a little bit of effort to set up, but are worth it. They permit large amounts of data to be organized in multiple ways that allow for efficient and rapid exploration via powerful declarative query languages; durable and reliable storage, via transactional features that maintain data consistency; scaling to large data sizes; and intuitive analysis, both within the DBMS itself and via bridges to other data analysis packages and tools when specialized analyses are required. DBMSs have become a critical component of a great variety of applications, from handling transactions in financial systems to delivering data as a service to power websites, dashboards, and applications. If you are using a production-level enterprise system, chances are there is a database in the back end. They are multipurpose and well suited for organizing social science data and for supporting analytics for data exploration.

DBMSs make many easy things trivial, and many hard things easy. They are easy to use but can appear daunting to those unfamiliar with their concepts and workings. A basic understanding of databases and of when and how to use DBMSs is an important element of the social data scientist's knowledge base. We therefore provide in this chapter an introduction to databases and how to use them. We describe different types of databases and their various features, and how different types can be applied in different contexts. We describe basic features like how to get started, set up a database schema, ingest data, query data within a database, and get results out. We also discuss how to link from databases to other tools, such as Python, R, and Stata (if you really have to). Chapter 5 describes how to apply parallel computing methods when needed.

4.2 DBMS: When and why

Consider the following three data sets:

1. 10,000 records describing research grants, each specifying the principal investigator, institution, research area, proposal title, award date, and funding amount in comma-separated-value (CSV) format.

2. 10 million records in a variety of formats from funding agencies, web APIs, and institutional sources describing people, grants, funding agencies, and patents.

3. 10 billion Twitter messages and associated metadata—around 10 terabytes (10^{13} bytes) in total, and increasing at a terabyte a month.

Which tools should you use to manage and analyze these data sets? The answer depends on the specifics of the data, the analyses that you want to perform, and the life cycle within which data and analyses are embedded. Table 4.1 summarizes relevant factors, which we now discuss.

Table 4.1. When to use different data management and analysis technologies

Text files, spreadsheets, and scripting language
• Your data are small • Your analysis is simple • You do not expect to repeat analyses over time
Statistical packages
• Your data are modest in size • Your analysis maps well to your chosen statistical package
Relational database
• Your data are structured • Your data are large • You will be analyzing changed versions of your data over time • You want to share your data and analyses with others
NoSQL database
• Your data are unstructured • Your data are extremely large

In the case of data set 1 (10,000 records describing research grants), it may be feasible to leave the data in their original file, use spreadsheets, pivot tables, or write programs in scripting languages* such as Python or R to ask questions of those files. For example, someone familiar with such languages can quickly create a script to extract from data set 1 all grants awarded to one investigator, compute average grant size, and count grants made each year in different areas.

★ A scripting language is a programming language used to automate tasks that could otherwise be performed one by one be the user.

However, this approach also has disadvantages. Scripts do not provide inherent control over the file structure. This means that if you obtain new data in a different format, your scripts need to be updated. You cannot just run them over the newly acquired file. Scripts can also easily become unreasonably slow as data volumes grow. A Python or R script will not take long to search a list of 1,000 grants to find those that pertain to a particular institution. But what if you have information about 1 million grants, and for each grant you want to search a list of 100,000 investigators, and for each investigator, you want to search a list of 10 million papers to see whether that investigator is listed as an author of each paper? You now have $1{,}000{,}000 \times 100{,}000 \times 10{,}000{,}000 = 10^{18}$ comparisons to perform. Your simple script may now run for hours or even days. You can speed up the search process by constructing indices, so that, for example, when given a grant, you can find the associated investigators in constant time rather than in time proportional to the number of investigators. However, the construction of such indices is itself a time-consuming and error-prone process.

For these reasons, the use of scripting languages alone for data analysis is rarely to be recommended. This is not to say that all analysis computations can be performed in database systems. A programming language will also often be needed. But many data access and manipulation computations are best handled in a database.

Researchers in the social sciences frequently use statistical packages* such as R, SAS, SPSS, and Stata for data analysis. Because these systems integrate some crude data management, statistical analysis, and graphics capabilities in a single package, a researcher can often carry out a data analysis project of modest size within the same environment. However, each of these systems has limitations that hinder its use for modern social science research, especially as data grow in size and complexity.

★ A statistical package is a specialized compute program for analysis in statistics and economics.

Take Stata, for example. Stata always loads the entire data set into the computer's working memory, and thus you would have no problems loading data set 1. However, depending on your computer's memory, it could have problems dealing with with data set 2 and certainly would not be able to handle data set 3. In addition, you would need to perform this data loading step each time you start working on the project, and your analyses would be limited to what Stata can do. SAS can deal with larger data sets, but is renowned for being hard to learn and use. Of course there are workarounds in statistical packages. For example, in Stata you can deal with larger file sizes by choosing to only load the variables or cases that you need for the analysis [211]. Likewise, you can deal with more complex data by creating a system of files that each can be linked as needed for a particular analysis through a common identifier variable.

▶ For example, the Panel Study of Income Dynamics [181] has a series of files that are related and can be combined through common identifier variables [182].

Those solutions essentially mimic core functions of a DBMS, and you would be well advised to set up such system, especially if you find yourself in a situation where the data set is constantly updated through different users, if groups of users have different rights to use your data or should only have access to subsets of the data, and if the analysis takes place on a server that sends results to a client (browser). Statistics packages also have difficulty working with more than one data source at a time—something that DBMSs are designed to do well.

These considerations bring us to the topic of this chapter, namely database management systems. A DBMS* handles all of the issues listed above, and more. As we will see below when we look at concrete examples, a DBMS allows the programmer to define a logical design that fits the structure of their data. The DBMS then

★ DBMS is a system that interacts with users, other applications, and the database itself to capture and analyze data.

implements a *data model* (more on this below) that allows these data to be stored, queried, and updated efficiently and reliably on disk, thus providing independence from underlying physical storage. It supports efficient access to data through *query languages* and automatic optimization of those queries to permit fast analysis. Importantly, it also support concurrent access by multiple users, which is not an option for file-based data storage. It supports *transactions*, meaning that any update to a database is performed in its entirety or not at all, even in the face of computer failures or multiple concurrent updates. And it reduces the time spent both by analysts, by making it easy to express complex analytical queries concisely, and on data administration, by providing simple and uniform data administration interfaces.

A *database* is a structured collection of data about entities and their relationships. It models real-world objects—both entities (e.g., grants, investigators, universities) and relationships (e.g., "Steven Weinberg" works at "University of Texas at Austin")—and captures structure in ways that allow these entities and relationships to be queried for analysis. A *database management system* is a software suite designed to safely store and efficiently manage databases, and to assist with the maintenance and discovery of the relationships that database represents. In general, a DBMS encompasses three key components, as shown in Table 4.2: its *data model* (which defines how data are represented: see Box 4.1), its *query language* (which defines how the user interacts with the data), and support for *transactions and crash recovery* (to ensure reliable execution despite system failures).*

★ Some key DBMS features are often lacking in standard statistical packages: a standard query language (with commands that allow analyses or data manipulation on a subgroup of cases defined during the analysis, for example "group by ...," "order by ..."), keys (for speed improvement), and an explicit model of a relational data structure.

Box 4.1: Data model

A *data model* specifies the data elements associated with a problem domain, the properties of those data elements, and how those data elements relate to one another. In developing a data model, we commonly first identity the entities that are to be modeled and then define their properties and relationships. For example, when working on the science of science policy (see Figure 1.2), the entities include people, products, institutions, and funding, each of which has various properties (e.g., for a person, their name, address, employer); relationships include "is employed by" and "is funded by." This conceptual data model can then be translated into relational tables or some other database representation, as we describe next.

Table 4.2. Key components of a DBMS

	Data model	**Query language**	**Transactions, crash recovery**
User-facing	For example: relational, semi-structured	For example: SQL (for relational), XPath (for semi-structured)	Transactions
Internal	Mapping data to storage systems; creating and maintaining indices	Query optimization and evaluation; consistency	Locking, concurrency control, recovery

Literally hundreds of different open source, commercial, and cloud-hosted versions DBMSs are available. However, you only need to understand a relatively small number of concepts and major database types to make sense of this diversity. Table 4.3 defines the major classes of DBMSs that we will consider in this book. We consider only a few of these in any detail.

Relational DBMSs are the most widely used and mature systems, and will be the optimal solution for many social science data analysis purposes. We describe relational DBMSs in detail below, but in brief, they allow for the efficient storage, organization, and analysis of large quantities of *tabular* data: data organized as tables, in which rows represent entities (e.g., research grants) and columns represent attributes of those entities (e.g., principal investigator, institution, funding level). The associated Structured Query Language (SQL) can then be used to perform a wide range of analyses, which are executed with high efficiency due to sophisticated indexing and query planning techniques.

▶ Sometimes, as discussed in Chapter 3, the links are one to one and sometimes one to many.

While relational DBMSs have dominated the database world for decades, other database technologies have become popular for various classes of applications in recent years. As we will see, these alternative NoSQL DBMSs have typically been motivated by a desire to scale the quantities of data and/or number of users that can be supported and/or to deal with unstructured data that are not easily represented in tabular form. For example, a key–value store can organize large numbers of records, each of which associates an arbitrary key with an arbitrary value. These stores, and in particular variants called *document stores* that permit text search on the stored values, are widely used to organize and process the billions of records that can be obtained from web crawlers. We review below some of these alternatives and the factors that may motivate their use.

Table 4.3. Types of databases: relational (first row) and various types of NoSQL (other rows)

Type	Examples	Advantages	Disadvantages	Uses
Relational database	MySQL, PostgreSQL, Oracle, SQL Server, Teradata	Consistency (ACID)	Fixed schema; typically harder to scale	Transactional systems: order processing, retail, hospitals, etc.
Key–value store	Dynamo, Redis	Dynamic schema; easy scaling; high throughput	Not immediately consistent; no higher-level queries	Web applications
Column store	Cassandra, HBase	Same as key–value; distributed; better compression at column level	Not immediately consistent; using all columns is inefficient	Large-scale analysis
Document store	CouchDB, MongoDB	Index entire document (JSON)	Not immediately consistent; no higher-level queries	Web applications
Graph database	Neo4j, InfiniteGraph	Graph queries are fast	Difficult to do non-graph analysis	Recommendation systems, networks, routing

Relational and NoSQL databases (and indeed other solutions, such as statistical packages) can also be used together. Consider, for example, Figure 4.1, which depicts data flows commonly encountered in large research projects. Diverse data are being collected from different sources: JSON documents from web APIs, web pages from web scraping, tabular data from various administrative databases, Twitter data, and newspaper articles. There may be hundreds or even thousands of data sets in total, some of which may be extremely large. We initially have no idea of what schema* to use for the different data sets, and indeed it may not be feasible to define a unified set of schema, so diverse are the data and so rapidly are new data sets being acquired. Furthermore, the way we organize the data may vary according to our intended purpose. Are we interested in geographic, temporal, or thematic relationships among different entities? Each type of analysis may require a different organization.

★ A schema defines the structure of a database in a formal language defined by the DBMS. See Section 4.3.3.

For these reasons, a common storage solution is to first load all data into a large NoSQL database. This approach makes all data available via a common (albeit limited) query interface. Researchers can then extract from this database the specific elements that are of interest for their work, loading those elements into a re-

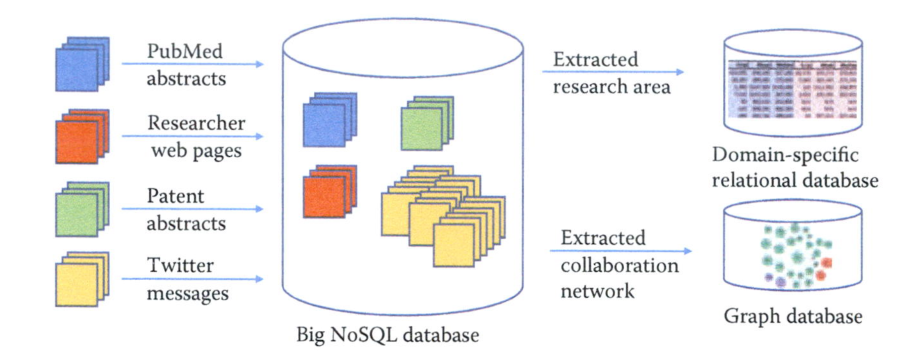

Figure 4.1. A research project may use a NoSQL database to accumulate large amounts of data from many different sources, and then extract selected subsets to a relational or other database for more structured processing

lational DBMS, another specialized DBMS (e.g., a graph database), or a statistical package for more detailed analysis. As part of the process of loading data from the NoSQL database into a relational database, the researcher will necessarily define schemas, relationships between entities, and so forth. Analysis results can be stored in a relational database or back into the NoSQL store.

4.3 Relational DBMSs

We now provide a more detailed description of relational DBMSs. Relational DBMSs implement the relational data model, in which data are represented as sets of records organized in tables. This model is particularly well suited for the structured, regular data with which we frequently deal in the social sciences; we discuss in Section 4.5 alternative data models, such as those used in NoSQL databases.

We use the data shown in Figure 4.2 to introduce key concepts. These two CSV format files describe grants made by the US National Science Foundation (NSF). One file contains information about grants, the other information about investigators. How should you proceed to manipulate and analyze these data?

The main concept underlying the relational data model is a *table* (also referred to as a *relation*): a set of rows (also referred to as tuples, records, or observations), each with the same columns (also referred to as fields, attributes or variables). A database consists of multiple tables. For example, we show in Figure 4.3 how the data

The file **grants.csv**

```
# Identifier,Person,Funding,Program
1316033,Steven Weinberg,666000,Elem. Particle Physics/Theory
1336199,Howard Weinberg,323194,ENVIRONMENTAL ENGINEERING
1500194,Irving Weinberg,200000,Accelerating Innovation Rsrch
1211853,Irving Weinberg,261437,GALACTIC ASTRONOMY PROGRAM
```

The file **investigators.csv**

```
# Name,Institution,Email
Steven Weinberg,University of Texas at Austin,weinberg@utexas.edu
Howard Weinberg,University of North Carolina Chapel Hill,
Irving Weinberg,University of Maryland College Park,irving@ucmc.edu
```

Figure 4.2. CSV files representing grants and investigators. Each line in the first table specifies a grant number, investigator name, total funding amount, and NSF program name; each line in the second gives an investigator name, institution name, and investigator email address

contained in the two CSV files of Figure 4.2 may be represented as two tables. The `Grants` table contains one tuple for each row in grants.csv, with columns `GrantID`, `Person`, `Funding`, and `Program`. The `Investigators` table contains one tuple for each row in investigators.csv, with columns `ID`, `Name`, `Institution`, and `Email`. The CSV files and tables contain essentially the same information, albeit with important differences (the addition of an `ID` field in the `Investigators` table, the substitution of an `ID` column for the `Person` column in the `Grants` table) that we will explain below.

The use of the relational data model provides for physical independence: a given table can be stored in many different ways. SQL queries are written in terms of the logical representation of tables (i.e., their schema definition). Consequently, even if the physical organization of the data changes (e.g., a different layout is used to store the data on disk, or a new index is created to speed up access for some queries), the queries need not change. Another advantage of the relational data model is that, since a table is a *set*, in a mathematical sense, simple and intuitive set operations (e.g., union, intersection) can be used to manipulate the data, as we discuss below. We can easily, for example, determine the intersection of two relations (e.g., grants that are awarded to a specific institution), as we describe in the following. The database further ensures that the data comply with the model (e.g., data types, key uniqueness, entity relationships), essentially providing core quality assurance.

Number	Person	Funding	Program
1316033	1	660,000	Elem. Particle Physics/Theory
1336199	2	323,194	ENVIRONMENTAL ENGINEERING
1500194	3	200,000	Accelerating Innovation Rsrch
1211853	3	261,437	GALACTIC ASTRONOMY PROGRAM

ID	Name	Institution	Email
1	Steven Weinberg	University of Texas at Austin	weinberg@utexas.edu
2	Howard Weinberg	University of North Carolina Chapel Hill	
3	Irving Weinberg	University of Maryland College Park	irving@ucmc.edu

Figure 4.3. Relational tables `Grants` and `Investigators` corresponding to the grants.csv and investigators.csv data in Figure 4.2, respectively. The only differences are the representation in a tabular form, the introduction of a unique numerical investigator identifier (`ID`) in the `Investigators` table, and the substitution of that identifier for the investigator name in the `Grants` table

4.3.1 Structured Query Language (SQL)

We use query languages to manipulate data in a database (e.g., to add, update, or delete data elements) and to retrieve (raw and aggregated) data from a database (e.g., data elements that certain properties). Relational DBMSs support SQL, a simple, powerful query language with a strong formal foundation based on logic, a foundation that allows relational DBMSs to perform a wide variety of sophisticated optimizations. SQL is used for three main purposes:

- Data definition: e.g., creation of new tables,
- Data manipulation: queries and updates,
- Control: creation of assertions to protect data integrity.

We introduce each of these features in the following, although not in that order, and certainly not completely. Our goal here is to give enough information to provide the reader with insights into how relational databases work and what they do well; an in-depth SQL tutorial is beyond the scope of this book but is something we highly recommend readers seek elsewhere.

4.3.2 Manipulating and querying data

SQL and other query languages used in DBMSs support the concise, declarative specification of complex queries. Because we are eager to show you something immediately useful, we cover these features first, before talking about how to define data models.

Example: Identifying grants of more than $200,000

Here is an SQL query to identify all grants with total funding of at most $200,000:

```
select * from Grants
where Funding <= 200,000;
```

(Here and elsewhere in this chapter, we show SQL key words in blue.)

Notice SQL's declarative nature: this query can be read almost as the English language statement, "select all rows from the `Grants` table for which the `Funding` column has value less than or equal 200,000." This query is evaluated as follows:

1. The input table specified by the `from` clause, `Grants`, is selected.
2. The condition in the `where` clause, `Funding <= 200,000`, is checked against all rows in the input table to identify those rows that match.
3. The `select` clause specifies which columns to keep from the matching rows, that is, which columns make the schema of the output table. (The "*" indicates that all columns should be kept.)

The answer, given the data in Figure 4.3, is the following single-row table. (The fact that an SQL query returns a table is important when it comes to creating more complex queries: the result of a query can be stored into the database as a new table, or passed to another query as input.)

Number	Person	Funding	Program
1500194	3	200,000	Accelerating Innovation Rsrch

DBMSs automatically optimize declarative queries such as the example that we just presented, translating them into a set of low-level data manipulations (an imperative *query plan*) that can be evaluated efficiently. This feature allows users to write queries without having to worry too much about performance issues—the database does the worrying for you. For example, a DBMS need not consider every row in the `Grants` table in order to identify those with funding less than $200,000, a strategy that would be slow if the `Grants` table were large: it can instead use an index to retrieve the relevant records much more quickly. We discuss indices in more detail in Section 4.3.6.

The querying component of SQL supports a wide variety of manipulations on tables, whether referred to explicitly by a table name (as in the example just shown) or constructed by another query. We just saw how to use the `select` operator to both pick certain rows (what is termed *selection*) and certain columns (what is called *projection*) from a table.

Example: Finding grants awarded to an investigator

★ In statistical packages, the term *merge* or *append* is often used when data sets are combined.

We want to find all grants awarded to the investigator with name "Irving Weinberg." The information required to answer this question is distributed over two tables, Grants and Investigators, and so we *join*★ the two tables to combine tuples from both:

```
select Number, Name, Funding, Program
from Grants, Investigators
where Grants.Person = Investigators.ID
and Name = "Irving Weinberg";
```

This query combines tuples from the Grants and Investigators tables for which the Person and ID fields match. It is evaluated in a similar fashion to the query presented above, except for the from clause: when multiple tables are listed, as here, the conditions in the where clause are checked for all different combinations of tuples from the tables defined in the from clause (i.e., the cartesian product of these tables)—in this case, a total of $3 \times 4 = 12$ combinations. We thus determine that Irving Weinberg has two grants. The query further selects the Number, Name, Funding, and Program fields from the result, giving the following:

Number	Name	Funding	Program
1500194	Irving Weinberg	200,000	Accelerating Innovation Rsrch
1211853	Irving Weinberg	261,437	GALACTIC ASTRONOMY PROGRAM

This ability to join two tables in a query is one example of how SQL permits concise specifications of complex computations. This joining of tables via a cartesian product operation is formally called a *cross join*. Other types of join are also supported. We describe one such, the *inner join*, in Section 4.6.

SQL aggregate functions allow for the computation of aggregate statistics over tables. For example, we can use the following query to determine the total number of grants and their total and average funding levels:

```
select count(*) as 'Number', sum(Funding) as 'Total',
       avg(Funding) as 'Average'
from Grants;
```

This yields the following:

Number	Total	Average
4	1444631	361158

The group by operator can be used in conjunction with the aggregate functions to group the result set by one or more columns.

For example, we can use the following query to create a table with three columns: investigator name, the number of grants associated with the investigator, and the aggregate funding:

```
select Name, count(*) as 'Number',
       avg(Funding) as 'Average funding'
from Grants, Investigators
where Grants.Person = Investigators.ID
group by Name;
```

We obtain the following:

Name	Number	Average Funding
Steven Weinberg	1	666000
Howard Weinberg	1	323194
Irving Weinberg	2	230719

4.3.3 Schema design and definition

We have seen that a relational database comprises a set of tables. The task of specifying the structure of the data to be stored in a database is called *logical design*. This task may be performed by a database administrator, in the case of a database to be shared by many people, or directly by users, if they are creating databases themselves. More specifically, the logical design process involves defining a *schema*. A schema comprises a set of tables (including, for each table, its columns and their types), their relationships, and integrity constraints.

The first step in the logical design process is to identify the entities that need to be modeled. In our example, we identified two important classes of entity: "grants" and "investigators." We thus define a table for each; each row in these two tables will correspond to a unique grant or investigator, respectively. (In a more complete and realistic design, we would likely also identify other entities, such as institutions and research products.) During this step, we will often find ourselves breaking information up into multiple tables, so as to avoid duplicating information.

For example, imagine that we were provided grant information in the form of one CSV file rather than two, with each line providing a grant number, investigator, funding, program, institution, and email. In this file, the name, institution, and email address for Irving Weinberg would then appear twice, as he has two grants, which can lead to errors when updating values and make it difficult to represent certain information. (For example, if we want to add an investigator who does not yet have a grant, we will need to create a tuple (row) with empty slots for all columns (variables) associated with grants.) Thus we would want to break up the single big table into the two tables that we defined here. This breaking up of information across different tables to avoid repetition of information is referred to as normalization.*

★ Normalization involves organizing columns and tables of a relational database to minimize data redundancy.

The second step in the design process is to define the columns that are to be associated with each entity. For each table, we define a set of columns. For example, given the data in Figure 4.2, those columns will likely include, for a grant, an award identifier, title, investigator, and award amount; for an investigator, a name, university, and email address. In general, we will want to ensure that each row in our table has a key: a set of columns that uniquely identifies that row. In our example tables, grants are uniquely identified by `Number` and investigators by `ID`.

► Normalization can be done in statistical packages as well. For example, as noted above, PSID splits its data into different files linked through ID variables. The difference here is that the DBMS makes creating, navigating, and querying the resulting data particularly easy.

The third step in the design process is to capture relationships between entities. In our example, we are concerned with just one relationship, namely that between grants and investigators: each grant has an investigator. We represent this relationship between tables by introducing a `Person` column in the `Grants` table, as shown in Figure 4.3. Note that we do not simply duplicate the investigator names in the two tables, as was the case in the two CSV files shown in Figure 4.2: these names might not be unique, and the duplication of data across tables can lead to later inconsistencies if a name is updated in one table but not the other.

The final step in the design process is to represent integrity constraints (or rules) that must hold for the data. In our example, we may want to specify that each grant must be awarded to an investigator; that each value of the grant identifier column must be unique (i.e., there cannot be two grants with the same number); and total funding can never be negative. Such restrictions can be achieved by specifying appropriate constraints at the time of schema creation, as we show in Listing 4.1, which contains the code used to create the two tables that make up our schema.

Listing 4.1 contains four SQL statements. The first two statements, lines 1 and 2, simply set up our new database. The `create table` statement in lines 4–10 creates our first table. It specifies the table name (`Investigators`) and, for each of the four columns, the column name and its type.* Relational DBMSs offer a rich set of types to choose from when designing a schema: for example, `int` or `integer` (synonyms); `real` or `float` (synonyms); `char(n)`, a fixed-length string of `n` characters; and `varchar(n)`, a variable-length string of up to `n` characters. Types are important for several reasons. First, they allow for more efficient encoding of data. For

★ These storage types will be familiar to many of you from statistical software packages.

```
create database grantdata;
use grantdata;

create table Investigators (
    ID int auto_increment,
    Name varchar(100) not null,
    Institution varchar(256) not null,
    Email varchar(100),
    primary key(ID)
);

create table Grants (
    Number int not null,
    Person int not null,
    Funding float unsigned not null,
    Program varchar(100),
    primary key(Number)
);
```

Listing 4.1. Code to create the `grantdata` database and its `Investigators` and `Grants` tables

example, the `Funding` field in the grants.csv file of Figure 4.2 could be represented as a string in the `Grants` table, `char(15)`, say, to allow for large grants. By representing it as a floating point number instead (line 15 in Listing 4.1), we reduce the space requirement per grant to just four bytes. Second, types allow for integrity checks on data as they are added to the database: for example, that same type declaration for `Funding` ensures that only valid numbers will be entered into the database. Third, types allow for type-specific operations on data, such as arithmetic operations on numbers (e.g., min, max, sum).

Other SQL features allow for the specification of additional constraints on the values that can be placed in the corresponding column. For example, the `not null` constraints for `Name` and `Institution` (lines 6, 7) indicate that each investigator must have a name and an institution, respectively. (The lack of such a constraint on the `Email` column shows that an investigator need not have an email address.)

4.3.4 Loading data

So far we have created a database and two tables. To complete our simple SQL program, we show in Listing 4.2 the two statements that load the data of Figure 4.2 into our two tables. (Here and elsewhere in this chapter, we use the MySQL DBMS. The SQL

```
load data local infile "investigators.csv"
    into table Investigators
    fields terminated by ","
    ignore 1 lines
    (Name, Institution, Email);

load data local infile "grants.csv" into table Grants
    fields terminated by ","
    ignore 1 lines
    (Number, @var, Funding, Program)
set Person = (select ID from Investigators
              where Investigators.Name=@var);
```

Listing 4.2. Code to load data into the `Investigators` and `Grants` tables

syntax used by different DBMSs differs in various, mostly minor ways.) Each statement specifies the name of the file from which data is to be read and the table into which it is to be loaded. The `fields terminated by ","` statement tells SQL that values are separated by columns, and `ignore 1 lines` tells SQL to skip the header. The list of column names is used to specify how values from the file are to be assigned to columns in the table.

For the `Investigators` table, the three values in each row of the investigators.csv file are assigned to the `Name`, `Institution`, and `Email` columns of the corresponding database row. Importantly, the `auto_increment` declaration on the `ID` column (line 5 in Listing 4.1) causes values for this column to be assigned automatically by the DBMS, as rows are created, starting at `1`. This feature allows us to assign a unique integer identifier to each investigator as its data are loaded.

For the `Grants` table, the `load data` call (lines 7–12) is somewhat more complex. Rather than loading the investigator name (the second column of each line in our data file, represented here by the variable `@var`) directly into the database, we use an SQL query (the `select` statement in lines 11–12) to retrieve from the `Investigators` table the `ID` corresponding to that name. By thus replacing the investigator name with the unique investigator identifier, we avoid replicating the name across the two tables.

4.3.5 Transactions and crash recovery

A DBMS protects the data that it stores from computer crashes: if your computer stops running suddenly (e.g., your operating system crashes or you unplug the power), the contents of your database are not corrupted. It does so by supporting *transactions*. A transaction is an atomic sequence of database actions. In general, every SQL statement is executed as a transaction. You can also specify sets of statements to be combined into a single transaction, but we do not cover that capability here. The DBMS ensures that each transaction is executed completely even in the case of failure or error: if the transaction succeeds, the results of all operations are recorded permanently ("persisted") in the database, and if it fails, all operations are "rolled back" and no changes are committed. For example, suppose we ran the following SQL statement to convert the funding amounts in the `Grants` table from dollars to euros, by scaling each number by 0.9. The `update` statement specifies the table to be updated and the operation to be performed, which in this case is to update the `Funding` column of each row. The DBMS will ensure that either no rows are altered or all are altered.

```
update Grants set Grants.Funding = Grants.Funding*0.9;
```

Transactions are also key to supporting multi-user access. The *concurrency control* mechanisms in a DBMS allow multiple users to operate on a database concurrently, as if they were the only users of the system: transactions from multiple users can be interleaved to ensure fast response times, while the DBMS ensures that the database remains consistent. While entire books could be (and have been) written on concurrency in databases, the key point is that read operations can proceed concurrently, while update operations are typically serialized.

4.3.6 Database optimizations

A relational DBMS applies query planning and optimization methods with the goal of evaluating queries as efficiently as possible. For example, if a query asks for rows that fit two conditions, one cheap to evaluate and one expensive, a relational DBMS may filter first on the basis of the first condition, and then apply the second conditions only to the rows identified by that first filter. These sorts of optimization are what distinguish SQL from other programming languages, as they allow the user to write queries declaratively and rely on the DBMS to come up with an efficient execution strategy.

Nevertheless, the user can help the DBMS to improve performance. The single most powerful performance improvement tool is the index, an internal data structure that the DBMS maintains to speed up queries. While various types of indices can be created, with different characteristics, the basic idea is simple. Consider the column `ID` in our `Investigators` table. Assume that there are N rows in the table. In the absence of an index, a query that refers to a column value (e.g., `where ID=3`) would require a linear scan of the table, taking on average $N/2$ comparisons and in the worst case N comparisons. A binary tree index allows the desired value to be found with just $\log_2 N$ comparisons.

Example: Using indices to improve database performance

Consider the following query:

```
select ID, Name, sum(Funding) as TotalFunding
  from Grants, Investigators
    where Investigators.ID=Grants.Person
  group by ID;
```

This query joins our two tables to link investigators with the grants that they hold, groups grants by investigator (using `group by`), and finally sums the funding associated with the grants held by each investigator. The result is the following:

ID	Name	TotalFunding
1	Steven Weinberg	666000
2	Howard Weinberg	323194
3	Irving Weinberg	461437

In the absence of indices, the DBMS must compare each row in `Investigators` with each row in `Grants`, checking for each pair whether `Investigators.ID = Grants.Person` holds. As the two tables in our sample database have only three and four rows, respectively, the total number of comparisons is only $3 \times 4 = 12$. But if we had, say, 1 million investigators and 1 million grants, then the DBMS would have to perform 1 trillion comparisons, which would take a long time. (More importantly in many cases, it would have to perform a large number of disk I/O operations if the tables did not fit in memory.) An index on the `ID` column of the `Investigators` table reduces the number of operations dramatically, as the DBMS can then take each of the 1 million rows in the `Grants` table and, for each row, identify the matching row(s) in `Investigators` via an index lookup rather than a linear scan.

In our example table, the `ID` column has been specified to be a `primary key`, and thus an index is created for it automatically. If it were not, we could easily create the desired index as follows:

```
alter table Investigators add index(ID);
```

It can be difficult for the user to determine when an index is required. A good rule of thumb is to create an index for any column that is queried often, that is, appears on the right-hand side of a `where` statement. However, the presence of indices makes updates more expensive, as every change to a column value requires that the index be rebuilt to reflect the change. Thus, if your data are highly dynamic, you should carefully select which indices to create. (For bulk load operations, a common practice is to drop indices prior to the data import, and re-create them once the load is completed.) Also, indices take disk space, so you need to consider the tradeoff between query efficiency and resources.

The `explain` command can be useful for determining when indices are required. For example, we show in the following some of the output produced when we apply `explain` to our query. (For this example, we have expanded the two tables to 1,000 rows each, as our original tables are too small for MySQL to consider the use of indices.) The output provides useful information such as the key(s) that could be used, if indices exist (`Person` in the `Grants` table, and the primary key, `ID`, for the `Investigators` table); the key(s) that are actually used (the primary key, `ID`, in the `Investigators` table); the column(s) that are compared to the index (`Investigators.ID` is compared with `Grants.Person`); and the number of rows that must be considered (each of the 1,000 rows in `Grants` is compared with one row in `Investigators`, for a total of 1,000 comparisons).

```
mysql> explain select ID, Name, sum(Funding) as TotalFunding
       from Grants, Investigators
       where Investigators.ID=Grants.Person group by ID;
+---------------+---------------+---------+---------------+------+
| table         | possible_keys | key     | ref           | rows |
+---------------+---------------+---------+---------------+------+
| Grants        | Person        | NULL    | NULL          | 1000 |
| Investigators | PRIMARY       | PRIMARY | Grants.Person |    1 |
+---------------+---------------+---------+---------------+------+
```

Contrast this output with the output obtained for equivalent tables in which `ID` is not a primary key. In this case, no keys are used and thus 1,000 × 1,000 = 1,000,000 comparisons and the associated disk reads must be performed.

```
+---------------+---------------+------+------+------+
| table         | possible_keys | key  | ref  | rows |
+---------------+---------------+------+------+------+
| Grants        | Person        | NULL | NULL | 1000 |
| Investigators | ID            | NULL | NULL | 1000 |
+---------------+---------------+------+------+------+
```

A second way in which the user can contribute to performance improvement is by using appropriate table definitions and data types. Most DBMSs store data on disk. Data must be read from disk into memory before it can be manipulated. Memory accesses are fast, but loading data into memory is expensive: accesses to main memory can be a million times faster than accesses to disk. Therefore, to ensure queries are efficient, it is important to minimize the number of disk accesses. A relational DBMS automatically optimizes queries: based on how the data are stored, it transforms a SQL query into a query plan that can be executed efficiently, and chooses an execution strategy that minimizes disk accesses. But users can contribute to making queries efficient. As discussed above, the choice of types made when defining schemas can make a big difference. As a rule of thumb, only use as much space as needed for your data: the smaller your records, the more records can be transferred to main memory using a single disk access. The design of relational tables is also important. If you put all columns in a single table (do not normalize), more data will come into memory than is required.

4.3.7 Caveats and challenges

It is important to keep the following caveats and challenges in mind when using SQL technology with social science data.

Data cleaning Data created outside an SQL database, such as data in files, are not always subject to strict constraints: data types may not be correct or consistent (e.g., numeric data stored as text) and consistency or integrity may not be enforced (e.g., absence of primary keys, missing foreign keys). Indeed, as the reader probably knows well from experience, data are rarely perfect. As a result, the data may fail to comply with strict SQL schema requirements and fail to load, in which case either data must be cleaned before or during loading, or the SQL schema must be relaxed.

Missing values Care must be taken when loading data in which some values may be missing or blank. SQL engines represent and refer to a missing or blank value as the built-in constant `null`. Counterintuitively, when loading data from text files (e.g., CSV), many SQL engines require that missing values be represented explicitly by the term `null`; if a data value is simply omitted, it may fail to load or be incorrectly represented, for example as zero or the empty string (`" "`) instead of `null`. Thus, for example, the second row in the investigators.csv file of Figure 4.2:

```
Howard Weinberg,University of North Carolina Chapel Hill,
```

may need to be rewritten as:

```
Howard Weinberg,University of North Carolina Chapel Hill,null
```

Metadata for categorical variables SQL engines are metadata poor: they do not allow extra information to be stored about a variable (field) beyond its base name and type (`int`, `char`, etc., as introduced in Section 4.3.3). They cannot, for example, record directly the fact that the column `class` can only take one of three values, `animal`, `vegetable`, or `mineral`, or what these values mean. Common practice is thus to store information about possible values in another table (commonly referred to as a *dimension table*) that can be used as a lookup and constraint, as in the following:

Table **class_values**

Value	**Description**
`animal`	Is alive
`vegetable`	Grows
`mineral`	Isn't alive and doesn't grow

A related concept is that a column or list of columns may be declared `primary key` or `unique`. Either says that no two tuples of the table may agree in all the column(s) on the list. There can be only one `primary key` for a table, but several `unique` columns. No column of a `primary key` can ever be `null` in any tuple. But columns declared `unique` may have `null`s, and there may be several tuples with `null`.

4.4 Linking DBMSs and other tools

Query languages such as SQL are not general-purpose programming languages; they support easy, efficient access to large data sets, but are not intended to be used for complex calculations. When complex computations are required, one can embed query language statements into a programming language or statistical package. For example, we might want to calculate the interquartile range of funding for all grants. While this calculation can be accomplished in SQL, the resulting SQL code will be complicated. Languages like Python make such statistical calculations straightforward, so it is natural to write a Python (or R, SAS, Stata, etc.) program that connects to the DBMS that contains our data, fetches the required data from the DBMS, and then calculates the interquartile range of those data. The program can then, if desired, store the result of this calculation back into the database.

Many relational DBMSs also have built-in analytical functions or often now embed the R engine, providing significant in-database

```
from mysql.connector import MySQLConnection, Error
from python_mysql_dbconfig import read_db_config

def retrieve_and_analyze_data():
    try:
        # Open connection to the MySQL database
        dbconfig = read_db_config()
        conn = MySQLConnection(**dbconfig)
        cursor = conn.cursor()
        # Transmit the SQL query to the database
        cursor.execute('select Funding from Grants;')
        # Fetch all rows of the query response
        rows = [row for row in cur.fetchall()]
        calculate_inter_quartile_range(rows)
    except Error as e:
        print(e)
    finally:
        cursor.close()
        conn.close()

if __name__ == '__main__':
    retrieve_and_analyze_data()
```

Listing 4.3. Embedding SQL in Python

statistical and analytical capabilities and alleviating the need for external processing.

Example: Embedding database queries in Python

The Python script in Listing 4.3 shows how this embedding of database queries in Python is done. This script establishes a connection to the database (lines 7–9), transmits the desired SQL query to the database (line 11), retrieves the query results into a Python array (line 13), and calls a Python procedure (not given) to perform the desired computation (line 14). A similar program could be used to load the results of a Python (or R, SAS, Stata, etc.) computation into a database.

Example: Loading other structured data

We saw in Listing 4.2 how to load data from CSV files into SQL tables. Data in other formats, such as the commonly used JSON, can also be loaded into a relational DBMS. Consider, for example, the following JSON format data, a simplified version of data shown in Chapter 2.

```
1  [
2    {
3      institute : Janelia Campus,
4      name : Laurence Abbott,
5      role : Senior Fellow,
6      state : VA,
7      town : Ashburn
8    },
9    {
10     institute : Jackson Lab,
11     name : Susan Ackerman,
12     role : Investigator,
13     state : ME,
14     town : Bar Harbor
15   }
16 ]
```

While some relational DBMSs provide built-in support for JSON objects, we assume here that we want to convert these data into normal SQL tables. Using one of the many utilities for converting JSON into CSV, we can construct the following CSV file, which we can load into an SQL table using the method shown earlier.

```
institute,name,role,state,town
Janelia Campus,Laurence Abbott,Senior Fellow,VA,Ashburn
Jackson Lab,Susan Ackerman,Investigator,ME,Bar Harbor
```

But into what table? The two records each combine information about a person with information about an institute. Following the schema design rules given in Section 4.3.3, we should *normalize* the data by reorganizing them into two tables, one describing people and one describing institutes. Similar problems arise when JSON documents contain nested structures. For example, consider the following alternative JSON representation of the data above. Here, the need for normalization is yet more apparent.

```
1  [
2    {
3      name : Laurence Abbott,
4      role : Senior Fellow,
5      employer : { institute : Janelia Campus,
6                   state : VA,
7                   town : Ashburn}
8    },
9    {
10     name : Susan Ackerman,
11     role : Investigator,
12     employer: { institute : Jackson Lab,
13                 state : ME,
14                 town : Bar Harbor}
15   }
16 ]
```

Thus, the loading of JSON data into a relational database usually requires both work on schema design (Section 4.3.3) and data preparation.

4.5 NoSQL databases

While relational DBMSs have dominated the database world for several decades, other database technologies exist and indeed have become popular for various classes of applications in recent years. As we will see, these alternative technologies have typically been motivated by a desire to scale the quantities of data and/or number of users that can be supported, and/or to support specialized data types (e.g., unstructured data, graphs). Here we review some of these alternatives and the factors that may motivate their use.

4.5.1 Challenges of scale: The CAP theorem

For many years, the big relational database vendors (Oracle, IBM, Sybase, and to a lesser extent Microsoft) have been the mainstay of how data were stored. During the Internet boom, startups looking for low-cost alternatives to commercial relational DBMSs turned to MySQL and PostgreSQL. However, these systems proved inadequate for big sites as they could not cope well with large traffic spikes, for example when many customers all suddenly wanted to order the same item. That is, they did not *scale*.

An obvious solution to scaling databases is to partition and/or replicate data across multiple computers, for example by distributing different tables, or different rows from the same table, over multiple computers. However, partitioning and replication also introduce challenges, as we now explain. Let us first define some terms. In a system that comprises multiple computers:

- Consistency indicates that all computers see the same data at the same time.
- Availability indicates that every request receives a response about whether it succeeded or failed.
- Partition tolerance indicates that the system continues to operate even if a network failure prevents computers from communicating.

An important result in distributed systems (the so-called "CAP theorem" [51]) observes that it is not possible to create a distributed system with all three properties. This situation creates a challenge with large transactional data sets. Partitioning is needed in order to achieve high performance, but as the number of computers grows, so too does the likelihood of network disruption among pair(s) of computers. As strict consistency cannot be achieved at the same time as availability and partition tolerance, the DBMS designer must choose between high consistency and high availability for a particular system.

The right combination of availability and consistency will depend on the needs of the service. For example, in an e-commerce setting, it makes sense to choose high availability for a checkout process, in order to ensure that requests to add items to a shopping cart (a revenue-producing process) can be honored. Errors can be hidden from the customer and sorted out later. However, for order submission—when a customer submits an order—it makes sense to favor consistency because several services (credit card processing, shipping and handling, reporting) need to access the data simultaneously. However, in almost all cases, availability is chosen over consistency.

4.5.2 NoSQL and key–value stores

Relational DBMSs were traditionally motivated by the need for transaction processing and analysis, which led them to put a premium on consistency and availability. This led the designers of these systems to provide a set of properties summarized by the acronym ACID [137, 347]:

- Atomic: All work in a transaction completes (i.e., is committed to stable storage) or none of it completes.
- Consistent: A transaction transforms the database from one consistent state to another consistent state.
- Isolated: The results of any changes made during a transaction are not visible until the transaction has committed.
- Durable: The results of a committed transaction survive failures.

The need to support extremely large quantities of data and numbers of concurrent clients has led to the development of a range of alternative database technologies that relax consistency and thus these ACID properties in order to increase scalability and/or availability. These systems are commonly referred to as NoSQL (for "not SQL"—or, more recently, "not only SQL," to communicate that they may support SQL-like query languages) because they usually do not require a fixed table schema nor support joins and other SQL features. Such systems are sometimes referred to as BASE [127]: Basically Available (the system seems to work all the time), Soft state (it does not have to be consistent all the time), and Eventually consistent (it becomes consistent at some later time). The data systems used in essentially all large Internet companies (Google, Yahoo!, Facebook, Amazon, eBay) are BASE.

Dozens of different NoSQL DBMSs exist, with widely varying characteristics as summarized in Table 4.3. The simplest are *key-value stores* such as Redis, Amazon Dynamo, Apache Cassandra, and Project Voldemort. We can think of a key–value store as a relational database with a single table that has just two columns, key and value, and that supports just two operations: store (or update) a key–value pair, and retrieve the value for a given key.

Example: Representing investigator data in a NoSQL database

We might represent the contents of the investigators.csv file of Figure 4.2 (in a NoSQL database) as follows.

Key	Value
Investigator_StevenWeinberg_Institution	University of Texas at Austin
Investigator_StevenWeinberg_Email	weinberg@utexas.edu
Investigator_HowardWeinberg_Institution	University of North Carolina Chapel Hill
Investigator_IrvingWeinberg_Institution	University of Maryland College Park
Investigator_IrvingWeinberg_Email	irving@ucmc.edu

A client can then read and write the value associated with a given *key* by using operations such as the following:

- **Get**(*key*) returns the value associated with *key*.
- **Put**(*key*, *value*) associates the supplied *value* with *key*.
- **Delete**(*key*) removes the entry for *key* from the data store.

Key–value stores are thus particularly easy to use. Furthermore, because there is no schema, there are no constraints on what values can be associated with a key. This lack of constraints can be useful if we want to store arbitrary data. For example, it is trivial to add the following records to a key–value store; adding this information to a relational table would require schema modifications.

Key	Value
Investigator_StevenWeinberg_FavoriteColor	Blue
Investigator_StevenWeinberg_Awards	Nobel

Another advantage is that if a given key would have no value (e.g., Investigator_HowardWeinberg_Email), we need not create a record. Thus, a key-value store can achieve a more compact representation of sparse data, which would have many empty fields if expressed in relational form.

A third advantage of the key-value approach is that a key-value store is easily partitioned and thus can scale to extremely large sizes. A key-value DBMS can partition the space of keys (e.g., via a hash on the key) across different computers for scalability. It can also replicate key-value pairs across multiple computers for availability. Adding, updating, or querying a key-value pair requires simply sending an appropriate message to the computer(s) that hold that pair.

The key-value approach also has disadvantages. As we can see from the example, users must be careful in their choice of keys if they are to avoid name collisions. The lack of schema and constraints can also make it hard to detect erroneous keys and values. Key-value stores typically do not support join operations (e.g., "which investigators have the Nobel and live in Texas?"). Many key-value stores also relax consistency constraints and do not provide transactional semantics.

4.5.3 Other NoSQL databases

The simple structure of key-value stores allows for extremely fast and scalable implementations. However, as we have seen, many interesting data cannot be easily modeled as key-value pairs. Such concerns have motivated the development of a variety of other NoSQL systems that offer, for example, richer data models: document-based (CouchDB and MongoDB), graph-based (Neo4J), and column-based (Cassandra, HBase) databases.

In document-based databases, the value associated with a key can be a structured document: for example, a JSON document, permitting the following representation of our investigators.csv file plus the additional information that we just introduced.

Key	**Value**
Investigator_StevenWeinberg	{ institution : University of Texas at Austin, email : weinberg@utexas.edu, favcolor : Blue, award : Nobel }
Investigator_HowardWeinberg	{ institution : University of North Carolina Chapel Hill }
Investigator_IrvingWeinberg	{ institution : University of Maryland College Park, email : irving@ucmc.edu }

Associated query languages may permit queries within the document, such as regular expression searches, and retrieval of selected fields, providing a form of a relational DBMS's selection and projection capabilities (Section 4.3.2). For example, MongoDB allows us to ask for documents in a collection called `investigators` that have "University of Texas at Austin" as their institution and the Nobel as an award.

```
db.investigators.find(
   { institution: 'University of Texas at Austin',
     award: 'Nobel' }
)
```

A column-oriented DBMS stores data tables by columns rather than by rows, as is common practice in relational DBMSs. This approach has advantages in settings where aggregates must frequently be computed over many similar data items: for example, in clinical data analysis. Google Cloud BigTable and Amazon RedShift are two cloud-hosted column-oriented NoSQL databases. HBase and Cassandra are two open source systems with similar characteristics. (Confusingly, the term *column oriented* is also often used to refer to SQL database engines that store data in columns instead of rows: for example, Google BigQuery, HP Vertica, Terradata, and the open source MonetDB. Such systems are not to be confused with column-based NoSQL databases.)

Graph databases store information about graph structures in terms of nodes, edges that connect nodes, and attributes of nodes and edges. Proponents argue that they permit particularly straightforward navigation of such graphs, as when answering queries such as "find all the friends of the friends of my friends"—a task that would require multiple joins in a relational database.

4.6 Spatial databases

Social science research commonly involves spatial data. Socioeconomic data may be associated with census tracts, data about the distribution of research funding and associated jobs with cities and states, and crime reports with specific geographic locations. Furthermore, the quantity and diversity of such spatially resolved data are growing rapidly, as are the scale and sophistication of the systems that provide access to these data. For example, just one urban data store, Plenario, contains many hundreds of data sets about the city of Chicago [64].

Researchers who work with spatial data need methods for representing those data and then for performing various queries against them. Does crime correlate with weather? Does federal spending on research spur innovation within the locales where research occurs? These and many other questions require the ability to quickly determine such things as which points exist within which regions, the areas of regions, and the distance between two points. Spatial databases address these and many other related requirements.

Example: Spatial extensions to relational databases

Spatial extensions have been developed for many relational databases: for example, Oracle Spatial, DB2 Spatial, and SQL Server Spatial. We use the PostGIS extensions to the PostgreSQL relational database here. These extensions implement support for spatial data types such as `point`, `line`, and `polygon`, and operations such as `st_within` (returns `true` if one object is contained within another), `st_dwithin` (returns `true` if two objects are within a specified distance of each other), and `st_distance` (returns the distance between two objects). Thus, for example, given two tables with rows for schools and hospitals in Illinois (`illinois_schools` and `illinois_hospitals`, respectively; in each case, the column `the_geom` is a polygon for the object in question) and a third table with a single row representing the city of Chicago (`chicago_citylimits`), we can easily find the names of all schools within the Chicago city limits:

```
select illinois_schools.name
  from illinois_schools, chicago_citylimits
  where st_within(illinois_schools.the_geom,
                  chicago_citylimits.the_geom);
```

We join the two tables `illinois_schools` and `chicago_citylimits`, with the `st_within` constraint constraining the selected rows to those representing schools within the city limits. Here we use the inner join introduced in Section 4.3.2. This query could also be written as:

```
select illinois_schools.name
  from illinois_schools left join chicago_citylimits
  on st_within(illinois_schools.the_geom,
                  chicago_citylimits.the_geom);
```

We can also determine the names of all schools that do *not* have a hospital within 3,000 meters:

```
select s.name as 'School Name'
  from illinois_schools as s
    left join illinois_hospitals as h
      on st_dwithin(s.the_geom, h.the_geom, 3000)
  where h.gid is null;
```

Here, we use an alternative form of the join operator, the *left join*—or, more precisely, the *left excluding join*. The expression

```
table1 left join table2 on constraint
```

Table 4.4. Three types of *join* illustrated: the inner join, as used in Section 4.3.2, the left join, and left excluding join

Inner join	Left join	Left excluding join
A B	A B	A B
select *columns* from Table_A A inner join Table_B B on A.Key = B.Key	select *columns* from Table_A A left join Table_B B on A.Key = B.Key	select *columns* from Table_A A left join Table_B B on A.Key = B.Key where B.Key is null

returns all rows from the left table (table1) with the matching rows in the right table (table2), with the result being null in the right side when there is no match. This selection is illustrated in the middle column of Table 4.4. The addition of the where h.gid is null then selects only those rows in the left table with no right-hand match, as illustrated in the right-hand column of Table 4.4. Note also the use of the as operator to rename the columns illinois_schools and illinois_hospitals. In this case, we rename them simply to make our query more compact.

4.7 Which database to use?

The question of which DBMS to use for a social sciences data management and analysis project depends on many factors. We introduced some relevant rules in Table 4.1. We expand on those considerations here.

4.7.1 Relational DBMSs

If your data are structured, then a relational DBMS is almost certainly the right technology to use. Many open source, commercial, and cloud-hosted relational DBMSs exist. Among the open source DBMSs, MySQL and PostgreSQL (often simply Postgres) are particularly widely used. MySQL is the most popular. It is particularly easy to install and use, but does not support all features of the SQL standard. PostgreSQL is fully standard compliant and supports useful features such as full text search and the PostGIS extensions mentioned in the previous section, but can be more complex to work with.

Popular commercial relational DBMSs include IBM DB2, Microsoft SQL Server, and Oracle RDMS. These systems are heavily used in commercial settings. There are free community editions, and some large science projects use enterprise features via academic licensing: for example, the Sloan Digital Sky Survey uses Microsoft SQL Server [367] and the CERN high-energy physics lab uses Oracle [132].

We also see increasing use being made of cloud-hosted relational DBMSs such as Amazon Relational Database Service (RDS; this supports MySQL, PostgreSQL, and various commercial DBMSs), Microsoft Azure, and Google Cloud SQL. These systems obviate the need to install local software, administer a DBMS, or acquire hardware to run and scale your database. Particularly if your database is bigger than can fit on your workstation, a cloud-hosted solution can be a good choice.

4.7.2 NoSQL DBMSs

Few social science problems have the scale that might motivate the use of a NoSQL DBMS. Furthermore, while defining and enforcing a schema can involve some effort, the benefits of so doing are considerable. Thus, the use of a relational DBMS is usually to be recommended.

Nevertheless, as noted in Section 4.2, there are occasions when a NoSQL DBMS can be a highly effective, such as when working with large quantities of unstructured data. For example, researchers analyzing large collections of Twitter messages frequently store the messages in a NoSQL document-oriented database such as MongoDB. NoSQL databases are also often used to organize large numbers of records from many different sources, as illustrated in Figure 4.1.

4.8 Summary

A key message of this book is that you should, whenever possible, use a database. Database management systems are one of the great achievements of information technology, permitting large amounts of data to be stored and organized so as to allow rapid and reliable exploration and analysis. They have become a central component of a great variety of applications, from handling transactions in financial systems to serving data published in websites. They are particularly well suited for organizing social science data and for supporting analytics for data exploration.

DBMSs provide an environment that greatly simplifies data management and manipulation. They make many easy things trivial, and many hard things easy. They automate many other error-prone, manual tasks associated with query optimization. While they can by daunting to those unfamiliar with their concepts and workings, they are in fact easy to use. A basic understanding of databases and of when and how to use DBMSs is an important element of the social data scientist's knowledge base.

4.9 Resources

The enormous popularity of DBMSs means that there are many good books to be found. Classic textbooks such as those by Silberschatz et al. [347] and Ramakrishnan and Gherke [316] provide a great deal of technical detail. The DB Engines website collects information on DBMSs [351]. There are also many also useful online tutorials, and of course StackExchange and other online forums often have answers to your technical questions.

Turning to specific technologies, the *SQL Cookbook* [260] provides a wonderful introduction to SQL. We also recommend the SQL Cheatsheet [22] and a useful visual depiction of different SQL join operators [259]. Two good books on the PostGIS geospatial extensions to the PostgreSQL database are the *PostGIS Cookbook* [85] and *PostGIS in Action* [285]. The online documentation is also excellent [306]. The monograph *NoSQL Databases* [364] provides much useful technical detail.

We did not consider in this chapter the native extensible Markup Language (XML) and Resource Description Framework (RDF) triple stores, as these are not typically used for data management. However, they do play a fundamental role in metadata and knowledge management. See, for example, Sesame [53, 336]

If you are interested in the state of database and data management research, the recent Beckman Report [2] provides a useful perspective.

Chapter 5

Programming with Big Data

Huy Vo and Claudio Silva

Big data is sometimes defined as data that are too big to fit onto the analyst's computer. This chapter provides an overview of clever programming techniques that facilitate the use of data (often using parallel computing). While the focus is on one of the most widely used big data programming paradigms and its most popular implementation, Apache Hadoop, the goal of the chapter is to provide a conceptual framework to the key challenges that the approach is designed to address.

5.1 Introduction

There are many definitions of big data, but perhaps the most popular one is from Laney's report [229] in 2001 that defines big data as data with the three Vs: *volume* (large data sets), *velocity* (real-time streaming data), and *variety* (various data forms). Other authors have since proposed additional Vs for various purposes: *veracity*, *value*, *variability*, and *visualization*. It is common nowadays to see big data being associated with five or even seven Vs—the big data definition itself is also getting "big" and getting difficult to keep track. A simpler but also broader definition of big data is data sets that are "so large or complex that traditional data processing applications are inadequate," as described by Wikipedia. We note that in this case, the definition adapts to the task at hand, and it is also tool dependent.

For example, we could consider a problem to involve big data if a spreadsheet gets so large that Excel can no longer load the entire data set into memory for analysis—or if there are so many features in a data set that a machine learning classifier would take an unreasonable amount of time (say days) to finish instead of a few seconds

▶ See Chapter 6, in particular Section 6.5.2.

or minutes. In such cases, the analyst has to develop customized solutions to interrogate the data, often by taking advantage of parallel computing. In particular, one may have to get a larger computer, with more memory and/or stronger processors, to cope with expensive computations; or get a cluster of machines to speed up the computing time by distributing the workload among them. While the former solution would not scale well due to the limited amount of processors a single computer can have, the latter solution needs to deal with nontraditional programming infrastructures. Big data technologies, in a nutshell, are the technologies that make these infrastructures more usable by users without a computer science background.

Parallel computing and big data are hardly new ideas for dealing with computational challenges. Scientists have routinely been working on data sets much larger than a single machine can handle for several decades, especially at the DOE National Laboratories [87, 337] where high-performance computing has been a major technology trend. This is also demonstrated by the history of research in distributed computing and data management going back to the 1980s [401].

There are major technological differences between this type of big data technology and what is covered in this chapter. True parallel computing involves designing clever parallel algorithms, often from scratch, taking into account machine-dependent constraints to achieve maximum performance from the particular architecture. They are often implemented using message passing libraries, such as implementations of the Message Passing Interface (MPI) standard [142] (available since the mid-1990s). Often the expectation is that the code developed will be used for substantial computations. In that scenario it makes sense to optimize a code for maximum performance knowing the same code will be used repeatedly. Contrast that with big data for exploration, where we might need to keep changing the analysis code very often. In this case, what we are trying to optimize is the analyst's time, not computing time.

With storage and networking getting significantly cheaper and faster, big data sets could easily become available to data enthusiasts with just a few mouse clicks, e.g., the Amazon Web Service Public Data Sets [11]. These enthusiasts may be policymakers, government employees, or managers who would like to draw insights and (business) value from big data. Thus, it is crucial for big data to be made available to nonexpert users in such a way that they can process the data without the need for a supercomputing expert. One such approach is to build big data frameworks within which commands can be implemented just as they would in a small data framework. Also, such a framework should be as simple as possible, even if not as efficient as custom-designed parallel solutions. Users should expect that if their code works within these frameworks for small data, it will also work for big data.

In order to achieve scalability for big data, many frameworks only implement a small subset of data operations and fully automate the parallelism of these operations. Users are expected to develop their codes using only the subset if they expect their code to scale to large data sets. MapReduce, one of the most widely used big data programming paradigms, is no exception to this rule. As its name suggests, the framework only supports two operations: map and reduce. The next sections will provide an overview of MapReduce and its most popular implementation, Apache Hadoop.

5.2 The MapReduce programming model

The MapReduce framework was proposed by Jeffrey Dean and Sanjay Ghemawat at Google in 2004 [91]. Its origins date back to conceptually similar approaches first described in the early 1980s. MapReduce was indeed inspired by the *map* and *reduce* functions of functional programming, though its `reduce` function is more of a *group-by-key* function, producing a list of values, instead of the traditional reduce, which outputs only a single value. MapReduce is a record-oriented model, where each record is treated as a *key-value* pair; thus, both `map` and `reduce` functions operate on key-value pair data.

A typical MapReduce job is composed of three phases—map, shuffle, and reduce—taking a list of key-value pairs `[(k$_1$,v$_2$), (k$_2$,v$_2$), ..., (k$_n$,v$_n$)]` as input. In the map phase, each input key-value pair is run through the `map` function, and zero or more new key-value pairs are output. In the shuffle phase, the framework sorts the outputs of the map phase, grouping pairs by keys before sending each of them to the `reduce` function. In the reduce phase, each grouping of values are processed by the `reduce` function, and the result is a list of new values that are collected for the job output.

In brief, a MapReduce job just takes a list of key-value pairs as input and produces a list of values as output. Users only need to implement interfaces of the `map` and `reduce` functions (the shuffle phase is not very customizable) and can leave it up to the system that implements MapReduce to handle all data communications and parallel computing. We can summarize the MapReduce logic as follows:

map: $(k_i,\ v_i) \rightarrow [f(k'_{i1},\ v'_{i1}), f(k'_{i2},\ v'_{i2}), ..]$
for a user-defined function f.
reduce: $(k''_i,\ [v''_{i1}, v''_{i2}, ..]) \rightarrow [v'''_1, v'''_2, ..]$
where $\{k'_i\} \equiv \{k''_j\}$ and $v''' = g(v'')$ for a user-defined function g.

Example: Counting NSF awards

To gain a better understanding of these MapReduce operators, imagine that we have a list of NSF principal investigators, along with their email information and award IDs as below. Our task is to count the number of awards for each institution. For example, given the four records below, we will discover that the Berkeley Geochronology Center has two awards, while New York University and the University of Utah each have one.

```
AwardId,FirstName,LastName,EmailAddress
0958723,Roland,Mundil,rmundil@bgc.org
0958915,Randall,Irmis,irmis@umnh.utah.edu
1301647,Zaher,Hani,zh8@nyu.edu
1316375,David,Shuster,dshuster@bgc.org
```

We observe that institutions can be distinguished by their email address domain name. Thus, we adopt of a strategy of first grouping all award IDs by domain names, and then counting the number of distinct award within each group. In order to do this, we first set the `map` function to scan input lines and extract institution information and award IDs. Then, in the `reduce` function, we simply count unique IDs on the data, since everything is already grouped by institution. Python pseudo-code is provided in Listing 5.1.

In the map phase, the input will be transformed into tuples of institutions and award ids:

```
"0958723,Roland,Mundil,rmundil@bgc.org"       →  ("bgc.org", 0958723)
"0958915,Randall,Irmis,irmis@umnh.utah.edu"   →  ("utah.edu", 0958915)
"1301647,Zaher,Hani,zh8@nyu.edu"              →  ("nyu.edu", 1301647)
"1316375,David,Shuster,dshuster@bgc.org"      →  ("bgc.org", 1316375)
```

Then the tuples will be grouped by institutions and be counted by the `reduce` function.

```
("bgc.org", [0958723,1316375])  →  ("bgc.org", 2)
("utah.edu", [0958915])         →  ("utah.edu", 1)
("nyu.edu", [1301647])          →  ("nyu.edu", 1)
```

```
# Input  : a list of text lines
# Output : a list of domain name and award ids
def MAP(lines):
    for line in lines:
        fields     = line.strip('\n').split(',')
        awardId    = fields[0]
        domainName = fields[3].split('@')[-1].split('.')[-2:]
        yield (domainName, awardId)

# Input  : a list of domain name and award ids
# Output : a list of domain name and award count
def REDUCE(pairs):
    for (domainName, awardIds) in pairs:
        count = len(set(awardIds))
        yield (domainName, count)
```

Listing 5.1. Python pseudo-code for the `map` and `reduce` functions to count the number of awards per institution

As we have seen so far, the MapReduce programming model is quite simple and straightforward, yet it supports a simple parallelization model. In fact, it has been said to be *too* simple and criticized as "a major step backwards" [94] for large-scale, data-intensive applications. It is hard to argue that MapReduce is offering something truly innovative when MPI has been offering similar scatter and reduce operations since 1995, and Python has had high-order functions (`map`, `reduce`, `filter`, and `lambda`) since its 2.2 release in 1994. However, the biggest strength of MapReduce is its simplicity. Its simple programming model has brought many nonexpert users to big data analysis. Its simple architecture has also inspired many developers to develop advanced capabilities, such as support for distributed computing, data partitioning, and streaming processing. (A downside of this diversity of interest is that available features and capabilities can vary considerably, depending on the specific implementation of MapReduce that is being used.)

We next describe two specific implementations of the MapReduce model: Hadoop and Spark.

5.3 Apache Hadoop MapReduce

The names MapReduce and Apache Hadoop (or Hadoop)* are often used interchangeably, but they are conceptually different. MapReduce is simply a programming paradigm with a layer of abstraction that allows a set of data processing pipelines to be expressed without too much tailoring for how it will be executed exactly. The

★ The term *Hadoop* refers to the creator's son's toy elephant.

MapReduce model tells us which class of data structure (key-value pairs) and data transformations (map and reduce) it is supporting; however, it does not specifically state how the framework should be implemented; for example, it does not specify how data should be stored or how the computation be executed (in particular, parallelized). Hadoop [399], on the other hand, is a specific implementation of MapReduce with exact specifications of how data and computation are handled inside the system.

Hadoop was originally designed for batch data processing at scale, with the target of being able to run in environments with thousands of machines. Supporting such a large computing environment puts several constraints on the system; for instance, with so many machines, the system had to assume computing nodes would fail. Hadoop is an enhanced MapReduce implementation with the support for fault tolerance, distributed storage, and data parallelism through two added key design features: (1) a distributed file system called the Hadoop Distributed File System (HDFS); and (2) a data distribution strategy that allows computation to be moved to the data during execution.

5.3.1 The Hadoop Distributed File System

The Hadoop Distributed File System [345] is arguably the most important component that Hadoop added to the MapReduce framework. In a nutshell, it is a distributed file system that stripes data across all the nodes of a Hadoop cluster. HDFS splits large data files into smaller blocks which are managed by different nodes in the cluster. Each block is also replicated across several nodes as an attempt to ensure that a full copy of the data is still available even in the case of computing node failures. The block size as well as the number of replications per block are fully customized by users when they create files on HDFS. By default, the block size is set to 64 MB with a replication factor of 3, meaning that the system may encounter at least two concurrent node failures without losing any data. HDFS also actively monitors failures and re-replicates blocks on failed nodes to make sure that the number of replications for each block always stays at the user-defined settings. Thus, if a node fails, and only two copies of some data exist, the system will quickly copy those data to a working node, thus raising the number of copies to three again. This dynamic replication the primary mechanism for fault tolerance in Hadoop.

Note that data blocks are replicated and distributed across several machines. This could create a problem for users, because if they had to manage the data manually, they might, for example, have to access more than one machine to fetch a large data file. Fortunately, Hadoop provides infrastructure for managing this complexity, including command line programs as well as an API that users can employ to interact with HDFS as if it were a local file system.

This is one example that reinforces our discussion on big data technology being all about making things work seamlessly regardless of the computational environment; it should be possible for the user to use the system as though they are using their local workstation. Hadoop and HDFS are great examples of this approach. For example, one can run `ls` and `mkdir` to list and create a directory on HDFS, or even use `tail` to inspect file contents as one would expect in a Linux file system. The following code shows some examples of interacting with HDFS.

```
# Creating a temporary folder
hadoop dfs -mkdir /tmp/mytmp

# Upload a CSV file from our local machine to HDFS
hadoop dfs -put myfile.csv /tmp/mytmp

# Listing all files under mytmp folder
hadoop dfs -ls /tmp/mytmp

# Upload another file with five replications and 128MB per block
hadoop -D dfs.replication=5 -D dfs.block.size=128M \
       dfs -put mylargefile.csv /tmp/mytmp

# Download a file to our local machine
hadoop dfs -get /tmp/mytmp/myfile.csv .
```

5.3.2 Hadoop: Bringing compute to the data

The configuration of parallel computing infrastructure is a fairly complex task. At the risk of oversimplifying, we consider the computing environment as comprising a *compute cluster* with substantial computing power (e.g., thousands of computing cores), and a *storage cluster* with petabytes of disk space, capable of storing and serving data quickly to the compute cluster. These two clusters have quite different hardware specifications: the first is optimized for CPU performance and the second for storage occupancy. The two systems are typically configured as separate physical hardware.

Running compute jobs on such hardware often goes like this. When a user requests to run an intensive task on a particular data

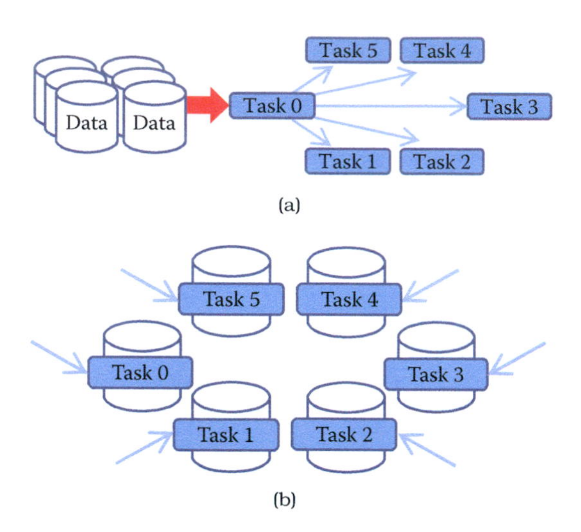

Figure 5.1. (a) The traditional parallel computing model where data are brought to the computing nodes. (b) Hadoop's parallel computing model: bringing compute to the data [242]

set, the system will first reserve a set of computing nodes. Then the data are partitioned and copied from the storage server into these computing nodes before the task is executed. This process is illustrated in Figure 5.1(a). This computing model will be referred to as *bringing data to computation*. In this model, if a data set is being analyzed in multiple iterations, it is very likely that the data will be copied multiple times from the storage cluster to the compute nodes without reusability. This is because the compute node scheduler normally does not have or keep knowledge of where data have previously been held. The need to copy data multiple times tends to make such a computation model inefficient, and I/O becomes the bottleneck when all tasks constantly pull data from the storage cluster (the green arrow). This in turn leads to poor scalability; adding more nodes to the computing cluster would not increase its performance.

To solve this problem, Hadoop implements a *bring compute to the data* strategy that combines both computing and storage at each node of the cluster. In this setup, each node offers both computing power and storage capacity. As shown in Figure 5.1(b), when users submit a task to be run on a data set, the scheduler will first look for nodes that contain the data, and if the nodes are available, it will schedule the task to run directly on those nodes. If a node is busy with another task, data will still be copied to available nodes, but the scheduler will maintain records of the copy for subsequent use of the data. In addition, data copying can be minimized by increasing the data duplication in the cluster, which also increases the potential for parallelism, since the scheduler has more choices to allocate computing without copying. Since both the compute and data storage are closely coupled for this model, it is best suited for data-intensive applications.

Given that Hadoop was designed for batch data processing at scale, this model fits the system nicely, especially with the support of HDFS. However, in an environment where tasks are more compute intensive, a traditional high-performance computing environment is probably best since it tends to spend more resources on CPU cores. It should be clear now that the Hadoop model has hardware implications, and computer architects have optimized systems for data-intensive computing.

Now that we are equipped with the knowledge that Hadoop is a MapReduce implementation that runs on HDFS and a bring-compute-to-the-data model, we can go over the design of a Hadoop MapReduce job. A MapReduce job is still composed of three phases: map, shuffle, and reduce. However, Hadoop divides the map and reduce phases into smaller tasks.

Each map phase in Hadoop is divided into five tasks: input format, record reader, mapper, combiner, and partitioner. An input format task is in charge of talking to the input data presumably sitting on HDFS, and splitting it into partitions (e.g., by breaking lines at line breaks). Then a record reader task is responsible for translating the split data into the key–value pair records so that they can be processed by the mapper. By default, Hadoop parses files into key–value pairs of line numbers and line contents. However, both input formats and record readers are fully customizable and can be programmed to read custom data including binary files. It is important to note that input formats and record readers only provide data partitioning; they do not move data around computing nodes.

After the records are generated, mappers are spawned—typically on nodes containing the blocks—to run through these records and output zero or more new key–value pairs. A mapper in Hadoop is equivalent to the `map` function of the MapReduce model that we discussed earlier. The selection of the key to be output from the mapper will heavily depend on the data processing pipeline and could greatly affect the performance of the framework. Mappers are executed concurrently in Hadoop as long as resources permit.

A combiner task in Hadoop is similar to a `reduce` function in the MapReduce framework, but it only works locally at each node: it takes output from mappers executed on the same node and produces aggregated values. Combiners are optional but can be used to greatly reduce the amount of data exchange in the shuffle phase; thus, users are encouraged to implement this whenever possible. A common practice is when a `reduce` function is both commutative and associative, and has the same input and output format, one can just use the `reduce` function as the combiner. Nevertheless, combiners are not guaranteed to be executed by Hadoop, so this should only be treated as a hint. Its execution must not affect the correctness of the program.

A partitioner task is the last process taking place in the map phase on each mapper node, where it hashes the key of each key-value pair output from the mappers or the combiners into bins. By default, the partitioner uses object hash codes and modulus operations to direct a designated reducer to pull data from a map node. Though it is possible to customize the partitioner, it is only advisable to do so when one fully understands the intermediate data distribution as well as the specifications of the cluster. In general, it is better to leave this job to Hadoop.

Each reduce phase in Hadoop is divided into three tasks: reducer, output format, and record writer. The `reducer` task is equivalent to the `reduce` function of the MapReduce model. It basically groups the data produced by the mappers by keys and runs a `reduce` function on each list of grouping values. It outputs zero or more key–value pairs for the output format task, which then translates them into a writable format for the record writer task to serialize on HDFS. By default, Hadoop will separate the key and value with a tab and write separate records on separate lines. However, this behavior is fully customizable. Similarly, the map phase reducers are also executed concurrently in Hadoop.

5.3.3 Hardware provisioning

Hadoop requires a distributed cluster of machines to operate efficiently. (It can be set up to run entirely on a single computer, but this should only be done for technology demonstration purposes.) This is mostly because the MapReduce performance heavily depends on the total I/O throughput (i.e., disk read and write) of the entire system. Having a distributed cluster, where each machine has its own set of hard drives, is one of the most efficient ways to maximize this throughput.

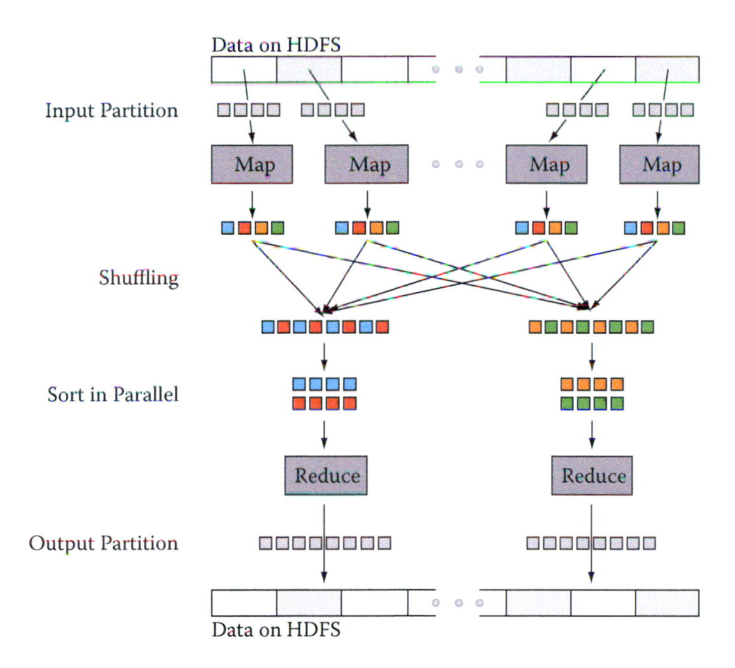

Figure 5.2. Data transfer and communication of a MapReduce job in Hadoop. Data blocks are assigned to several maps, which emit key–value pairs that are shuffled and sorted in parallel. The reduce step emits one or more pairs, with results stored on the HDFS

A typical Hadoop cluster consists of two types of machine: masters and workers. Master machines are those exclusively reserved for running services that are critical to the framework operations. Some examples are the NameNode and the JobTracker services, which are tasked to manage how data and tasks are distributed among the machines, respectively. The worker machines are reserved for data storage and for running actual computation tasks (i.e., map and reduce). It is normal to have worker machines that can be included or removed from an operational cluster on demand. This ability to vary the number of worker nodes makes the overall system more tolerant of failure. However, master machines are usually required to be running uninterrupted.

Provisioning and configuring the hardware for Hadoop, like any other parallel computing, are some of the most important and complex tasks in setting up a cluster, and they often require a lot of experience and careful consideration. Major big data vendors provide guidelines and tools to facilitate the process [18, 23, 81].

```python
#!/usr/bin/env python
import sys

def parseInput():
    for line in sys.stdin:
        yield line

if __name__=='__main__':
    for line in parseInput():
        fields     = line.strip('\n').split(',')
        awardId    = fields[0]
        domainName = fields[3].split('@')[-1].split('.')[-2:]
        print '%s\t%s' % (domainName,awardId)
```

Listing 5.2. A Hadoop streaming mapper in Python

Nevertheless, most decisions will be based on the types of analysis to be run on the cluster, for which only you, as the user, can provide the best input.

5.3.4 Programming language support

Hadoop is written entirely in Java, thus it is best supporting applications written in Java. However, Hadoop also provides a *streaming API* that allows arbitrary code to be run inside the Hadoop MapReduce framework through the use of UNIX pipes. This means that we can supply a mapper program written in Python or C++ to Hadoop as long as that program reads from the standard input and writes to the standard output. The same mechanism also applies for the combiner and reducer. For example, we can develop from the Python pseudo-code in Listing 5.1 to a complete Hadoop streaming mapper (Listing 5.2) and reducer (Listing 5.3).

```python
#!/usr/bin/env python
import sys

def parseInput():
    for line in sys.stdin:
        yield line

if __name__=='__main__':
    for line in parseInput():
        (domainName, awardIds) = line.split('\t')
        count = len(set(awardIds))
        print '%s\t%s' % (domainName, count)
```

Listing 5.3. A Hadoop streaming reducer in Python

It should be noted that in Hadoop streaming, intermediate key-value pairs (the data flowing between mappers and reducers) must be in tab-delimited format, thus we replace the original `yield` command with a `print` formatted with tabs. Though the input format and record reader are still customizable in Hadoop streaming, they must be supplied as Java classes. This is one of the biggest limitations of Hadoop for Python developers. They not only have to split their code into separate mapper and reducer programs, but also need to learn Java if they want to work with nontextual data.

5.3.5 Fault tolerance

By default, HDFS uses checksums to enforce data integrity on its file system use data replication for recovery of potential data losses. Taking advantage of this, Hadoop also maintains fault tolerance of MapReduce jobs by storing data at every step of a MapReduce job to HDFS, including intermediate data from the combiner. Then the system checks whether a task fails by either looking at its heartbeats (data activities) or whether it has been taking too long. If a task is deemed to have failed, Hadoop will kill it and run it again on a different node. The time limit for the heartbeats and task running duration may also be customized for each job. Though the mechanism is simple, it works well on thousands of machines. It is indeed highly robust because of the simplicity of the model.

5.3.6 Limitations of Hadoop

Hadoop is a great system, and probably the most widely used MapReduce implementation. Nevertheless, it has important limitations, as we now describe.

- Performance: Hadoop has proven to be a scalable implementation that can run on thousands of cores. However, it is also known for having a relatively high job setup overheads and suboptimal running time. An empty task in Hadoop (i.e., with no mapper or reducer) can take roughly 30 seconds to complete even on a modern cluster. This overhead makes it unsuitable for real-time data or interactive jobs. The problem comes mostly from the fact that Hadoop monitoring processes only lives within a job, thus it needs to start and stop these processes each time a job is submitted, which in turns results in this major overhead. Moreover, the brute force approach of maintaining fault tolerance by storing everything on HDFS is expensive, especially when for large data sets.

- Hadoop streaming support for non-Java applications: As mentioned previously, non-Java applications may only be integrated with Hadoop through the Hadoop streaming API. However, this API is far from optimal. First, input formats and record readers can only be written in Java, making it impossible to write advanced MapReduce jobs entirely in a different language. Second, Hadoop streaming only communicates with Hadoop through Unix pipes, and there is no support for data passing within the application using native data structure (e.g., it is necessary to convert Python tuples into strings in the mappers and convert them back into tuples again in reducers).

- Real-time applications: With the current setup, Hadoop only supports batch data processing jobs. This is by design, so it is not exactly a limitation of Hadoop. However, given that more and more applications are dealing with real-time massive data sets, the community using MapReduce for real-time processing is constantly growing. Not having support for streaming or real-time data is clearly a disadvantage of Hadoop over other implementations.

- Limited data transformation operations: This is more of a limitation of MapReduce than Hadoop per se. MapReduce only supports two operations, map and reduce, and while these operations are sufficient to describe a variety of data processing pipelines, there are classes of applications that MapReduce is not suitable for. Beyond that, developers often find themselves rewriting simple data operations such as data set joins, finding a min or max, and so on. Sometime, these tasks require more than one map-and-reduce operation, resulting in multiple MapReduce jobs. This is both cumbersome and inefficient. There are tools to automate this process for Hadoop; however, they are only a layer above, and it is not easy to integrate with existing customized Hadoop applications.

5.4 Apache Spark

In addition to Apache Hadoop, other notable MapReduce implementations include MongoDB, GreenplumDB, Disco, Riak, and Spark.

▶ See Chapter 4.

MongoDB, Riak, and Greenplum DB are all database systems and thus their MapReduce implementations focus more on the interoperability of MapReduce and the core components such as MongoDB's aggregation framework, and leave it up to users to customize the MapReduce functionalities for broader tasks. Some of these systems, such as Riak, only parallelize the map phase, and run the reduce phase on the local machine that request the tasks. The main advantage of the three implementations is the ease with which they connect to specific data stores. However, their support for general data processing pipelines is not as extensive as that of Hadoop.

Disco, similar to Hadoop, is designed to support MapReduce in a distributed computing environment, but it written in Erlang with a Python interface. Thus, for Python developers, Disco might be a better fit. However, it has significantly fewer supporting applications, such as access control and workflow integration, as well as a smaller developing community. This is why the top three big data platforms, Cloudera, Hortonworks, and MapR, still build primarily on Hadoop.

Apache Spark is another implementation that aims to support beyond MapReduce. The framework is centered around the concept of resilient distributed data sets and data transformations that can operate on these objects. An innovation in Spark is that the fault tolerance of resilient distributed data sets can be maintained without flushing data onto disks, thus significantly improving the system performance (with a claim of being 100 times faster than Hadoop). Instead, the fault-recovery process is done by replaying a log of data transformations on check-point data. Though this process could take longer than reading data straight from HDFS, it does not occur often and is a fair tradeoff between processing performance and recovery performance.

Beyond map and reduce, Spark also supports various other transformations [147], including filter, data join, and aggregation. Streaming computation can also be done in Spark by asking Spark to reserve resources on a cluster to constantly stream data to/from the cluster. However, this streaming method might be resource intensive (still consuming resources when there is no data coming). Additionally, Spark plays well with the Hadoop ecosystem, particularly with the distributed file system (HDFS) and resource manager (YARN), making it possible to be built on top of current Hadoop applications.

Another advantage of Spark is that it supports Python natively; thus, developers can run Spark in a fraction of the time required for Hadoop. Listing 5.4 provides the full code for the previous example

```
import sys
from pyspark import SparkContext
def mapper(lines):
    for line in lines:
        fields     = line.strip('\n').split(',')
        awardId    = fields[0]
        domainName = fields[3].split('@')[-1].split('.')[-2:]
        yield (domainName, awardId)

def reducer(pairs):
    for (domainName, awardIds) in pairs:
        count = len(set(awardIds))
        yield (domainName, count)

if __name__=='__main__':
    hdfsInputPath  = sys.argv[1]
    hdfsOutputFile =  sys.argv[2]
    sc = SparkContext(appName="Counting Awards")
    output = sc.textFile(hdfsInputPath) \
        .mapPartitions(mapper) \
        .groupByKey() \
        .mapPartitions(reducer)

    output.saveAsTextFile(hdfsInputPath)
```

Listing 5.4. Python code for a Spark program that counts the number of awards per institution using MapReduce

written entirely in Spark. It should be noted that Spark's concept of the `reduceByKey` operator is not the same as Hadoop's, as it is designed to aggregate all elements of a data set into a single element. The closest simulation of Hadoop's MapReduce pattern is a combination of `mapPartitions`, `groupByKey` and `mapPartitions`, as shown in the next example.

Example: Analyzing home mortgage disclosure application data

We use a financial services analysis problem to illustrate the use of Apache Spark.

Mortgage origination data provided by the Consumer Protection Financial Bureau provide insightful details of the financial health of the real estate market. The data [84], which are a product of the Home Mortgage Disclosure Act (HMDA), highlight key attributes that function as strong indicators of health and lending patterns.

Lending institutions, as defined by section 1813 in Title 12 of the HMDA, decide on whether to originate or deny mortgage applications based on credit risk. In order to determine this credit risk, lenders must evaluate certain features relative

Table 5.1. Home Mortgage Disclosure Act data size

Year	Records	File Size (Gigabytes)
2007	26,605,696	18
2008	17,391,571	12
2009	19,493,492	13
2010	16,348,558	11
2011	14,873,416	9.4
2012	18,691,552	12
2013	17,016,160	11
Total	**130,420,445**	**86.4**

Table 5.2. Home Mortgage Disclosure Act data fields

Index	Attribute	Type
0	Year	Integer
1	State	String
2	County	String
3	Census Tract	String
4	Loan Amount	Float
5	Applicant Income	Float
6	Loan Originated	Boolean
...	...	...

to the applicant, the underlying property, and the location. We want to determine whether census tract clusters could be created based on mortgage application data and whether lending institutions' perception of risk is held constant across the entire USA.

For the first step of this process, we study the debt-income ratio for loans originating in different census tracts. This could be achieved simply by computing the debt-income ratio for each loan application and aggregating them for each year by census tract number. A challenge, however, is that the data set provided by HMDA is quite extensive. In total, HMDA data contain approximately 130 million loan applications between 2007 and 2013. As each record contains 47 attributes, varying in types from continuous variables such as loan amounts and applicant income to categorical variables such as applicant gender, race, loan type, and owner occupancy, the entire data set results in about 86 GB of information. Parsing the data alone could take up to hours on a single machine if using a naïve approach that scans through the data sequentially. Tables 5.1 and 5.2 highlight the breakdown in size per year and data fields of interest.

Observing the transactional nature of the data, where the aggregation process could be distributed and merged across multiple partitions of the data, we could complete this task in much less time by using Spark. Using a cluster consisting of 1,200 cores, the Spark program in Listing 5.5 took under a minute to complete. The substantial performance gain comes not so much from the large number of processors available, but mostly from the large I/O bandwidth available on the cluster thanks to the 200 distributed hard disks and fast network interconnects.

5.5 Summary

Big data means that it is necessary to both store very large collections of data and perform aggregate computations on those data. This chapter spelled out an important data storage approach (the Hadoop Distributed File System) and a way of processing large-scale

```
import ast
import sys
from pyspark import SparkContext

def mapper(lines):
    for line in lines:
        fields = ast.literal_eval('(%s)' % line)
        (year, state, county, tract) = fields[:4]
        (amount, income, originated) = fields[4:]

        key = (year, state, county, tract)
        value = (amount, income)

        # Only count originated loans
        if originated:
            yeild (key, value)

def sumDebtIncome(debtIncome1, debtIncome2):
    return (debtIncome1[0] + debtIncome2[0], debtIncome1[1] +
    debtIncome2[1])

if __name__=='__main__':
    hdfsInputPath  = sys.argv[1]
    hdfsOutputFile =  sys.argv[2]
    sc = SparkContext(appName="Counting Awards")
    sumValues = sc.textFile(hdfsInputPath) \
        .mapPartitions(mapper) \
        .reduceByKey(sumDebtIncome)

    # Actually compute the aggregated debt income
    output = sumValues.mapValues(lambda debtIncome: debtIncome[0]/
    debtIncome[1])

    output.saveAsTextFile(hdfsInputPath)
```

Listing 5.5. Python code for a Spark program to aggregate the debt–income ratio for loans originated in different census tracts

data sets (the MapReduce model, as implemented in both Hadoop and Spark). This in-database processing model enables not only high-performance analytics but also the provision of more flexibility for analysts to work with data. Instead of going through a database administrator for every data gathering, data ingestion, or data transformation task similar to a traditional data warehouse approach, the analysts rather "own" the data in the big data environment. This increases the analytic throughput as well as the time to insight, speeding up the decision-making process and thus increasing the business impact, which is one of the main drivers for big data analytics.

5.6 Resources

There are a wealth of online resources describing both Hadoop and Spark. See, for example, the tutorials on the Apache Hadoop [19] and Spark [20] websites. Albanese describes how to use Hadoop for social science [9], and Lin and Dyer discuss the use of MapReduce for text analysis [238].

We have not discussed here how to deal with data that are located remotely from your computer. If such data are large, then moving them can be an arduous task. The Globus transfer service [72] is commonly used for the transfer and sharing of such data. It is available on many research research systems and is easy to install on your own computer.

Part II
Modeling and Analysis

Chapter 6

Machine Learning

Rayid Ghani and Malte Schierholz

This chapter introduces you to the value of machine learning in the social sciences, particularly focusing on the overall machine learning process as well as clustering and classification methods. You will get an overview of the machine learning pipeline and methods and how those methods are applied to solve social science problems. The goal is to give an intuitive explanation for the methods and to provide practical tips on how to use them in practice.

6.1 Introduction

You have probably heard of "machine learning" but are not sure exactly what it is, how it differs from traditional statistics, and what you can do with it. In this chapter, we will demystify machine learning, draw connections to what you already know from statistics and data analysis, and go deeper into some of the unique concepts and methods that have been developed in this field. Although the field originates from computer science (specifically, artificial intelligence), it has been influenced quite heavily by statistics in the past 15 years. As you will see, many of the concepts you will learn are not entirely new, but are simply called something else. For example, you already are familiar with logistic regression (a classification method that falls under the supervised learning framework in machine learning) and cluster analysis (a form of unsupervised learning). You will also learn about new methods that are more exclusively used in machine learning, such as random forests and support vector machines. We will keep formalisms to a minimum and focus on getting the intuition across, as well as providing practical tips. Our hope is this chapter will make you comfortable and familiar with machine learning vocabulary, concepts, and processes, and allow you to further explore and use these methods and tools in your own research and practice.

6.2 What is machine learning?

When humans improve their skills with experience, they are said to learn. Is it also possible to program computers to do the same? Arthur Samuel, who coined the term *machine learning* in 1959 [323], was a pioneer in this area, programming a computer to play checkers. The computer played against itself and human opponents, improving its performance with every game. Eventually, after sufficient *training* (and experience), the computer became a better player than the human programmer. Today, machine learning has grown significantly beyond learning to play checkers. Machine learning systems have learned to drive (and park) autonomous cars, are embedded inside robots, can recommend books, products, and movies we are (sometimes) interested in, identify drugs, proteins, and genes that should be investigated further to cure diseases, detect cancer and other diseases in medical imaging, help us understand how the human brain learns language, help identify which voters are persuadable in elections, detect which students are likely to need extra support to graduate high school on time, and help solve many more problems. Over the past 20 years, machine learning has become an interdisciplinary field spanning computer science, artificial intelligence, databases, and statistics. At its core, machine learning seeks to design computer systems that improve over time with more experience. In one of the earlier books on machine learning, Tom Mitchell gives a more operational definition, stating that: "A computer program is said to learn from experience E with respect to some class of tasks T and performance measure P, if its performance at tasks in T, as measured by P, improves with experience E" [258].

Machine learning grew from the need to build systems that were adaptive, scalable, and cost-effective to build and maintain. A lot of tasks now being done using machine learning used to be done by rule-based systems, where experts would spend considerable time and effort developing and maintaining the rules. The problem with those systems was that they were rigid, not adaptive, hard to scale, and expensive to maintain. Machine learning systems started becoming popular because they could improve the system along all of these dimensions. Box 6.1 mentions several examples where machine learning is being used in commercial applications today. Social scientists are uniquely placed today to take advantage of the same advances in machine learning by having better methods to solve several key problems they are tackling. We will give concrete examples later in this chapter.

▸ See Chapter 3.

Box 6.1: Commercial machine learning examples

- Speech recognition: Speech recognition software uses machine learning algorithms that are built on large amounts of initial training data. Machine learning allows these systems to be tuned and adapt to individual variations in speaking as well as across different domains.

- Autonomous cars: The ongoing development of self-driving cars applies techniques from machine learning. An onboard computer continuously analyzes the incoming video and sensor streams in order to monitor the surroundings. Incoming data are matched with annotated images to recognize objects like pedestrians, traffic lights, and potholes. In order to assess the different objects, huge training data sets are required where similar objects already have been identified. This allows the autonomous car to decide on which actions to take next.

- Fraud detection: Many public and private organizations face the problem of fraud and abuse. Machine learning systems are widely used to take historical cases of fraud and flag fraudulent transactions as they take place. These systems have the benefit of being adaptive, and improving with more data over time.

- Personalized ads: Many online stores have personalized recommendations promoting possible products of interest. Based on individual shopping history and what other similar users bought in the past, the website predicts products a user may like and tailors recommendations. Netflix and Amazon are two examples of companies whose recommendation software predicts how a customer would rate a certain movie or product and then suggests items with the highest predicted ratings. Of course there are some caveats here, since they then adjust the recommendations to maximize profits.

- Face recognition: Surveillance systems, social networking platforms, and imaging software all use face detection and face recognition to first detect faces in images (or video) and then tag them with individuals for various tasks. These systems are trained by giving examples of faces to a machine learning system which then learns to detect new faces, and tag known individuals.

This chapter is not an exhaustive introduction to machine learning. There are many books that have done an excellent job of that [124, 159, 258]. Instead, we present a short and understandable introduction to machine learning for social scientists, give an overview of the overall machine learning process, provide an intuitive introduction to machine learning methods, give some practical tips that will be helpful in using these methods, and leave a lot of the statistical theory to machine learning textbooks. As you read more about machine learning in the research literature or the media, you will encounter names of other fields that are related (and practically the same for most social science audiences), such as statistical learning, data mining, and pattern recognition.

6.3 The machine learning process

▶ See Chapter 10.

When solving problems using machine learning methods, it is important to think of the larger data-driven problem-solving process of which these methods are a small part. A typical machine learning problem requires researchers and practitioners to take the following steps:

1. Understand the problem and goal: This sounds obvious but is often nontrivial. Problems typically start as vague descriptions of a goal—improving health outcomes, increasing graduation rates, understanding the effect of a variable X on an outcome Y, etc. It is really important to work with people who understand the domain being studied to dig deeper and define the problem more concretely. What is the analytical formulation of the metric that you are trying to optimize?

2. Formulate it as a machine learning problem: Is it a classification problem or a regression problem? Is the goal to build a model that generates a ranked list prioritized by risk, or is it to detect anomalies as new data come in? Knowing what kinds of tasks machine learning can solve will allow you to map the problem you are working on to one or more machine learning settings and give you access to a suite of methods.

3. Data exploration and preparation: Next, you need to carefully explore the data you have. What additional data do you need or have access to? What variable will you use to match records for integrating different data sources? What variables exist in the data set? Are they continuous or categorical? What about

missing values? Can you use the variables in their original form or do you need to alter them in some way?

4. Feature engineering: In machine learning language, what you might know as independent variables or predictors or factors or covariates are called "features." Creating good features is probably the most important step in the machine learning process. This involves doing transformations, creating interaction terms, or aggregating over data points or over time and space.

5. Method selection: Having formulated the problem and created your features, you now have a suite of methods to choose from. It would be great if there were a single method that always worked best for a specific type of problem, but that would make things too easy. Typically, in machine learning, you take a collection of methods and try them out to empirically validate which one works the best for your problem. We will give an overview of leading methods that are being used today in this chapter.

6. Evaluation: As you build a large number of possible models, you need a way to select the model that is the best. This part of the chapter will cover the validation methodology to first validate the models on historical data as well as discuss a variety of evaluation metrics. The next step is to validate using a field trial or experiment.

7. Deployment: Once you have selected the best model and validated it using historical data as well as a field trial, you are ready to put the model into practice. You still have to keep in mind that new data will be coming in, and the model might change over time. We will not cover too much of those aspects in this chapter, but they are important to keep in mind.

6.4 Problem formulation: Mapping a problem to machine learning methods

When working on a new problem, one of the first things we need to do is to map it to a class of machine learning methods. In general, the problems we will tackle, including the examples above, can be grouped into two major categories:

1. Supervised learning: These are problems where there exists a target variable (continuous or discrete) that we want to predict

or classify data into. Classification, prediction, and regression all fall into this category. More formally, supervised learning methods predict a value Y given input(s) X by learning (or estimating or fitting or training) a function F, where $F(X) = Y$. Here, X is the set of variables (known as *features* in machine learning, or in other fields as *predictors*) provided as input and Y is the target/dependent variable or a *label* (as it is known in machine learning).

The goal of supervised learning methods is to search for that function F that best predicts Y. When the output Y is categorical, this is known as *classification*. When Y is a continuous value, this is called *regression*. Sound familiar?

One key distinction in machine learning is that the goal is not just to find the best function F that can predict Y for observed outcomes (known Ys) but to find one that best generalizes to new, unseen data. This distinction makes methods more focused on generalization and less on just fitting the data we have as best as we can. It is important to note that you do that implicitly when performing regression by not adding more and more higher-order terms to get better fit statistics. By getting better fit statistics, we *overfit* to the data and the performance on new (unseen) data often goes down. Methods like the lasso [376] penalize the model for having too many terms by performing what is known as *regularization*.*

★ In statistical terms, regularization is an attempt to avoid overfitting the model.

2. Unsupervised learning: These are problems where there does not exist a target variable that we want to predict but we want to understand "natural" groupings or patterns in the data. Clustering is the most common example of this type of analysis where you are given X and want to group similar Xs together. Principal components analysis (PCA) and related methods also fall into the unsupervised learning category.

In between the two extremes of supervised and unsupervised learning, there is a spectrum of methods that have different levels of supervision involved (Figure 6.1). Supervision in this case is the presence of target variables (known in machine learning as *labels*). In unsupervised learning, none of the data points have labels. In supervised learning, all data points have labels. In between, either the percentage of examples with labels can vary or the types of labels can vary. We do not cover the weakly supervised and semi-supervised methods much in this chapter, but this is an active area of research in machine learning. Zhu [414] provides more details.

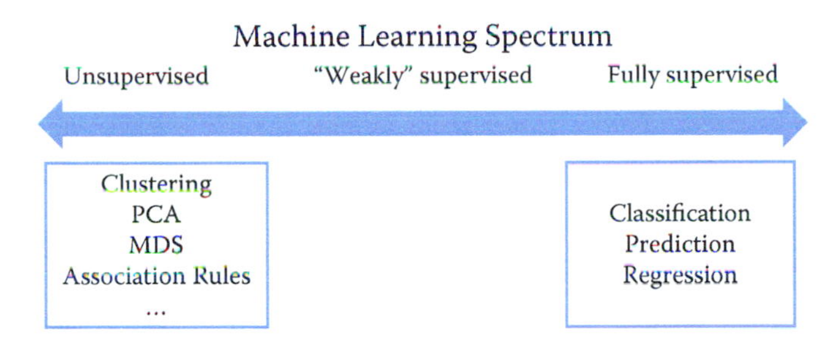

Figure 6.1. Spectrum of machine learning methods from unsupervised to supervised learning

6.5 Methods

We will start by describing unsupervised learning methods and then go on to supervised learning methods. We focus here on the intuition behind the methods and the algorithm, as well as practical tips, rather than on the statistical theory that underlies the methods. We encourage readers to refer to machine learning books listed in Section 6.11 for more details. Box 6.2 gives brief definitions of several terms we will use in this section.

6.5.1 Unsupervised learning methods

As mentioned earlier, unsupervised learning methods are used when we do not have a target variable to predict but want to understand "natural" clusters or patterns in the data. These methods are often used for initial data exploration, as in the following examples:

1. When faced with a large corpus of text data—for example, email records, congressional bills, speeches, or open-ended free-text survey responses—unsupervised learning methods are often used to understand and get a handle on what the data contain.

2. Given a data set about students and their behavior over time (academic performance, grades, test scores, attendance, etc.), one might want to understand typical behaviors as well as trajectories of these behaviors over time. Unsupervised learning methods (clustering) can be applied to these data to get student "segments" with similar behavior.

3. Given a data set about publications or patents in different fields, we can use unsupervised learning methods (association

Box 6.2: Machine learning vocabulary

- Learning: In machine learning, you will notice the term *learning* that will be used in the context of "learning" a model. This is what you probably know as *fitting* or *estimating* a function, or *training* or *building* a model. These terms are all synonyms and are used interchangeably in the machine learning literature.
- Examples: These are data points and instances.
- Features: These are independent variables, attributes, predictor variables, and explanatory variables.
- Labels: These include the response variable, dependent variable, and target variable.
- Underfitting: This happens when a model is too simple and does not capture the structure of the data well enough.
- Overfitting: This happens when a model is possibly too complex and models the noise in the data, which can result in poor generalization performance. Using in-sample measures to do model selection can result in that.
- Regularization: This is a general method to avoid overfitting by applying additional constraints to the model that is learned. A common approach is to make sure the model weights are, on average, small in magnitude. Two common regularizations are L_1 regularization (used by the lasso), which has a penalty term that encourages the sum of the absolute values of the parameters to be small; and L_2 regularization, which encourages the sum of the squares of the parameters to be small.

rules) to figure out which disciplines have the most collaboration and which fields have researchers who tend to publish across different fields.

Clustering Clustering is the most common unsupervised learning technique and is used to group data points together that are similar to each other. The goal of clustering methods is to produce clusters with high intra-cluster (within) similarity and low inter-cluster (between) similarity.

Clustering algorithms typically require a distance (or similarity) metric* to generate clusters. They take a data set and a distance metric (and sometimes additional parameters), and they generate clusters based on that distance metric. The most common distance metric used is Euclidean distance, but other commonly used metrics are Manhattan, Minkowski, Chebyshev, cosine, Hamming, Pearson, and Mahalanobis. Often, domain-specific similarity metrics can be designed for use in specific problems. For example, when performing the record linkage tasks discussed in Chapter 3, you can design a similarity metric that compares two first names and assigns them a high similarity (low distance) if they both map to the same canonical name, so that, for example, Sammy and Sam map to Samuel.

★ Distance metrics are mathematical formulas to calculate the distance between two objects.
For example, *Manhattan distance* is the distance a car would drive from one place to another place in a grid-based street system, whereas *Euclidian distance* (in two-dimensional space) is the "straight-line" distance between two points.

Most clustering algorithms also require the user to specify the number of clusters (or some other parameter that indirectly determines the number of clusters) in advance as a parameter. This is often difficult to do a priori and typically makes clustering an iterative and interactive task. Another aspect of clustering that makes it interactive is often the difficulty in automatically evaluating the quality of the clusters. While various analytical clustering metrics have been developed, the best clustering is task-dependent and thus must be evaluated by the user. There may be different clusterings that can be generated with the same data. You can imagine clustering similar news stories based on the topic content, based on the writing style or based on sentiment. The right set of clusters depends on the user and the task they have. Clustering is therefore typically used for exploring the data, generating clusters, exploring the clusters, and then rerunning the clustering method with different parameters or modifying the clusters (by splitting or merging the previous set of clusters). Interpreting a cluster can be nontrivial: you can look at the centroid of a cluster, look at frequency distributions of different features (and compare them to the prior distribution of each feature), or you can build a decision tree (a supervised learning method we will cover later in this chapter) where the target variable is the cluster ID that can describe the cluster using the features in your data. A good example of a tool that allows interactive clustering from text data is Ontogen [125].

k-means clustering The most commonly used clustering algorithm is called k-means, where k defines the number of clusters. The algorithm works as follows:

1. Select k (the number of clusters you want to generate).

2. Initialize by selecting k points as centroids of the k clusters. This is typically done by selecting k points uniformly at random.

3. Assign each point a cluster according to the nearest centroid.

4. Recalculate cluster centroids based on the assignment in (3) as the mean of all data points belonging to that cluster.

5. Repeat (3) and (4) until convergence.

The algorithm stops when the assignments do not change from one iteration to the next (Figure 6.2). The final set of clusters, however, depend on the starting points. If they are initialized differently, it is possible that different clusters are obtained. One common practical trick is to run k-means several times, each with different (random) starting points. The k-means algorithm is fast, simple, and easy to use, and is often a good first clustering algorithm to try and see if it fits your needs. When the data are of the form where the mean of the data points cannot be computed, a related method called K-medoids can be used [296].

Expectation-maximization (EM) clustering You may be familiar with the EM algorithm in the context of imputing missing data. EM is a general approach to maximum likelihood in the presence of incomplete data. However, it is also used as a clustering method where the missing data are the clusters a data point belongs to. Unlike k-means, where each data point gets assigned to only one cluster, EM does a soft assignment where each data point gets a probabilistic assignment to various clusters. The EM algorithm iterates until the estimates converge to some (locally) optimal solution.

The EM algorithm is fairly good at dealing with outliers as well as high-dimensional data, compared to k-means. It also has a few limitations. First, it does not work well with a large number of clusters or when a cluster contains few examples. Also, when the value of k is larger than the number of actual clusters in the data, EM may not give reasonable results.

Mean shift clustering Mean shift clustering works by finding dense regions in the data by defining a window around each data point and computing the mean of the data points in the window. Then it shifts the center of the window to the mean and repeats the algorithm till

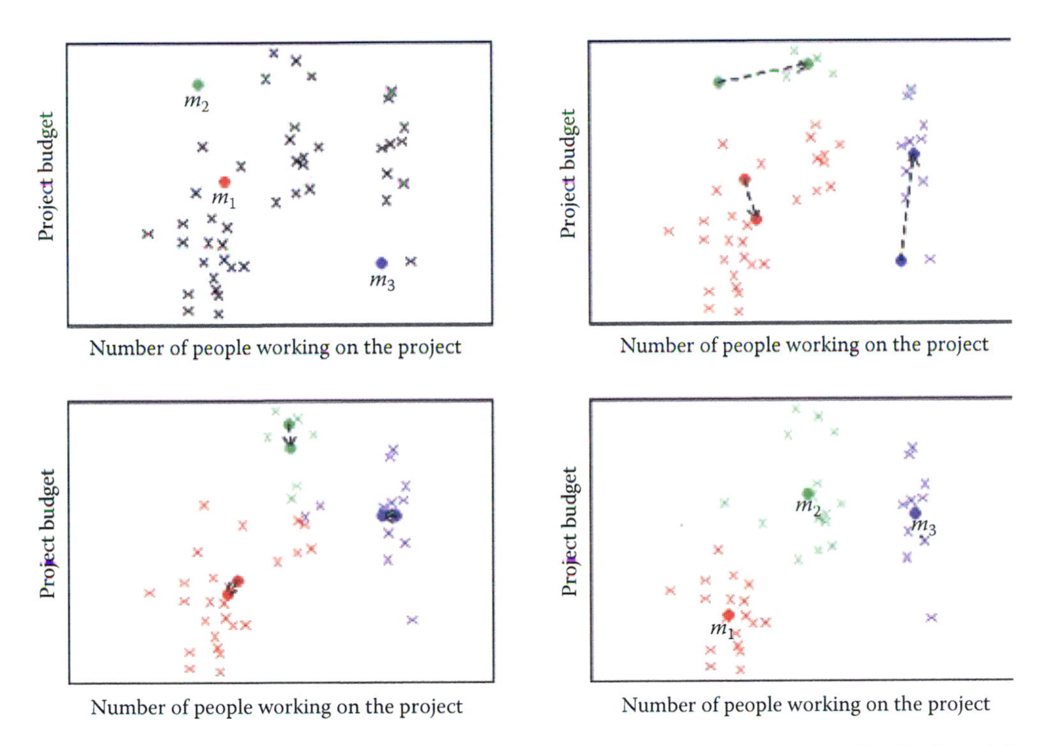

Figure 6.2. Example of k-means clustering with $k = 3$. The upper left panel shows the distribution of the data and the three starting points m_1, m_2, m_3 placed at random. On the upper right we see what happens in the first iteration. The cluster means move to more central positions in their respective clusters. The lower left panel shows the second iteration. After six iterations the cluster means have converged to their final destinations and the result is shown in the lower right panel

it converges. After each iteration, we can consider that the window shifts to a denser region of the data set. The algorithm proceeds as follows:

1. Fix a window around each data point (based on the bandwidth parameter that defines the size of the window).
2. Compute the mean of data within the window.
3. Shift the window to the mean and repeat till convergence.

Mean shift needs a bandwidth parameter h to be tuned, which influences the convergence rate and the number of clusters. A large h might result in merging distinct clusters. A small h might result in too many clusters. Mean shift might not work well in higher dimensions since the number of local maxima is pretty high and it might converge to a local optimum quickly.

One of the most important differences between mean shift and k-means is that k-means makes two broad assumptions: the number of clusters is already known and the clusters are shaped spherically (or elliptically). Mean shift does not assume anything about the number of clusters (but the value of h indirectly determines that). Also, it can handle arbitrarily shaped clusters.

The k-means algorithm is also sensitive to initializations, whereas mean shift is fairly robust to initializations. Typically, mean shift is run for each point, or sometimes points are selected uniformly randomly. Similarly, k-means is sensitive to outliers, while mean shift is less sensitive. On the other hand, the benefits of mean shift come at a cost—speed. The k-means procedure is fast, whereas classic mean shift is computationally slow but can be easily parallelized.

Hierarchical clustering The clustering methods that we have seen so far, often termed *partitioning* methods, produce a flat set of clusters with no hierarchy. Sometimes, we want to generate a hierarchy of clusters, and methods that can do that are of two types:

1. Agglomerative (bottom-up): Start with each point as its own cluster and iteratively merge the closest clusters. The iterations stop either when the clusters are too far apart to be merged (based on a predefined distance criterion) or when there is a sufficient number of clusters (based on a predefined threshold).

2. Divisive (top-down): Start with one cluster and create splits recursively.

Typically, agglomerative clustering is used more often than divisive clustering. One reason is that it is significantly faster, although both of them are typically slower than direct partition methods such as k-means and EM. Another disadvantage of these methods is that they are *greedy*, that is, a data point that is incorrectly assigned to the "wrong" cluster in an earlier split or merge cannot be reassigned again later on.

Spectral clustering Figure 6.3 shows the clusters that k-means would generate on the data set in the figure. It is obvious that the clusters produced are not the clusters you would want, and that is one drawback of methods such as k-means. Two points that are far

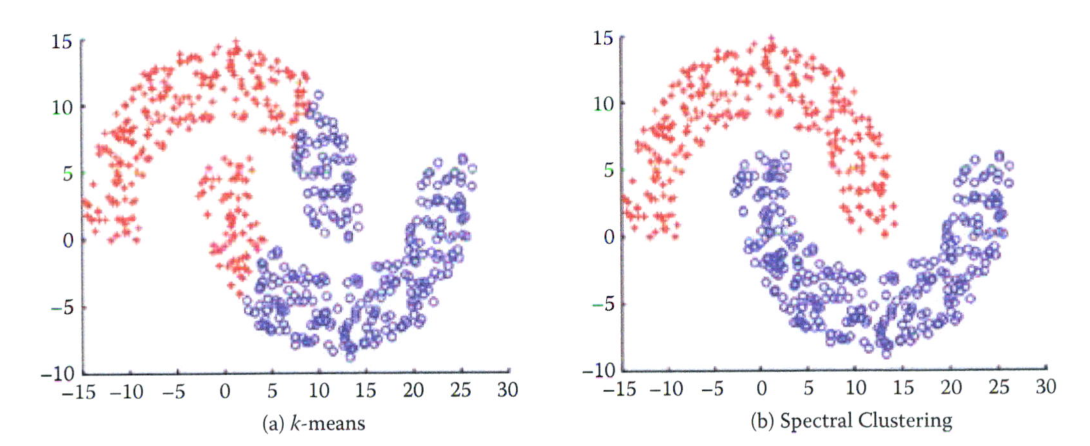

Figure 6.3. The same data set can produce drastically different clusters: (a) k-means; (b) spectral clustering

away from each other will be put in different clusters even if there are other data points that create a "path" between them. Spectral clustering fixes that problem by clustering data that are connected but not necessarily (what is called) compact or clustered within convex boundaries. Spectral clustering methods work by representing data as a graph (or network), where data points are nodes in the graph and the edges (connections between nodes) represent the similarity between the two data points.

The algorithm works as follows:

1. Compute a similarity matrix from the data. This involves determining a pairwise distance function (using one of the distance functions we described earlier).

2. With this matrix, we can now perform graph partitioning, where connected graph components are interpreted as clusters. The graph must be partitioned such that edges connecting different clusters have low weights and edges within the same cluster have high values.

3. We can now partition these data represented by the similarity matrix in a variety of ways. One common way is to use the normalized cuts method. Another way is to compute a graph Laplacian from the similarity matrix.

4. Compute the eigenvectors and eigenvalues of the Laplacian.

5. The k eigenvectors are used as proxy data for the original data set, and they are fed into k-means clustering to produce cluster assignments for each original data point.

Spectral clustering is in general much better than k-means in clustering performance but much slower to run in practice. For large-scale problems, k-means is a preferred clustering algorithm to run because of efficiency and speed.

Principal components analysis Principal components analysis is another unsupervised method used for finding patterns and structure in data. In contrast to clustering methods, the output is not a set of clusters but a set of *principal components* that are linear combinations of the original variables. PCA is typically used when you have a large number of variables and you want a reduced number that you can analyze. This approach is often called *dimensionality reduction*. It generates linearly uncorrelated dimensions that can be used to understand the underlying structure of the data. In mathematical terms, given a set of data on n dimensions, PCA aims to find a linear subspace of dimension d lower than n such that the data points lie mainly on this linear subspace.

PCA is related to several other methods you may already know about. Multidimensional scaling, factor analysis, and independent component analysis differ from PCA in the assumptions they make, but they are often used for similar purposes of dimensionality reduction and discovering the underlying structure in a data set.

Association rules Association rules are a different type of analysis method and originate from the data mining and database community, primarily focused on finding frequent co-occurring associations among a collection of items. This methods is sometimes referred to as "market basket analysis," since that was the original application area of association rules. The goal is to find associations of items that occur together more often than you would randomly expect. The classic example (probably a myth) is "men who go to the store to buy diapers will also tend to buy beer at the same time." This type of analysis would be performed by applying association rules to a set of supermarket purchase data.

Association rules take the form $X_1, X_2, X_3 \Rightarrow Y$ with support S and confidence C, implying that when a transaction contains items $\{X_1, X_2, X_3\}$ C% of the time, they also contain item Y and there are at least S% of transactions where the antecedent is true. This is useful in cases where we want to find patterns that are both *frequent* and *statistically significant*, by specifying thresholds for support S and confidence C.

Support and confidence are useful metrics to generate rules but are often not enough. Another important metric used to generate rules (or reduce the number of spurious patterns generated) is *lift*. Lift is simply estimated by the ratio of the joint probability of two items, x and y, to the product of their individual probabilities: $P(x, y)/[P(x)P(y)]$. If the two items are statistically independent, then $P(x, y) = P(x)P(y)$, corresponding to a lift of 1. Note that anti-correlation yields lift values less than 1, which is also an interesting pattern, corresponding to mutually exclusive items that rarely occur together.

Association rule algorithms work as follows: Given a set of transactions (rows) and items for that transaction:

1. Find all combinations of items in a set of transactions that occur with a specified minimum frequency. These combinations are called *frequent itemsets*.

2. Generate association rules that express co-occurrence of items within frequent itemsets.

For our purposes, association rule methods are an efficient way to take a *basket* of features (e.g., areas of publication of a researcher, different organizations an individual has worked at in their career, all the cities or neighborhoods someone may have lived in) and find co-occurrence patterns. This may sound trivial, but as data sets and number of features get larger, it becomes computationally expensive and association rule mining algorithms provide a fast and efficient way of doing it.

6.5.2 Supervised learning

We now turn to the problem of supervised learning, which typically involves methods for classification, prediction, and regression. We will mostly focus on classification methods in this chapter since many of the regression methods in machine learning are fairly similar to methods with which you are already familiar. Remember that classification means predicting a discrete (or categorical) variable. Some of the classification methods that we will cover can also be used for regression, a fact that we will mention when describing that method.

In general, supervised learning methods take as input pairs of data points (X, Y) where X are the predictor variables (features) and

Y is the target variable (label). The supervised learning method then uses these pairs as *training data* and learns a model F, where $F(X) \sim Y$. This model F is then used to predict Ys for new data points X. As mentioned earlier, the goal is not to build a model that best fits known data but a model that is useful for future predictions and minimizes future generalization error. This is the key goal that differentiates many of the methods that you know from the methods that we will describe next. In order to minimize future error, we want to build models that are not just *overfitting* on past data.

Another goal, often prioritized in the social sciences, that machine learning methods do not optimize for is getting a structural form of the model. Machine learning models for classification can take different structural forms (ranging from linear models, to sets of rules, to more complex forms), and it may not always be possible to write them down in a compact form as an equation. This does not, however, make them incomprehensible or uninterpretable. Another focus of machine learning models for supervised learning is prediction, and not causal inference. Some of these models can be used to help with causal inference, but they are typically optimized for prediction tasks. We believe that there are many social science and policy problems where better prediction methods can be extremely beneficial.

▶ The topic of causal inference is addressed in more detail in Chapter 10.

In this chapter, we mostly deal with binary classification problems: that is, problems in which the data points are to be classified into one of two categories. Several of the methods that we will cover can also be used for multiclass classification (classifying a data point into one of n categories) or for multi-label classification (classifying a data point into m of n categories where $m \geq 1$). There are also approaches to take multiclass problems and turn them into a set of binary problems that we will mention briefly at the end of the chapter.

Before we describe supervised learning methods, we want to recap a few principles as well as terms that we have used and will be using in the rest of the chapter.

Training a model Once we have finished data exploration, filled in missing values, created predictor variables (features), and decided what our target variable (label) is, we now have pairs of X, Y to start training (or building) the model.

Using the model to score new data We are building this model so we can predict Y for a new set of Xs—using the model means,

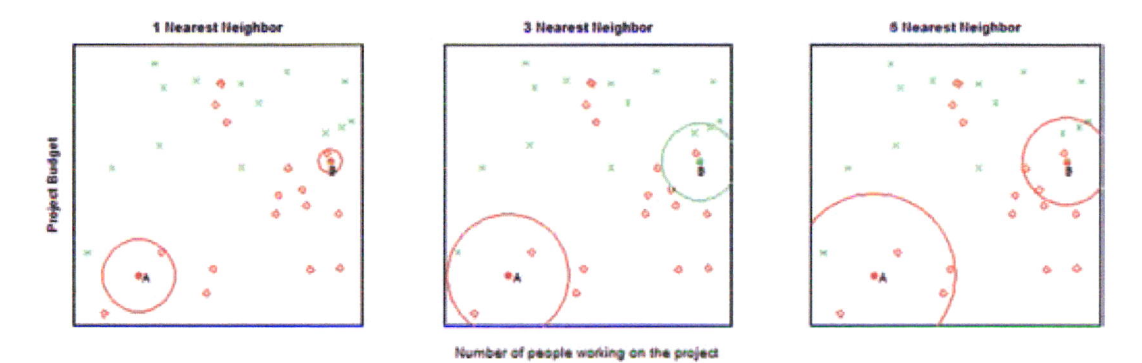

Figure 6.4. Example of k-nearest neighbor with $k = 1, 3, 5$ neighbors. We want to predict the points A and B. The 1-nearest neighbor for both points is red ("Patent not granted"), the 3-nearest neighbor predicts point A (B) to be red (green) with probability 2/3, and the 5-nearest neighbor predicts again both points to be red with probabilities 4/5 and 3/5, respectively.

getting new data, generating the same features to get the vector X, and then applying the model to produce Y.

One common technique for supervised learning is logistic regression, a method you will already be familiar with. We will give an overview of some of the other methods used in machine learning. It is important to remember that as you use increasingly powerful classification methods, you need more data to *train* the models.

k-nearest neighbor The method k-nearest neighbor (k-NN) is one of the simpler classification methods in machine learning. It belongs to a family of models sometimes known as *memory-based models* or *instance-based models*. An example is classified by finding its k nearest neighbors and taking majority vote (or some other aggregation function). We need two key things: a value for k and a distance metric with which to find the k nearest neighbors. Typically, different values of k are used to empirically find the best one. Small values of k lead to predictions having high variance but can capture the local structure of the data. Larger values of k build more global models that are lower in variance but may not capture local structure in the data as well.

Figure 6.4 provides an example for $k = 1, 3, 5$ nearest neighbors. The number of neighbors (k) is a parameter, and the prediction depends heavily on how it is determined. In this example, point B is classified differently if $k = 3$.

Training for k-NN just means storing the data, making this method useful in applications where data are coming in extremely quickly and a model needs to be updated frequently. All the work, however, gets pushed to scoring time, since all the distance calculations happen when a new data point needs to be classified. There are several optimized methods designed to make k-NN more efficient that are worth looking into if that is a situation that is applicable to your problem.

In addition to selecting k and an appropriate distance metric, we also have to be careful about the scaling of the features. When distances between two data points are large for one feature and small for a different feature, the method will rely almost exclusively on the first feature to find the closest points. The smaller distances on the second feature are nearly irrelevant to calculate the overall distance. A similar problem occurs when continuous and categorical predictors are used together. To resolve the scaling issues, various options for rescaling exist. For example, a common approach is to center all features at mean 0 and scale them to variance 1.

There are several variations of k-NN. One of these is weighted nearest neighbors, where different features are weighted differently or different examples are weighted based on the distance from the example being classified. The method k-NN also has issues when the data are sparse and has high dimensionality, which means that every point is far away from virtually every other point, and hence pairwise distances tend to be uninformative. This can also happen when a lot of features are irrelevant and drown out the relevant features' signal in the distance calculations.

Notice that the nearest-neighbor method can easily be applied to regression problems with a real-valued target variable. In fact, the method is completely oblivious to the type of target variable and can potentially be used to predict text documents, images, and videos, based on the aggregation function after the nearest neighbors are found.

Support vector machines Support vector machines are one of the most popular and best-performing classification methods in machine learning today. The mathematics behind SVMs has a lot of prerequisites that are beyond the scope of this book, but we will give you an intuition of how SVMs work, what they are good for, and how to use them.

We are all familiar with linear models that separate two classes by fitting a line in two dimensions (or a hyperplane in higher dimensions) in the middle (see Figure 6.5). An important decision that linear models have to make is which linear separator we should prefer when there are several we can build.

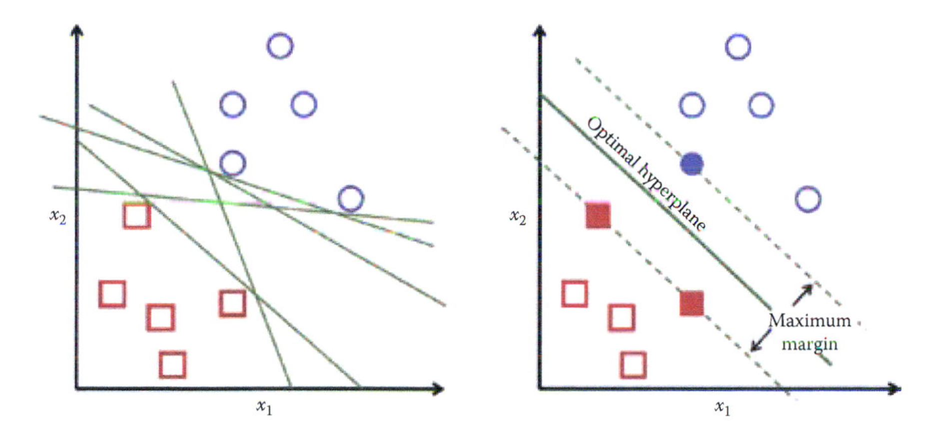

Figure 6.5. Support vector machines

You can see in Figure 6.5 that multiple lines offer a solution to the problem. Is any of them better than the others? We can intuitively define a criterion to estimate the worth of the lines: A line is bad if it passes too close to the points because it will be noise sensitive and it will not generalize correctly. Therefore, our goal should be to find the line passing as far as possible from all points.

The SVM algorithm is based on finding the hyperplane that maximizes the *margin* of the training data. The training examples that are closest to the hyperplane are called *support vectors* since they are *supporting* the margin (as the margin is only a function of the support vectors).

An important concept to learn when working with SVMs is *kernels*. SVMs are a specific instance of a class of methods called *kernel methods*. So far, we have only talked about SVMs as linear models. Linear works well in high-dimensional data but sometimes you need nonlinear models, often in cases of low-dimensional data or in image or video data. Unfortunately, traditional ways of generating nonlinear models get computationally expensive since you have to explicitly generate all the features such as squares, cubes, and all the interactions. Kernels are a way to keep the efficiency of the linear machinery but still build models that can capture nonlinearity in the data without creating all the nonlinear features.

You can essentially think of kernels as similarity functions and use them to create a linear separation of the data by (implicitly) mapping the data to a higher-dimensional space. Essentially, we take an n-dimensional input vector X, map it into a high-dimensional (possibly infinite-dimensional) feature space, and construct an optimal separating hyperplane in this space. We refer you to relevant papers for more detail on SVMs and nonlinear kernels [334, 339]. SVMs are also related to logistic regression, but use a different loss/penalty function [159].

When using SVMs, there are several parameters you have to optimize, ranging from the *regularization* parameter C, which determines the tradeoff between minimizing the training error and minimizing model complexity, to more kernel-specific parameters. It is often a good idea to do a grid search to find the optimal parameters. Another tip when using SVMs is to normalize the features; one common approach to doing that is to normalize each data point to be a vector of unit length.

Linear SVMs are effective in high-dimensional spaces, especially when the space is sparse such as text classification where the number of data points (perhaps tens of thousands) is often much less than the number of features (a hundred thousand to a million or more). SVMs are also fairly robust when the number of irrelevant features is large (unlike the k-NN approaches that we mentioned earlier) as well as when the class distribution is skewed, that is, when the class of interest is significantly less than 50% of the data.

One disadvantage of SVMs is that they do not directly provide probability estimates. They assign a score based on the distance from the margin. The farther a point is from the margin, the higher the magnitude of the score. This score is good for ranking examples, but getting accurate probability estimates takes more work and requires more labeled data to be used to perform probability calibrations.

In addition to classification, there are also variations of SVMs that can be used for regression [348] and ranking [70].

Decision trees Decision trees are yet another set of methods that are helpful for prediction. Typical decision trees learn a set of rules from training data represented as a tree. An exemplary decision tree is shown in Figure 6.6. Each level of a tree *splits* the tree to create a branch using a feature and a value (or range of values). In the example tree, the first split is made on the feature *number of visits in the past year* and the value 4. The second level of the tree now has two splits: one using *average length of visit* with value 2 days and the other using the value 10 days.

Various algorithms exist to build decision trees. C4.5, CHAID, and CART (Classification and Regression Trees) are the most popular.

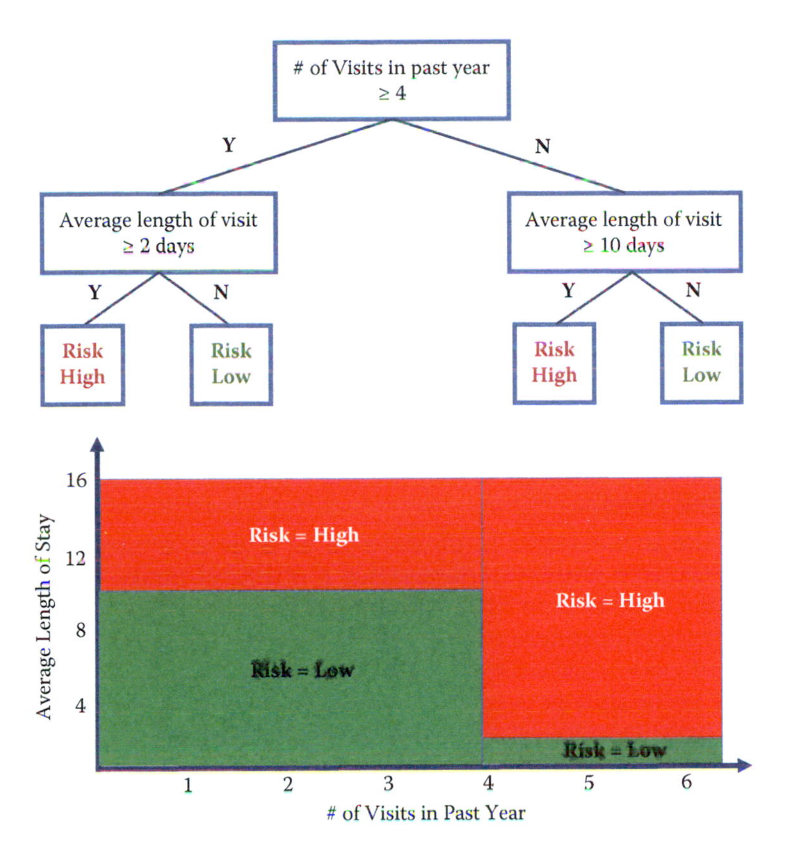

Figure 6.6. An exemplary decision tree. The top figure is the standard representation for trees. The bottom figure offers an alternative view of the same tree. The feature space is partitioned into numerous rectangles, which is another way to view a tree, representing its nonlinear character more explicitly

Each needs to determine the next best feature to split on. The goal is to find feature splits that can best reduce class impurity in the data, that is, a split that will ideally put all (or as many as possible) positive class examples on one side and all (or as many as possible) negative examples on the other side. One common measure of impurity that comes from information theory is *entropy*, and it is calculated as

$$H(X) = -\sum_{x} p(x) \log p(x).$$

Entropy is maximum (1) when both classes have equal numbers of examples in a node. It is minimum (0) when all examples are from the same class. At each node in the tree, we can evaluate all the possible features and select the one that most reduces the entropy given the tree so far. This expected change in entropy is known as *information gain* and is one of the most common criteria used to create decision trees. Other measures that are used instead of information gain are Gini and chi-squared.

If we keep constructing the tree in this manner, selecting the next best feature to split on, the tree ends up fairly deep and tends to overfit the data. To prevent overfitting, we can either have a stopping criterion or *prune* the tree after it is fully grown. Common stopping criteria include minimum number of data points to have before doing another feature split, maximum depth, and maximum purity. Typical pruning approaches use holdout data (or cross-validation, which will be discussed later in this chapter) to cut off parts of the tree.

Once the tree is built, a new data point is classified by running it through the tree and, once it reaches a terminal node, using some aggregation function to give a prediction (classification or regression). Typical approaches include performing maximum likelihood (if the leaf node contains 10 examples, 8 positive and 2 negative, any data point that gets into that node will get an 80% probability of being positive). Trees used for regression often build the tree as described above but then fit a linear regression model at each leaf node.

Decision trees have several advantages. The interpretation of a tree is straightforward as long as the tree is not too large. Trees can be turned into a set of rules that experts in a particular domain can possibly dig deeper into, validate, and modify. Trees also do not require too much feature engineering. There is no need to create interaction terms since trees can implicitly do that by splitting on two features, one after another.

Unfortunately, along with these benefits come a set of disadvantages. Decision trees, in general, do not perform well, compared to SVMs, random forests, or logistic regression. They are also unstable: small changes in data can result in very different trees. The lack of stability comes from the fact that small changes in the training data may lead to different splitting points. As a consequence, the whole tree may take a different structure. The suboptimal predictive performance can be seen from the fact that trees partition the predictor space into a few rectangular regions, each one predicting only a single value (see the bottom part of Figure 6.6).

Ensemble methods Combinations of models are generally known as model ensembles. They are among the most powerful techniques in machine learning, often outperforming other methods, although at the cost of increased algorithmic and model complexity.

The intuition behind building ensembles of models is to build several models, each somewhat different. This diversity can come from various sources such as: training models on subsets of the data; training models on subsets of the features; or a combination of these two.

Ensemble methods in machine learning have two things in common. First, they construct multiple, diverse predictive models from adapted versions of the training data (most often reweighted or resampled). Second, they combine the predictions of these models in some way, often by simple averaging or voting (possibly weighted).

Bagging Bagging stands for "bootstrap aggregation":* we first create bootstrap samples from the original data and then aggregate the predictions using models trained on each bootstrap sample. Given a data set of size N, the method works as follows:

★ Bootstrap is a general statistical procedure that draws random samples of the original data with replacement.

1. Create k bootstrap samples (with replacement), each of size N, resulting in k data sets. Only about 63% of the original training examples will be represented in any given bootstrapped set.

2. Train a model on each of the k data sets, resulting in k models.

3. For a new data point X, predict the output using each of the k models.

4. Aggregate the k predictions (typically using average or voting) to get the prediction for X.

A nice feature of this method is that any underlying model can be used, but decision trees are often the most commonly used base model. One reason for this is that decision tress are typically high variance and unstable, that is, they can change drastically given small changes in data, and bagging is effective at reducing the variance of the overall model. Another advantage of bagging is that each model can be trained in parallel, making it efficient to scale to large data sets.

Boosting Boosting is another popular ensemble technique, and it often results in improving the base classifier being used. In fact, if your only goal is improving accuracy, you will most likely find that boosting will achieve that. The basic idea is to keep training classifiers iteratively, each iteration focusing on examples that the previous one got wrong. At the end, you have a set of classifiers, each trained on smaller and smaller subsets of the training data. Given a new data point, all the classifiers predict the target, and a weighted average of those predictions is used to get the final prediction, where the weight is proportional to the accuracy of each classifier. The algorithm works as follows:

1. Assign equal weights to every example.
2. For each iteration:
   (a) Train classifier on the weighted examples.
   (b) Predict on the training data.
   (c) Calculate error of the classifier on the training data.
   (d) Calculate the new weighting on the examples based on the errors of the classifier.
   (e) Reweight examples.
3. Generate a weighted classifier based on the accuracy of each classifier.

One constraint on the classifier used within boosting is that it should be able to handle weighted examples (either directly or by replicating the examples that need to be overweighted). The most common classifiers used in boosting are decision stumps (single-level decision trees), but deeper trees can also work well.

Boosting is a common way to *boost* the performance of a classification method but comes with additional complexity, both in the training time and in interpreting the predictions. A disadvantage of boosting is that it is difficult to parallelize since the next iteration of boosting relies on the results of the previous iteration.

A nice property of boosting is its ability to identify outliers: examples that are either mislabeled in the training data, or are inherently ambiguous and hard to categorize. Because boosting focuses its weight on the examples that are more difficult to classify, the examples with the highest weight often turn out to be outliers. On the other hand, if the number of outliers is large (lots of noise in the data), these examples can hurt the performance of boosting by focusing too much on them.

Random forests Given a data set of size N and containing M features, the random forest training algorithm works as follows:

1. Create n bootstrap samples from the original data of size N. Remember, this is similar to the first step in bagging. Typically n ranges from 100 to a few thousand but is best determined empirically.

2. For each bootstrap sample, train a decision tree using m features (where m is typically much smaller than M) at each node of the tree. The m features are selected uniformly at random from the M features in the data set, and the decision tree will select the best split among the m features. The value of m is held constant during the forest growing.

3. A new test example/data point is classified by all the trees, and the final classification is done by majority vote (or another appropriate aggregation method).

Random forests are probably the most accurate classifiers being used today in machine learning. They can be easily parallelized, making them efficient to run on large data sets, and can handle a large number of features, even with a lot of missing values. Random forests can get complex, with hundreds or thousands of trees that are fairly deep, so it is difficult to interpret the learned model. At the same time, they provide a nice way to estimate feature importance, giving a sense of what features were important in building the classifier.

Another nice aspect of random forests is the ability to compute a proximity matrix that gives the similarity between every pair of data points. This is calculated by computing the number of times two examples land in the same terminal node. The more that happens, the closer the two examples are. We can use this proximity matrix for clustering, locating outliers, or explaining the predictions for a specific example.

Stacking Stacking is a technique that deals with the task of learning a meta-level classifier to combine the predictions of multiple base-level classifiers. This meta-algorithm is trained to combine the model predictions to form a final set of predictions. This can be used for both regression and classification. The algorithm works as follows:

1. Split the data set into n equal-sized sets: $set_1, set_2, \ldots, set_n$.

2. Train base models on all possible combinations of $n - 1$ sets and, for each model, use it to predict on set_i what was left out of the training set. This would give us a set of predictions on every data point in the original data set.

3. Now train a second-stage stacker model on the predicted classes or the predicted probability distribution over the classes from the first-stage (base) model(s).

By using the first-stage predictions as features, a stacker model gets more information on the problem space than if it were trained in isolation. The technique is similar to cross-validation, an evaluation methodology that we will cover later in this chapter.

Neural networks and deep learning Neural networks are a set of multi-layer classifiers where the outputs of one layer feed into the inputs of the next layer. The layers between the input and output layers are called *hidden layers*, and the more hidden layers a neural network has, the more complex functions it can learn. Neural networks were popular in the 1980s and early 1990s, but then fell out of fashion because they were slow and expensive to train, even with only one or two hidden layers. Since 2006, a set of techniques has been developed that enable learning in deeper neural networks. These techniques have enabled much deeper (and larger) networks to be trained—people now routinely train networks with five to ten hidden layers. And it turns out that these perform far better on many problems than shallow neural networks (with just a single hidden layer). The reason for the better performance is the ability of deep nets to build up a complex hierarchy of concepts, learning multiple levels of representation and abstraction that help to make sense of data such as images, sound, and text.

Usually, with a supervised neural network you try to predict a target vector, Y, from a matrix of inputs, X. But when you train a deep neural network, it uses a combination of supervised and unsupervised learning. In an unsupervised neural network, you try to predict the matrix X using the same matrix X as the input. In doing this, the network can learn something intrinsic about the data without the help of a separate target or label. The learned information is stored as the weights of the network.

Currently, deep neural networks are trendy and a lot of research is being done on them. It is, however, important to keep in mind that they are applicable for a narrow class of problems with which social scientists would deal and that they often require a lot more data than are available in most problems. Training deep neural networks also requires a lot of computational power, but that is less likely to be an issue for most people. Typical cases where deep learning has been shown to be effective involve lots of images, video, and text data. We are still in the early stages of development of this class of methods, and the next few years will give us a much better understanding of why they are effective and the problems for which they are well suited.

6.6 Evaluation

The previous section introduced us to a variety of methods, all with certain pros and cons, and no single method guaranteed to outperforms others for a given problem. This section focuses on evaluation methods, with three primary goals:

1. Model selection: How do we select a method to use? What parameters should we select for that method?

2. Performance estimation: How well will our model do once it is deployed and applied to new data?

3. A deeper understanding of the model can point to inaccuracies of existing methods and provide a better understanding of the data and the problem we are tackling.

This section will cover evaluation methodologies as well as metrics that are commonly used. We will start by describing common evaluation methodologies that use existing data and then move on to field trials. The methodologies we describe below apply both to regression and classification problems.

6.6.1 Methodology

In-sample evaluation As social scientists, you already evaluate methods on how well they perform in-sample (on the set that the model was trained on). As we mentioned earlier in the chapter, the goal of machine learning methods is to generalize to new data, and validating models in-sample does not allow us to do that. We focus here on evaluation methodologies that allow us to optimize (as best as we can) for generalization performance. The methods are illustrated in Figure 6.7.

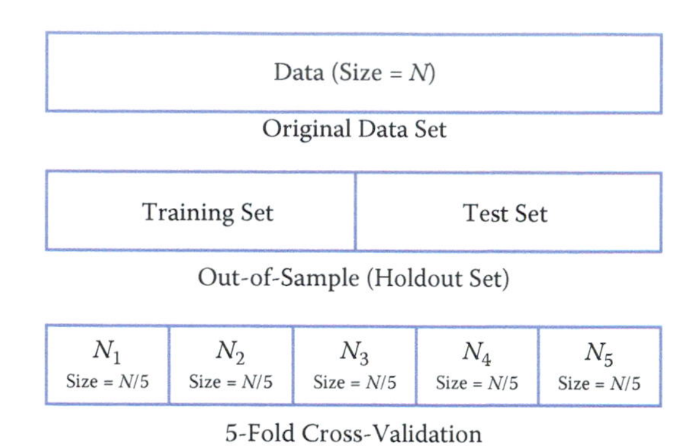

Figure 6.7. Validation methodologies: holdout set and cross-validation

Out-of-sample and holdout set The simplest way to focus on generalization is to *pretend* to generalize to new (unseen) data. One way to do that is to take the original data and randomly split them into two sets: a *training set* and a *test set* (sometimes also called the *holdout* or *validation set*). We can decide how much to keep in each set (typically the splits range from 50–50 to 80–20, depending on the size of the data set). We then train our models on the training set and classify the data in the test set, allowing us to get an estimate of the relative performance of the methods.

One drawback of this approach is that we may be extremely lucky or unlucky with our random split. One way to get around the problem that is to repeatedly create multiple training and test sets. We can then train on TR_1 and test on TE_1, train on TR_2 and test on TE_2, and so on. The performance measures on each test set can then give us an estimate of the performance of different methods and how much they vary across different random sets.

Cross-validation Cross-validation is a more sophisticated holdout training and testing procedure that takes away some of the shortcomings of the holdout set approach. Cross-validation begins by splitting a labeled data set into k partitions (called folds). Typically, k is set to 5 or 10. Cross-validation then proceeds by iterating k times. In each iteration, one of the k folds is held out as the test set, while the other $k - 1$ folds are combined and used to train the model. A nice property of cross-validation is that every example is used in one test set for testing the model. Each iteration of cross-validation gives us a performance estimate that can then be aggregated (typically averaged) to generate the overall estimate.

Figure 6.8. Temporal validation

An extreme case of cross-validation is called leave-one-out cross-validation, where given a data set of size N, we create N folds. That means iterating over each data point, holding it out as the test set, and training on the rest of the $N - 1$ examples. This illustrates the benefit of cross-validation by giving us good generalization estimates (by training on as much of the data set as possible) and making sure the model is tested on each data point.

Temporal validation The cross-validation and holdout set approaches described above assume that the data have no time dependencies and that the distribution is stationary over time. This assumption is almost always violated in practice and affects performance estimates for a model.

In most practical problems, we want to use a validation strategy that emulates the way in which our models will be used and provides an accurate performance estimate. We will call this *temporal validation*. For a given point in time t_i, we train our models only on information available to us before t_i to avoid training on data from the "future." We then predict and evaluate on data from t_i to $t_i + d$ and iterate, expanding the training window while keeping the test window size constant at d. Figure 6.8 shows this validation process with $t_i = 2010$ and $d = 1$ year. The test set window d depends on a few factors related to how the model will be deployed to best emulate reality:

1. How far out in the future do predictions need to be made? For example, if the set of students who need to be targeted for

interventions has to be finalized at the beginning of the school year for the entire year, then $d = 1$ year.

2. How often will the model be updated? If the model is being updated daily, then we can move the window by a day at a time to reflect the deployment scenario.

3. How often will the system get new data? If we are getting new data frequently, we can make predictions more frequently.

Temporal validation is similar to how time series models are evaluated and should be the validation approach used for most practical problems.

6.6.2 Metrics

The previous subsection focused on validation methodologies assuming we have a evaluation metric in mind. This section will go over commonly used evaluation metrics. You are probably familiar with using R^2, analysis of the residuals, and mean squared error (MSE) to evaluate the quality of regression models. For regression problems, the MSE calculates the average squared differences between predictions $\hat{y}_i$ and true values y_i. When prediction models have smaller MSE, they are better. However, the MSE itself is hard to interpret because it measures quadratic differences. Instead, the root mean squared error (RMSE) is more intuitive as it as measure of mean differences on the original scale of the response variable. Yet another alternative is the mean absolute error (MAE), which measures average absolute distances between predictions and true values.

We will now describe some additional evaluation metrics commonly used in machine learning for classification. Before we dive into metrics, it is important to highlight that machine learning models for classification typically do not predict 0/1 values directly. SVMs, random forests, and logistic regression all produce a score (which is sometimes a probability) that is then turned into 0 or 1 based on a user-specific threshold. You might find that certain tools (such as `sklearn`) use a default value for that threshold (often 0.5), but it is important to know that it is an arbitrary threshold and you should select the threshold based on the data, the model, and the problem you are solving. We will cover that a little later in this section.

Once we have turned the real-valued predictions into 0/1 classification, we can now create a *confusion matrix* from these pre-

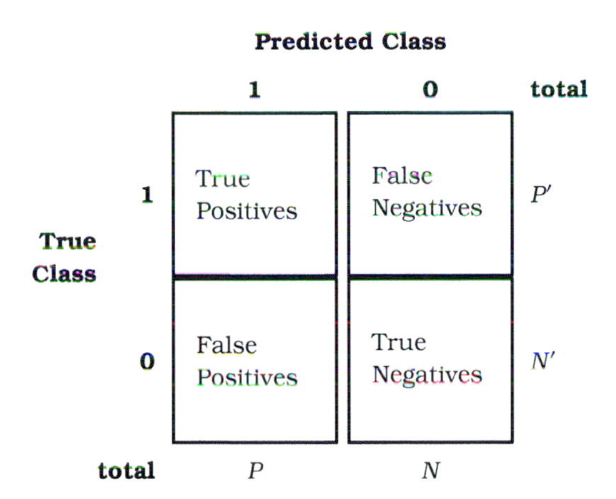

Figure 6.9. A *confusion matrix* created from real-valued predictions

dictions, shown in Figure 6.9. Each data point belongs to either the positive class or the negative class, and for each data point the prediction of the classifier is either correct or incorrect. This is what the four cells of the confusion matrix represent. We can use the confusion matrix to describe several commonly used evaluation metrics.

Accuracy is the ratio of correct predictions (both positive and negative) to all predictions:

$$\text{Accuracy} = \frac{TP + TN}{TP + TN + FP + FN} = \frac{TP + TN}{P + N} = \frac{TP + TN}{P' + N'},$$

where TP denotes true positives, TN true negatives, FP false positives, FN false negatives, and other symbols denote row or column totals as in Figure 6.9. Accuracy is the most commonly described evaluation metric for classification but is surprisingly the least useful in practical situations (at least by itself). One problem with accuracy is that it does not give us an idea of *lift* compared to baseline. For example, if we have a classification problem with 95% of the data as positive and 5% as negative, a classifier with 85% is performing worse than a dumb classifier that predicts positive all the time (and will have 95% accuracy).

Two additional metrics that are often used are precision and recall, which are defined as follows:

$$\text{Precision} = \frac{TP}{TP + FP} = \frac{TP}{P},$$

$$\text{Recall} = \frac{TP}{TP + FN} = \frac{TP}{P'}$$

(see also Box 7.3). Precision measures the accuracy of the classifier when it predicts an example to be positive. It is the ratio of correctly predicted positive examples (*TP*) to all examples predicted as positive (*TP* + *FP*). This measure is also called *positive predictive value* in other fields. Recall measures the ability of the classifier to find positive examples. It is the ratio of all the correctly predicted positive examples (*TP*) to all the positive examples in the data (*TP*+*FN*). This is also called *sensitivity* in other fields.

You might have encountered another metric called *specificity* in other fields. This measure is the true negative rate: the proportion of negatives that are correctly identified.

Another metric that is used is the F_1 score, which is the harmonic mean of precision and recall:

$$F_1 = \frac{2 * \text{Precision} * \text{Recall}}{\text{Precision} + \text{Recall}}$$

(see also equation (7.1)). This is often used when you want to balance both precision and recall.

There is often a tradeoff between precision and recall. By selecting different classification thresholds, we can vary and tune the precision and recall of a given classifier. A highly conservative classifier that only predicts a 1 when it is absolutely sure (say, a threshold of 0.9999) will most often be correct when it predicts a 1 (high precision) but will miss most 1s (low recall). At the other extreme, a classifier that says 1 to every data point (a threshold of 0.0001) will have perfect recall but low precision. Figure 6.10 show a precision-recall curve that is often used to represent the performance of a given classifier.

If we care about optimizing for the entire precision recall space, a useful metric is the *area under the curve* (AUC-PR), which is the area under the precision-recall curve. AUC-PR must not be confused with AUC-ROC, which is the area under the related receiver operating characteristic (ROC) curve. The ROC curve is created by plotting recall versus (1 - specificity). Both AUCs can be helpful metrics to compare the performance of different methods and the maximum value the AUC can take is 1. If, however, we care about a specific part on the precision-recall curve, we have to look at finer-grained metrics.

Let us consider an example from public health. Most public health agencies conduct inspections of various sorts to detect health hazard violations (lead hazards, for example). The number of possible places (homes or businesses) to inspect far exceeds the inspection resources typically available. Let us assume further that they

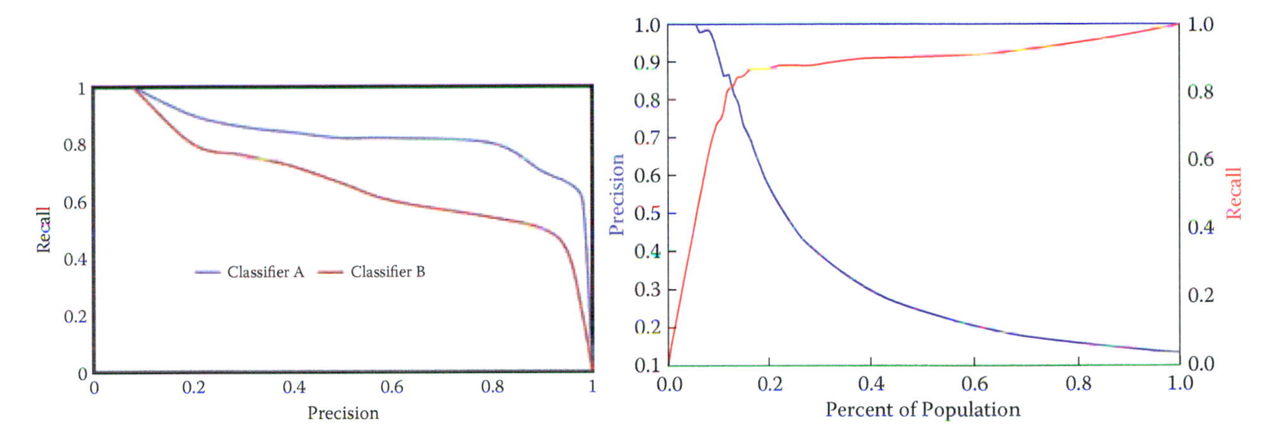

Figure 6.10. Precision–recall curve

Figure 6.11. Precision or recall at different thresholds

can only inspect 5% of all possible places; they would clearly want to prioritize the inspection of places that are most likely to contain the hazard. In this case, the model will score and rank all the possible inspection places in order of hazard risk. We would then want to know what percentage of the top 5% (the ones that will get inspected) are likely to be hazards, which translates to the precision in the top 5% of the most confidence predictions—precision at 5%, as it is commonly called (see Figure 6.11). *Precision at top k percent* is a common class of metrics widely used in information retrieval and search engine literature, where you want to make sure that the results retrieved at the top of the search results are accurate. More generally, this metric is often used in problems in which the class distribution is skewed and only a small percentage of the examples will be examined manually (inspections, investigations for fraud, etc.). The literature provides many case studies of such applications [219, 222, 307].

One last metric we want to mention is a class of cost-sensitive metrics where different costs (or benefits) can be associated with the different cells in the confusion matrix. So far, we have implicitly assumed that every correct prediction and every error, whether for the positive class or the negative class, has equal costs and benefits. In many practical problems, that is not the case. For example, we may want to predict whether a patient in a hospital emergency room is likely to go into cardiac arrest in the next six hours. The cost of a false positive in this case is the cost of the intervention (which may be a few extra minutes of a physician's time) while the cost of a false negative could be death. This type of analysis allows us to calculate the expected value of the predictions of a classifier and select the model that optimizes this cost-sensitive metric.

6.7 Practical tips

Here we highlight some practical tips that will be helpful when working with machine learning methods.

6.7.1 Features

So far in this chapter, we have focused a lot on methods and process, and we have not discussed features in detail. In social science, they are not called features but instead are known as variables or predictors. Good features are what makes machine learning systems effective. Feature generation (or engineering, as it is often called) is where the bulk of the time is spent in the machine learning process. As social science researchers or practitioners, you have spent a lot of time constructing features, using transformations, dummy variables, and interaction terms. All of that is still required and critical in the machine learning framework. One difference you will need to get comfortable with is that instead of carefully selecting a few predictors, machine learning systems tend to encourage the creation of lots of features and then empirically use holdout data to perform regularization and model selection. It is common to have models that are trained on thousands of features. Commonly used approaches to create features include:

- Transformations, such as log, square, and square root.
- Dummy (binary) variables: This is often done by taking categorical variables (such as city) and creating a binary variable for each value (one variable for each city in the data). These are also called indicator variables.
- Discretization: Several methods require features to be discrete instead of continuous. Several approaches exist to convert continuous variables into discrete ones, the most common of which is equal-width binning.
- Aggregation: Aggregate features often constitute the majority of features for a given problem. These aggregations use different aggregation functions (count, min, max, average, standard deviation, etc.), often over varying windows of time and space.

For example, given urban data, we would want to calculate the number (and min, max, mean, variance) of crimes within an m-mile radius of an address in the past t months for varying values of m and t, and then to use all of them as features in a classification problem.

In general, it is a good idea to have the complexity in features and use a simple model, rather than using more complex models with simple features. Keeping the model simple makes it faster to train and easier to understand.

6.7.2 Machine learning pipeline

When working on machine learning projects, it is a good idea to structure your code as a modular pipeline so you can easily try different approaches and methods without major restructuring. The Python workbooks supporting this book will give you an example of a machine learning pipeline. A good pipeline will contain modules for importing data, doing exploration, feature generation, classification, and evaluation. You can then instantiate a specific workflow by combining these modules.

An important component of the machine learning pipeline is comparing different methods. With all the methods out there and all the hyperparameters they come with, how do we know which model to use and which hyperparameters to select? And what happens when we add new features to the model or when the data have "temporal drift" and change over time? One simple approach is to have a nested set of `for` loops that loop over all the methods you have access to, then enumerate all the hyperparameters for that method, create a cross-product, and loop over all of them, comparing them across different evaluation metrics and selecting the best one to use going forward. You can even add different feature subsets and time slices to this `for` loop, as the example in the supporting workbooks will show.

6.7.3 Multiclass problems

In the supervised learning section, we framed classification problems as binary classification problems with a 0 or 1 output. There are many problems where we have multiple classes, such as classifying companies into their industry codes or predicting whether a student will drop out, transfer, or graduate. Several solutions have been designed to deal with the multiclass classification problem:

- Direct multiclass: Use methods that can directly perform multiclass classification. Examples of such methods are K-nearest neighbor, decision trees, and random forests. There are extensions of support vector machines that exist for multiclass classification as well [86], but they can often be slow to train.

- Convert to one versus all (OVA): This is a common approach to solve multiclass classification problems using binary classifiers. Any problem with n classes can be turned into n binary classification problems, where each classifier is trained to distinguish between one versus all the other classes. A new example can be classified by combining the predictions from all the n classifiers and selecting the class with the highest score. This is a simple and efficient approach, and one that is commonly used, but it suffers from each classification problem possibly having an imbalanced class distribution (due to the negative class being a collection of multiple classes). Another limitation of this approach is that it requires the scores of each classifier to be calibrated so that they are comparable across all of them.

- Convert to pairwise: In this approach, we can create binary classifiers to distinguish between each pair of classes, resulting in $\binom{n}{2}$ binary classifiers. This results in a large number of classifiers, but each classifier usually has a balanced classification problem. A new example is classified by taking the predictions of all the binary classifiers and using majority voting.

6.7.4 Skewed or imbalanced classification problems

A lot of problems you will deal with will not have uniform (balanced) distributions for both classes. This is often the case with problems in fraud detection, network security, and medical diagnosis where the class of interest is not very common. The same is true in many social science and public policy problems around behavior prediction, such as predicting which students will not graduate on time, which children may be at risk of getting lead poisoning, or which homes are likely to be abandoned in a given city. You will notice that applying standard machine learning methods may result in all the predictions being for the most frequent category in such situations, making it problematic to detect the infrequent classes. There has been a lot of work in machine learning research on dealing with such problems [73, 217] that we will not cover in detail here. Common approaches to deal with class imbalance include oversampling from the minority class and undersampling from the majority class. It is important to keep in mind that the sampling approaches do not need to result in a 1 : 1 ratio. Many supervised learning methods described in this chapter (such as SVMs) can work well even with a 10 : 1 imbalance. Also, it is critical to make sure that you only resample the training set; keep the distribution of the test set the same as that of the original data since you will not know the class labels of new data in practice and will not be able to resample.

6.8 How can social scientists benefit from machine learning?

In this chapter, we have introduced you to some new methods (both unsupervised and supervised), validation methodologies, and evaluation metrics. All of these can benefit social scientists as they tackle problems in research and practice. In this section, we will give a few concrete examples where what you have learned so far can be used to improve some social science tasks:

- Use of better prediction methods and methodology: Traditional statistics and social sciences have not focused much on methods for prediction. Machine learning researchers have spent the past 30 years developing and adapting methods focusing on that task. We believe that there is a lot of value for social science researchers and practitioners in learning more about those methods, applying them, and even augmenting them [210]. Two common tasks that can be improved using better prediction methods are generating counterfactuals (essentially a prediction problem) and matching. In addition, holdout sets and cross-validation can be used as a model selection methodology with any existing regression and classification methods, resulting in improved model selection and error estimates.

- Model misspecification: Linear and logistic regressions are common techniques for data analysis in the social sciences. One fundamental assumption within both is that they are additive over parameters. Machine learning provides tools when this assumption is too limiting. Hainmueller and Hazlett [148], for example, reanalyze data that were originally analyzed with logistic regression and come to substantially different conclusions. They argue that their analysis, which is more flexible and based on supervised learning methodology, provides three additional insights when compared to the original model. First, predictive performance is similar or better, although they do not need an extensive search to find the final model specification as it was done in the original analysis. Second, their model allows them to calculate average marginal effects that are mostly similar to the original analysis. However, for one covariate they find a substantially different result, which is due to model misspecification in the original model. Finally, the reanalysis also discovers interactions that were missed in the original publication.

- Better text analysis: Text is everywhere, but unfortunately humans are slow and expensive in analyzing text data. Thus, computers are needed to analyze large collections of text. Machine learning methods can help make this process more efficient. Feldman and Sanger [117] provide an overview of different automatic methods for text analysis. Grimmer and Stewart [141] give examples that are more specific for social scientists, and Chapter 7 provides more details on this topic.

- Adaptive surveys: Some survey questions have a large number of possible answer categories. For example, international job classifications describe more than 500 occupational categories, and it is prohibitive to ask all categories during the survey. Instead, respondents answer an open-ended question about their job and machine learning algorithms can use the verbatim answers to suggest small sets of plausible answer options. The respondents can then select which option is the best description for their occupation, thus saving the costs for coding after the interview.

- Estimating heterogeneous treatment effects: A standard approach to causal inference is the assignment of different treatments (e.g., medicines) to the units of interest (e.g., patients). Researchers then usually calculate the average treatment effect—the average difference in outcomes for both groups. It is also of interest if treatment effects differ for various subgroups (e.g., is a medicine more effective for younger people?). Traditional subgroup analysis has been criticized and challenged by various machine learning techniques [138, 178].

- Variable selection: Although there are many methods for variable selection, regularized methods such as the lasso are highly effective and efficient when faced with large amounts of data. Varian [386] goes into more detail and gives other methods from machine learning that can be useful for variable selection. We can also find interactions between pairs of variables (to feed into other models) using random forests, by looking at variables that co-occur in the same tree, and by calculating the strength of the interaction as a function of how many trees they co-occur in, how high they occur in the trees, and how far apart they are in a given tree.

6.9 Advanced topics

This has been a short but intense introduction to machine learning, and we have left out several important topics that are useful and interesting for you to know about and that are being actively researched in the machine learning community. We mention them here so you know what they are, but will not describe them in detail. These include:

- Semi-supervised learning,
- Active learning,
- Reinforcement learning,
- Streaming data,
- Anomaly detection,
- Recommender systems.

6.10 Summary

Machine learning is a active research field, and in this chapter we have given you an overview of how the work developed in this field can be used by social scientists. We covered the overall machine learning process, methods, evaluation approaches and metrics, and some practical tips, as well as how all of this can benefit social scientists. The material described in this chapter is a snapshot of a fast-changing field, and as we are seeing increasing collaborations between machine learning researchers and social scientists, the hope and expectation is that the next few years will bring advances that will allow us to tackle social and policy problems much more effectively using new types of data and improved methods.

6.11 Resources

Literature for further reading that also explains most topics from this chapter in greater depth:

- Hastie et al.'s *The Elements of Statistical Learning* [159] is a classic and is available online for free.
- James et al.'s *An Introduction to Statistical Learning* [187], from the same authors, includes less mathematics and is more approachable. It is also available online.
- Mitchell's *Machine Learning* [258] is a good introduction to some of the methods and gives a good motivation underlying them.
- Provost and Fawcett's *Data Science for Business* [311] is a good practical handbook for using machine learning to solve real-world problems.
- Wu et al.'s "Top 10 Algorithms in Data Mining" [409].

Software:

- Python (with libraries like `scikit-learn`, `pandas`, and more).
- R has many relevant packages [168].
- Cloud-based: AzureML, Amazon ML.
- Free: KNIME, Rapidminer, Weka (mostly for research use).
- Commercial: IBM Modeler, SAS Enterprise Miner, Matlab.

Many excellent courses are available online [412], including Hastie and Tibshirani's *Statistical Learning* [158].

Major conferences in this area include the International Conference on Machine Learning [177], the Annual Conference on Neural Information Processing Systems (NIPS) [282], and the ACM International Conference on Knowledge Discovery and Data Mining (KDD) [204].

Chapter 7

Text Analysis

Evgeny Klochikhin and Jordan Boyd-Graber

This chapter provides an overview of how social scientists can make use of one of the most exciting advances in big data—text analysis. Vast amounts of data that are stored in documents can now be analyzed and searched so that different types of information can be retrieved. Documents (and the underlying activities of the entities that generated the documents) can be categorized into topics or fields as well as summarized. In addition, machine translation can be used to compare documents in different languages.

7.1 Understanding what people write

You wake up and read the newspaper, a Facebook post, or an academic article a colleague sent you. You, like other humans, can digest and understand rich information, but an increasingly central challenge for humans is to cope with the deluge of information we are supposed to read and understand. In our use case of science, even Aristotle struggled with categorizing areas of science; the vast increase in the scope of written research has only made the challenge greater.

One approach is to use rule-based methods to tag documents for categorization. Businesses used to employ human beings to read the news and tag documents on topics of interest for senior management. The rules on how to assign these topics and tags were developed and communicated to these human beings beforehand. Such a manual categorization process is still common in multiple applications, e.g., systematic literature reviews [52].

However, as anyone who has used a search engine knows, newer approaches exist to categorize text and help humans cope with overload: computer-aided *text analysis*. Text data can be used to enrich "conventional" data sources, such as surveys and administrative data, since the words spoken or written by individuals often provide more nuanced and unanticipated insights. Chapter 3 discusses how to link data to create larger, more diverse data sets. The linkage data sets need not just be numeric, but can also include data sets consisting of text data.

▶ See Chapter 3.

Example: Using text to categorize scientific fields

The National Center for Science and Engineering Statistics, the US statistical agency charged with collecting statistics on science and engineering, uses a rule-based system to manually create categories of science; these are then used to categorize research as "physics" or "economics" [262, 288]. In a rule-based system there is no ready response to the question "how much do we spend on climate change, food safety, or biofuels?" because existing rules have not created such categories. Text analysis techniques can be used to provide such detail without manual collation. For example, data about research awards from public sources and about people funded on research grants from UMETRICS can be linked with data about their subsequent publications and related student dissertations from ProQuest. Both award and dissertation data are text documents that can be used to characterize what research has been done, provide information about which projects are similar within or across institutions, and potentially identify new fields of study [368].

Overall, text analysis can help with specific tasks that define application-specific subfields including the following:

- Searches and information retrieval: Text analysis tools can help find relevant information in large databases. For example, we used these techniques in systematic literature reviews to facilitate the discovery and retrieval of relevant publications related to early grade reading in Latin America and the Caribbean.

- Clustering and text categorization: Tools like topic modeling can provide a big picture of the contents of thousands of documents in a comprehensible format by discovering only the most important words and phrases in those documents.

- Text summarization: Similar to clustering, text summarization can provide value in processing large documents and text corpora. For example, Wang et al. [393] use topic modeling to produce category-sensitive text summaries and annotations on large-scale document collections.

- Machine translation: Machine translation is an example of a text analysis method that provides quick insights into documents written in other languages.

7.2 How to analyze text

Human language is complex and nuanced, which makes analysis difficult. We often make simplifying assumptions: we assume our input is perfect text; we ignore humor [149] and deception [280, 292]; and we assume "standard" English [212].

► See Chapter 6 for a discussion of speech recognition, which can turn spoken language into text.

Recognizing this complexity, the goal of text mining is to reduce the complexity of text and extract important messages in a comprehensible and meaningful way. This objective is usually achieved through text categorization or automatic classification. These tools can be used in multiple applications to gain salient insights into the relationships between words and documents. Examples include using machine learning to analyze the flow and topic segmentation of political debates and behaviors [276, 279] and to assign automated tags to documents [379].

► Classification, a machine learning method, is discussed in Chapter 6.

Information retrieval has a similar objective of extracting the most important messages from textual data that would answer a particular query. The process analyzes the full text or metadata related to documents and allows only relevant knowledge to be discovered and returned to the query maker. Typical information retrieval tasks include knowledge discovery [263], word sense disambiguation [269], and sentiment analysis [295].

The choice of appropriate tools to address specific tasks significantly depends on the context and application. For example, document classification techniques can be used to gain insights into the general contents of a large corpus of documents [368], or to discover a particular knowledge area, or to link corpora based on implicit semantic relationships [54].

In practical terms, some of the questions can be: How much does the US government invest in climate change research and nanotechnology? Or what are the main topics in the political debate on guns in the United States? Or how can we build a salient and dynamic taxonomy of all scientific research?

We begin with a review of established techniques to begin the process of analyzing text. Section 7.3 provides an overview of topic modeling, information retrieval and clustering, and other approaches accompanied by practical examples and applications. Section 7.4 reviews key evaluation techniques used to assess the validity, robustness and utility of derived results.

7.2.1 Processing text data

▶ Cleaning and processing are discussed extensively in Chapter 3.

The first important step in working with text data is cleaning and processing. Textual data are often messy and unstructured, which makes many researchers and practitioners overlook their value. Depending on the source, cleaning and processing these data can require varying amounts of effort but typically involve a set of established techniques.

Text corpora A set of multiple similar documents is called a *corpus*. For example, the Brown University Standard Corpus of Present-Day American English, or just the Brown Corpus [128], is a collection of processed documents from works published in the United States in 1961. The Brown Corpus was a historical milestone: it was a machine-readable collection of a million words across 15 balanced genres with each word tagged with its part of speech (e.g., noun, verb, preposition). The British National Corpus [383] repeated the same process for British English at a larger scale. The Penn Treebank [251] provides additional information: in addition to part-of-speech annotation, it provides *syntactic* annotation. For example, what is the object of the sentence "The man bought the hat"? These standard corpora serve as training data to train the classifiers and machine learning techniques to automatically analyze text [149].

However, not every corpus is effective for every purpose: the number and scope of documents determine the range of questions that you can ask and the quality of the answers you will get back: too few documents result in a lack of coverage, too many of the wrong kind of documents invite confusing noise.

Tokenization The first step in processing text is deciding what terms and phrases are meaningful. Tokenization separates sentences and terms from each other. The Natural Language Toolkit (NLTK) [39] provides simple reference implementations of standard natural language processing algorithms such as tokenization—for example, sentences are separated from each other using punctuation such as period, question mark, or exclamation mark. However, this does not cover all cases such as quotes, abbreviations, or informal communication on social media. While separating sentences in a single language is hard enough, some documents "code-switch," combining multiple languages in a single document. These complexities are best addressed through data-driven machine learning frameworks [209].

Stop words Once the tokens are clearly separated, it is possible to perform further text processing at a more granular, token level. Stop words are a category of words that have limited semantic meaning regardless of the document contents. Such words can be prepositions, articles, common nouns, etc. For example, the word "the" accounts for about 7% of all words in the Brown Corpus, and "to" and "of" are more than 3% each [247].

Hapax legomena are rarely occurring words that might have only one instance in the entire corpus. These words—names, misspellings, or rare technical terms—are also unlikely to bear significant contextual meaning. Similar to stop words, these tokens are often disregarded in further modeling either by the design of the method or by manual removal from the corpus before the actual analysis.

N-grams However, individual words are sometimes not the correct unit of analysis. For example, blindly removing stop words can obscure important phrases such as "systems of innovation," "cease and desist," or "commander in chief." Identifying these N-grams requires looking for statistical patterns to discover phrases that often appear together in fixed patterns [102]. These combinations of phrases are often called *collocations*, as their overall meaning is more than the sum of their parts.

Stemming and lemmatization Text normalization is another important aspect of preprocessing textual data. Given the complexity of natural language, words can take multiple forms dependent on the syntactic structure with limited change of their original meaning. For example, the word "system" morphologically has a plural "systems" or an adjective "systematic." All these words are semantically similar and—for many tasks—should be treated the same. For example, if a document has the word "system" occurring three times, "systems" once, and "systematic" twice, one can assume that the word "system" with similar meaning and morphological structure can cover all instances and that variance should be reduced to "system" with six instances.

The process for text normalization is often implemented using established lemmatization and stemming algorithms. A *lemma* is the original dictionary form of a word. For example, "go," "went," and "goes" will all have the lemma "go." The stem is a central part of a given word bearing its primary semantic meaning and uniting a group of similar lexical units. For example, the words "order" and "ordering" will have the same stem "ord." Morphy (a lemmatizer provided by the electronic dictionary WordNet), Lancaster Stemmer, and Snowball Stemmer are common tools used to derive lemmas and stems for tokens, and all have implementations in the NLTK [39].

All text-processing steps are critical to successful analysis. Some of them bear more importance than others, depending on the specific application, research questions, and properties of the corpus. Having all these tools ready is imperative to producing a clean input for subsequent modeling and analysis. Some simple rules should be followed to prevent typical errors. For example, stop words should not be removed before performing *n*-gram indexing, and a stemmer should not be used where data are complex and require accounting for all possible forms and meanings of words. Reviewing interim results at every stage of the process can be helpful.

7.2.2 How much is a word worth?

▸ Term weighting is an example of feature engineering discussed in Chapter 6.

Not all words are worth the same; in an article about electronics, "capacitor" is more important than "aspect." Appropriately weighting and calibrating words is important for both human and machine consumers of text data: humans do not want to see "the" as the most frequent word of every document in summaries, and classification algorithms benefit from knowing which features are actually important to making a decision.

Weighting words requires balancing how often a word appears in a local context (such as a document) with how much it appears overall in the document collection. Term frequency–inverse document frequency (TFIDF) [322] is a weighting scheme to explicitly balance these factors and prioritize the most meaningful words. The TFIDF model takes into account both the term frequency of a given token and its document frequency (Box 7.1) so that if a highly frequent word also appears in almost all documents, its meaning for the specific context of the corpus is negligible. Stop words are a good example when highly frequent words also bear limited meaning since they appear in virtually all documents of a given corpus.

Box 7.1: TFIDF

For every token t and every document d in the corpus D, TFIDF is calculated as

$$tfidf(t, d, D) = tf(t, d) \times idf(t, D),$$

where term frequency is either a simple count,

$$tf(t, d) = f(t, d),$$

or a more balanced quantity,

$$tf(t, d) = 0.5 + \frac{0.5 \times f(t, d)}{\max\{f(t, d) : t \in d\}},$$

and inverse document frequency is

$$idf(t, D) = \log \frac{N}{|\{d \in D : t \in d\}|}.$$

7.3 Approaches and applications

In this section, we discuss several approaches that allow users to perform an unsupervised analysis of large text corpora. That is, approaches that do not require extensive investment of time from experts or programmers to begin to understand large text corpora. The ease of using these approaches provides additional opportunities for social scientists and policymakers to gain insights into policy and research questions through text analysis.

First, we discuss topic modeling, an approach that discovers *topics* that constitute the high-level themes of a corpus. Topic modeling is often described as an *information discovery* process: describing what concepts are present in a corpus. Second, we discuss information retrieval, which finds the closest documents to a particular concept a user wants to discover. In contrast to topic modeling (exposing the primary concepts the corpus, heretofore unknown), information retrieval finds documents that express already known concepts. Other approaches can be used for document classification, sentiment analysis, and part-of-speech tagging.

7.3.1 Topic modeling

As topic modeling is a broad subfield of natural language processing and machine learning, we will restrict our focus to a single exemplar:

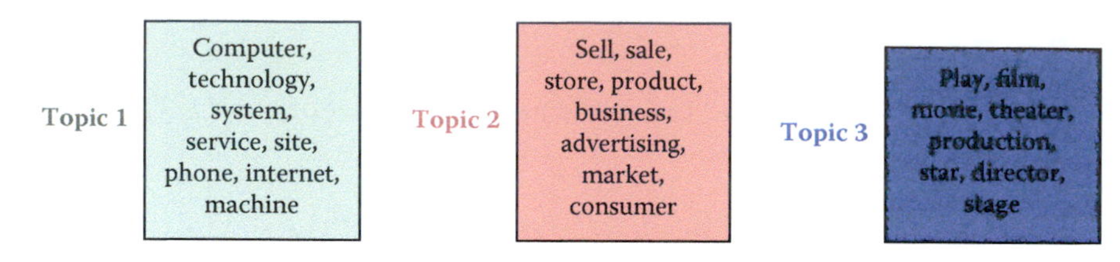

Figure 7.1. Topics are distributions over words. Here are three example topics learned by latent Dirichlet allocation from a model with 50 topics discovered from the *New York Times* [324]. Topic 1 seems to be about technology, Topic 2 about business, and Topic 3 about the arts

latent Dirichlet allocation (LDA) [43]. LDA is a fully Bayesian extension of probabilistic latent semantic indexing [167], itself a probabilistic extension of latent semantic analysis [224]. Blei and Lafferty [42] provide a more detailed discussion of the history of topic models.

LDA, like all topic models, assumes that there are topics that form the building blocks of a corpus. Topics are distributions over words and are often shown as a ranked list of words, with the highest probability words at the top of the list (Figure 7.1). However, we do not know what the topics are a priori; the challenge is to discover what they are (more on this shortly).

In addition to assuming that there exist some number of topics that explain a corpus, LDA also assumes that each document in a corpus can be explained by a small number of topics. For example, taking the example topics from Figure 7.1, a document titled "Red Light, Green Light: A Two-Tone LED to Simplify Screens" would be about Topic 1, which appears to be about technology. However, a document like "Forget the Bootleg, Just Download the Movie Legally" would require all three of the topics. The set of topics that are used by a document is called the document's *allocation* (Figure 7.2). This terminology explains the name *latent Dirichlet allocation*: each document has an allocation over latent topics governed by a Dirichlet distribution.

7.3.1.1 Inferring topics from raw text

Algorithmically, the problem can be viewed as a black box. Given a corpus and an integer K as input, provide the topics that best describe the document collection: a process called *posterior inference*. The most common algorithm for solving this problem is a technique called *Gibbs sampling* [131].

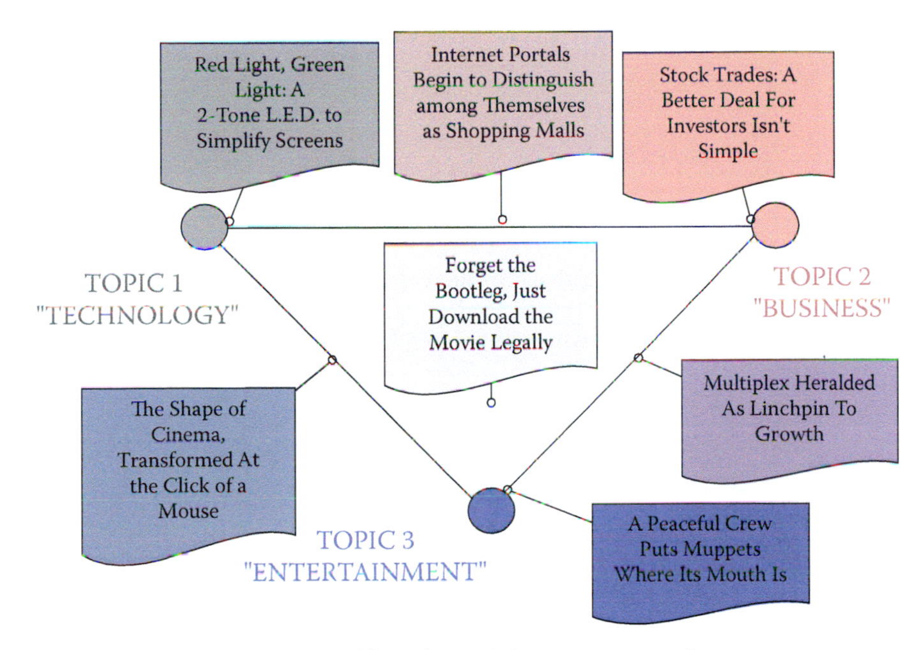

Figure 7.2. Allocations of documents to topics

Gibbs sampling works at the word level to discover the topics that best describe a document collection. Each word is associated with a single topic, explaining why that word appeared in a document. For example, consider the sentence "Hollywood studios are preparing to let people download and buy electronic copies of movies over the Internet." Each word in this sentence is associated with a topic: "Hollywood" might be associated with an arts topic; "buy" with a business topic; and "Internet" with a technology topic (Figure 7.3).

This is where we should eventually get. However, we do not know this to start. So we can initially assign words to topics randomly. This will result in poor topics, but we can make those topics better. We improve these topics by taking each word, pretending that we do not know the topic, and selecting a new topic for the word.

A topic model wants to do two things: it does not want to use many topics in a document, and it does not want to use many words in a topic. So the algorithm will keep track of how many times a document d has used a topic k, $N_{d,k}$, and how many times a topic k has used a word w, $V_{k,w}$. For notational convenience, it will also be useful to keep track of marginal counts of how many words are in a document,

$$N_{d,\cdot} \equiv \sum_k N_{d,k},$$

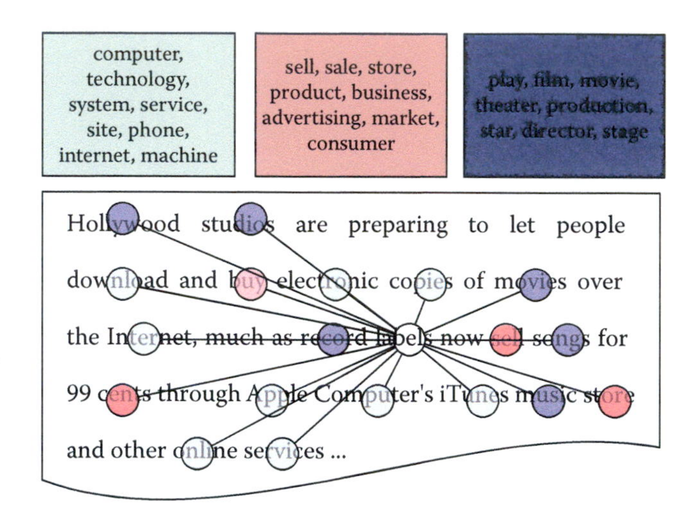

Figure 7.3. Each word is associated with a topic. Gibbs sampling inference iteratively resamples the topic assignments for each word to discover the most likely topic assignments that explain the document collection

and how many words are associated with a topic,

$$V_{k,\cdot} \equiv \sum_{w} V_{k,w}.$$

The algorithm removes the counts for a word from $N_{d,k}$ and $V_{k,w}$ and then changes the topic of a word (hopefully to a better topic than the one it had before). Through many thousands of iterations of this process, the algorithm can find topics that are coherent, useful, and characterize the data well.

The two goals of topic modeling—balancing document allocations to topics and topics' distribution over words—come together in an equation that multiplies them together. A good topic will be both common in a document and explain a word's appearance well.

Example: Gibbs sampling for topic models

The topic assignment $z_{d,n}$ of word n in document d is proportional to

$$p(z_{d,n} = k) \propto \left(\underbrace{\frac{N_{d,k} + \alpha}{N_{d,\cdot} + K\alpha}}_{\text{how much doc likes the topic}} \right) \left(\underbrace{\frac{V_{k,w_{d,n}} + \beta}{V_{k,\cdot} + V\beta}}_{\text{how much topic likes the word}} \right),$$

where α and β are smoothing factors that prevent a topic from having zero probability if a topic does not use a word or a document does not use a topic [390]. Recall that we do not include the token that we are sampling in the counts for N or V.

For the sake of concreteness, assume that we have three documents with the following topic assignments:

Document 1: ${}^{A}\text{dog}_3$ ${}^{B}\text{cat}_2$ ${}^{C}\text{cat}_3$ ${}^{D}\text{pig}_1$

Document 2: ${}^{E}\text{hamburger}_2$ ${}^{F}\text{dog}_3$ ${}^{G}\text{hamburger}_1$

Document 3: ${}^{H}\text{iron}_1$ ${}^{I}\text{iron}_3$ ${}^{J}\text{pig}_2$ ${}^{K}\text{iron}_2$

If we want to sample token B (the first instance of of "cat" in document 1), we compute the conditional probability for each of the three topics ($z = 1, 2, 3$):

$$p(z_B = 1) = \frac{1 + 1.000}{3 + 3.000} \times \frac{0 + 1.000}{3 + 5.000} = 0.333 \times 0.125 = 0.042,$$

$$p(z_B = 2) = \frac{0 + 1.000}{3 + 3.000} \times \frac{0 + 1.000}{3 + 5.000} = 0.167 \times 0.125 = 0.021, \text{ and}$$

$$p(z_B = 3) = \frac{2 + 1.000}{3 + 3.000} \times \frac{1 + 1.000}{4 + 5.000} = 0.500 \times 0.222 = 0.111.$$

To reiterate, we do not include token B in these counts: in computing these conditional probabilities, we consider topic 2 as never appearing in the document and "cat" as never appearing in topic 2. However, "cat" does appear in topic 3 (token C), so it has a higher probability than the other topics. After renormalizing, our conditional probabilities are $(0.24, 0.12, 0.64)$. We then sample the new assignment of token B to be topic 3 two times out of three. Griffiths and Steyvers [140] provide more details on the derivation of this equation.

Example code Listing 7.1 provides a function to compute the conditional probability of a single word and return the (unnormalized) probability to sample from.

7.3.1.2 Applications of topic models

Topic modeling is most often used for topic exploration, allowing users to understand the contents of large text corpora. Thus, topic models have been used, for example, to understand what the National Institutes of Health funds [368]; to compare and contrast what was discussed in the North and South in the Civil War [270]; and to understand how individuals code in large programming projects [253].

Topic models can also be used as features to more elaborate algorithms such as machine translation [170], detecting objects in images [392], or identifying political polarization [298].

```
def class_sample(docs, vocab, d, n, alpha,
               beta, theta, phi, num_topics):
  # Get the vocabulary ID of the word we are sampling
  type = docs[d][n]

  # Dictionary to store final result
  result = {}

  # Consider each topic possibility
  for kk in xrange(num_topics):
    # theta stores the number of times the document d uses
    # each topic kk; alpha is a smoothing parameter
    doc_contrib = (theta[d][kk] + alpha) / \
         (sum(theta[d].values()) + num_topics * alpha)

    # phi stores the number of times topic kk uses
    # this word type; beta is a smoothing parameter
    topic_contrib = (phi[kk][type] + beta) / \
            (sum(phi[kk].values()) + len(vocab) * beta)

    result[kk] = doc_contrib * topic_contrib
  return result
```

Listing 7.1. Python code to compute conditional probability of a single word and return the probability from which to sample

7.3.2 Information retrieval and clustering

Information retrieval is a large subdiscipline that encompasses a variety of methods and approaches. Its main advantage is using large-scale empirical data to make analytical inferences and class assignments. Compared to topic modeling, discussed above, information retrieval techniques can use external knowledge repositories to categorize given corpora as well as discover smaller and emerging areas within a large database.

A major concept of information retrieval is a search query that is usually a short phrase presented by a human or machine to retrieve a relevant answer to a question or discover relevant knowledge. A good example of a large-scale information retrieval system is a search engine, such as Google or Yahoo!, that provides the user with an opportunity to search the entire Internet almost instantaneously. Such fast searches are achieved by complex techniques that are linguistic (set-theoretic), algebraic, probabilistic, or feature-based.

Set-theoretic operations and Boolean logic Set-theoretic operations proceed from the assumption that any query is a set of linked components all of which need to be present in the returned result for it to be relevant. Boolean logic serves as the basis for such queries; it uses Boolean operators such as AND, OR, and NOT to combine query components. For example, the query

```
induction AND (physics OR logic)
```

will retrieve all documents in which the word "induction" is used, whether in a physical or logical sense. The *extended Boolean model* and *fuzzy retrieval* are enhanced approaches to calculating the relevance of retrieved documents based on such queries [232].

Search queries can be also enriched by wildcards and other connectors. For example, the character "*" typically substitutes for any possible character or characters depending on the settings of the query engine. (In some instances, search queries can run in *nongreedy* mode, in which case, for example, the phrase `inform*` might retrieve only text up to the end of the sentence. On the other hand, a *greedy* query might retrieve full text following the word or part of word, denoted as `inform*`, up to the end of the document, which would essentially mean the same as `inform*$`.) The wildcard "?" expects either one or no character in its place, and the wildcard "." expects exactly one character. Search queries enhanced with such symbols and Boolean operators are referred to as *regular expressions*.

Various databases and search engines can interpret Boolean operators and wildcards differently depending on their settings and therefore are prone to return rather different results. This behavior should be expected and controlled for while running searches on different data sources.

Example: Discover food safety awards

Food safety is an interdisciplinary research area that spans multiple scientific disciplines including biological sciences, agriculture, and food science. To retrieve food safety-related awards, we have to construct a Boolean-based search string that would look for terms and phrases in those documents and return only relevant results.

An example of such a string would typically be subdivided by category or search group connected to each other by the AND or OR operator:

1. General terms: `(food safety OR food securit* OR food insecurit*)`
2. Food pathogens: `(food*) AND (acanthamoeba OR actinobacteri* OR (anaerobic organ*) OR DDT OR ...)`
3. Biochemistry and toxicology: `(food*) AND (toxicolog* OR (activated carbon*) OR (acid-hydrol?zed vegetable protein*) OR aflatoxin* OR ...)`

4. Food processing and preservation: (food*) AND (process* OR preserv* OR fortif* OR extrac* OR ...)
5. Food quality and quality control: (food*) AND (qualit* OR (danger zon*) OR test* OR (risk analys*) OR ...)
6. Food-related diseases: (food* OR foodbo?rn* OR food-rela*) AND (diseas* OR hygien* OR allerg* OR diarrh?ea* OR nutrit* OR ...)

Different websites and databases use different search functions to return most relevant results given a query (e.g., "food safety"). Ideally, a user has access to a full database and can apply the same Python code based on regular expressions to all textual data. However, this is not always possible (e.g., when using proprietary databases, such as Web of Science). In those cases, it is important to follow the conventions of the information retrieval system. For example, one source might need phrases to be embedded in parentheses (i.e., (x-ray crystallograph.*)) while another database interface would require such phrases to be contained within quotation marks (i.e., ``x-ray crystallograph.*''). It is then critical to explore the search tips and rules on those databases to ensure that the most complete information is gathered and further analyzed.

Example code Python's built-in re package provides all the capability needed to construct and search with complex regular expressions. Listing 7.2 provides an example.

```
def fs_regex(nsf_award_abstracts,outfilename):

  # Construct simple search string divided by search groups
  food = "food.*"
  general = "safety|secur.*|insecur.*"
  pathogens = "toxicolog.*|acid-hydrolyzed vegetable protein.*|
    activated carbon.*"
  process = "process.*|preserv.*|fortif.*" #and so on

  # Open csv table with all NSF award abstracts in 2000-2014
  inpfile = open(nsf_award_abstracts,'rb')
  inpdata = csv.reader(infile)
  outfile = open(outfilename,'wb')
  output = csv.writer(outfile)
  for line in inpdata:
    award_id = line[0]
    title = line[1]
    abstract = line[2]
    if re.search(food,abstract) and (re.search(general,abstract)
      or re.search(pathogens,abstract) or re.search(process,
    abstract)):
      output.writerow(i+['food safety award'])
```

Listing 7.2. Python code to identify food safety related NSF awards with regular expressions

Algebraic models Algebraic models turn text into numbers to run mathematical operations and discover inherent interdependencies between terms and phrases, also defining the most important and meaningful among them. The vector space representation is a typical way of converting words into numbers, wherein every token is assigned with a sequential ID and a respective weight, be it a simple term frequency, TFIDF value, or any other assigned number.

Latent Dirichlet allocation, discussed in the preceding section, is a good example of a probabilistic model, while unsupervised machine learning techniques, such as random forest, can be used for feature-based modeling and information retrieval.

▶ Random forests are discussed for record linkages in Chapter 3 and described in more detail in Chapter 10.

Similarity measures and approaches Based on algebraic models, the user can either compare documents between each other or train a model that can be further inferred on a different corpus. Typical metrics involved in this process include cosine similarity and Kullback-Leibler divergence [218].

Cosine similarity is a popular measure in document classification. Given two documents d_a and d_b presented as term vectors $\overrightarrow{t_a}$ and $\overrightarrow{t_b}$, the cosine similarity is

$$SIM_C(\overrightarrow{t_a}, \overrightarrow{t_b}) = \frac{\overrightarrow{t_a} \cdot \overrightarrow{t_b}}{|\overrightarrow{t_a}| * |\overrightarrow{t_b}|}.$$

Example: Measuring cosine similarity between documents

NSF awards are not labeled by scientific field—they are labeled by program. This administrative classification is not always useful to assess the effects of certain funding mechanisms on disciplines and scientific communities. One approach is to understand how awards align with each other even if they were funded by different programs. Cosine similarity allows us to do just that.

Example code The Python `numpy` module is a powerful library of tools for efficient linear algebra computation. Among other things, it can be used to compute the cosine similarity of two documents represented by numeric vectors, as described above. The `gensim` module that is often used as a Python-based topic modeling implementation can be used to produce vector space representations of textual data. Listing 7.3 provides an example of measuring cosine similarity using these modules.

```
# Define cosine similarity function
def coss(v1,v2):
  return np.dot(v1,v2) /
         (np.sqrt(np.sum(np.square(v1))) *
          np.sqrt(np.sum(np.square(v2))))

def coss_nsf(nsf_climate_change,nsf_earth_science,outfile):

  # Open the source and compared to documents
  source = csv.reader(file(nsf_climate_change,'rb'))
  comparison = csv.reader(file(nsf_earth_science,'rb'))

  # Create an output file
  output = csv.writer(open(outfile,'wb'))

  # Read through the source and store value in static data container
  data = {}
  for row in source:
    award_id = row[0]
    abstract = row[1]
    data[award_id] = abstract

  # Read through the comparison file and compute similarity
  for row in comparison:
    award_id = row[0]
    # Assuming that abstract is cleaned, processed, tokenized, and
    # stored as a space-separated string of tokens
    abstract = row[1]
    abstract_for_dict = abstract.split(" ")
    # Construct dictionary of tokens and IDs
    dict_abstract = corpora.dictionary.Dictionary(abstract_for_dict)
    # Construct vector from dictionary
    # of all tokens and IDs in abstract
    abstr_vector = dict(dict_abstract.doc2bow(abstract))
    # Iterate through all stored abstracts in source corpus
    # and assign same token IDs using dictionary
    for key,value in data.items():
      source_id = key
      # Get all tokens from source abstract, assuming it is
      # tokenized and space-separated
      source_abstr = value.split(" ")
      source_vector = dict(dict_abstract.doc2bow(source_abstr))
      # Cosine similarity requires having same shape vectors.
      # Thus impute zeros for any missing tokens in source
      # abstract as compared to the target one
      add = { n:0 for n in abstr_vector.keys()
              if n not in source_dict.keys() }
      # Update source vector
      source_vector.update(add)
      source_vector = sorted(source_vector.items())
      abstr_vector = sorted(abstr_vector.items())
      # Compute cosine similarity
      similarity = coss(np.array([item[1] for item in abstr_vector]),
                        np.array([item[1] for item in source_dict])
      output.writerow([source_id,award_id,similarity])
```

Listing 7.3. Python code to measure cosine similarity between Climate Change and all other Earth Science NSF awards

Kullback–Leibler (KL) divergence is an asymmetric measure that is often enhanced by averaged calculations to ensure unbiased results when comparing documents between each other or running a classification task. Given two term vectors $\overrightarrow{t_a}$ and $\overrightarrow{t_b}$, the KL divergence from vector $\overrightarrow{t_a}$ to $\overrightarrow{t_b}$ is

$$D_{KL}(\overrightarrow{t_a}||\overrightarrow{t_b}) = \sum_{t=1}^{m} w_{t,a} \times \log\left(\frac{w_{t,a}}{w_{t,b}}\right),$$

where $w_{t,a}$ and $w_{t,b}$ are term weights in two vectors, respectively.

An averaged KL divergence metric is then defined as

$$D_{AvgKL}(\overrightarrow{t_a}||\overrightarrow{t_b}) = \sum_{t=1}^{m}(\pi_1 \times D(w_{t,a}||w_t) + \pi_2 \times D(w_{t,b}||w_t)),$$

where $\pi_1 = \frac{w_{t,a}}{w_{t,a}+w_{t,b}}$, $\pi_2 = \frac{w_{t,b}}{w_{t,a}+w_{t,b}}$, and $w_t = \pi_1 \times w_{t,a} + \pi_2 \times w_{t,b}$ [171].

A Python-based `scikit-learn` library provides an implementation of these measures as well as other machine learning models and approaches.

Knowledge repositories Information retrieval can be significantly enriched by the use of established knowledge repositories that can provide enormous amounts of organized empirical data for modeling and relevance calculations. Established corpora, such as the Brown Corpus and Lancaster–Oslo–Bergen Corpus, are one type of such preprocessed repositories.

Wikipedia and WordNet are examples of another type of lexical and semantic resources that are dynamic in nature and that can provide a valuable basis for consistent and salient information retrieval and clustering. These repositories have the innate hierarchy, or ontology, of words (and concepts) that are explicitly linked to each other either by inter-document links (Wikipedia) or by the inherent structure of the repository (WordNet). In Wikipedia, concepts thus can be considered as titles of individual Wikipedia pages and the contents of these pages can be considered as their extended semantic representation.

Information retrieval techniques build on these advantages of WordNet and Wikipedia. For example, Meij et al. [256] mapped search queries to the DBpedia ontology (derived from Wikipedia topics and their relationships), and found that this mapping enriches the search queries with additional context and concept relationships. One way of using these ontologies is to retrieve a predefined list of Wikipedia pages that would match a specific taxonomy. For example, scientific disciplines are an established way of tagging documents—some are in physics, others in chemistry, engineering, or computer science. If a user retrieves four Wikipedia pages on "Physics," "Chemistry," "Engineering," and "Computer Science," they can be further mapped to a given set of scientific documents to label and classify them, such as a corpus of award abstracts from the US National Science Foundation.

Personalized PageRank is a similarity system that can help with the task. This system uses WordNet to assess semantic relationships and relevance between a search query (document d) and possible results (the most similar Wikipedia article or articles). This system has been applied to text categorization [269] by comparing documents to *semantic model vectors* of Wikipedia pages constructed using WordNet. These vectors account for the term frequency and their relative importance given their place in the WordNet hierarchy, so that the overall *wiki* vector is defined as:

$$SMV_{wiki}(s) = \sum_{w \in Synonyms(s)} \frac{tf_{wiki}(w)}{|Synsets(w)|},$$

where w is a token within *wiki*, s is a WordNet synset that is associated with every token w in WordNet hierarchy, $Synonyms(s)$ is the set of words (i.e., synonyms) in the synset s, $tf_{wiki}(w)$ is the term frequency of the word w in the Wikipedia article *wiki*, and $Synsets(w)$ is the set of synsets for the word w.

The overall probability of a candidate document d (e.g., an NSF award abstract or a PhD dissertation abstract) matching the target query, or in our case a Wikipedia article *wiki*, is

$$wiki_{BEST} = \sum_{w_t \in doc} \max_{s \in Synsets(w_t)} SMV_{wiki}(s),$$

where $Synsets(w_t)$ is the set of synsets for the word w_t in the target document document (e.g., NSF award abstract) and $SMV_{wiki}(s)$ is the semantic model vector of a Wikipedia page, as defined above.

Applications Information retrieval can be used in a number of applications. Knowledge discovery, or information extraction, is perhaps its primary mission; in contrast, for users, the purpose of information retrieval applications is to retrieve the most relevant response to a query.

Document classification is another popular task where information retrieval methods can be helpful. Such systems, however, typically require a two-step process: The first phase defines all relevant information needed to answer the query. The second phase clusters the documents according to a set of rules or by allowing the machine to actively learn the patterns and classes. For example, one approach is to generate a taxonomy of concepts with associated Wikipedia pages and then map other documents to these pages through Personalized PageRank. In this case, disciplines, such as physics, chemistry, and engineering, can be used as the original labels, and NSF award abstracts can be mapped to these disciplinary categories through the similarity metrics (i.e., whichever of these disciplines scores the highest is the most likely to fit the disciplinary profile of an award abstract).

Another approach is to use the Wikipedia structure as a clustering mechanism in itself. For example, the article about "nanotechnology" links to a number of other Wikipedia pages as referenced in its content. "Quantum realm," "nanometer" or "National Nanotechnology Initiative" are among the meaningful concepts used in the description of nanotechnology that also have their own individual Wikipedia pages. Using these pages, we can assume that if a scientific document, such as an NSF award abstract, has enough similarity with any one of the articles associated with nanotechnology, it can be tagged as such in the classification exercise.

The process can also be turned around: if the user knows exactly the clusters of documents in a given corpus, these can be mapped to an external knowledge repository, such as Wikipedia, to discover yet unknown and emerging relationships between concepts that are not explicitly mentioned in the Wikipedia ontology at the current moment. This situation is likely given the time lag between the discovery of new phenomena, their introduction to the research community, and their adoption by the wider user community responsible for writing Wikipedia pages.

Examples Some examples from our recent work can demonstrate how Wikipedia-based labeling and labeled LDA [278, 315] cope with the task of document classification and labeling in the scientific domain. See Table 7.1.

7.3.3 Other approaches

Our focus in this chapter is on approaches that are language independent and require little (human) effort to analyze text data. In addition to topic modeling and information retrieval discussed above, natural language processing and computational linguistics

Table 7.1. Wikipedia articles as potential labels generated by n-gram indexing of NSF awards

Abstract excerpt	ProQuest subject category	Labeled LDA	Wikipedia-based labeling
Reconfigurable computing platform for small-scale resource-constrained robot. Specific applications often require robots of small size for reasons such as costs, access, and stealth. Small-scale robots impose constraints on resources such as power or space for modules ...	Engineering, Electronics and Electrical; Engineering, Robotics	Motor controller	Robotics, Robot, Field-programmable gate array
Genetic mechanisms of thalamic nuclei specification and the influence of thalamocortical axons in regulating neocortical area formation. Sensory information from the periphery is essential for all animal species to learn, adapt, and survive in their environment. The thalamus, a critical structure in the diencephalon, receives sensory information ...	Biology, Neurobiology	HSD2 neurons	Sonic hedgehog, Induced stem cell, Nervous system
Poetry 'n acts: The cultural politics of twentieth-century American poets' theater. This study focuses on the disciplinary blind spot that obscures the productive overlap between poetry and dramatic theater and prevents us from seeing the cultural work that this combination can perform ...	Literature, American; Theater	Audience	Counter-culture of the 1960s, Novel, Modernism

are rich, well-developed subdisciplines of computer science that can help analyze text data. While covering these subfields is beyond this chapter, we briefly discuss some of the most widely used approaches to process and understand natural language texts.

▶ Chapter 6 reviews supervised machine learning approaches.

In contrast to the *unsupervised* approaches discussed above, most techniques in natural language processing are *supervised* machine learning algorithms. Supervised machine learning produce labels y given inputs x—the algorithm's job is to learn how to automatically produce correct labels given automatic inputs x.

However, the algorithm must have access to many examples of x and y, often of the order of thousands of examples. This is expensive, as the labels often require linguistic expertise [251]. While it is possible to annotate data using crowdsourcing [350], this is not a panacea, as it often forces compromises in the complexity of the task or the quality of the labels.

In the sequel, we discuss how different definitions of x and y—both in the scope and structure of the examples and labels—define unique analyses of linguistic data.

Document classification If the examples x are documents and y are what these documents are about, the problem is called *document classification*. In contrast to the techniques in Section 7.3.1, document classification is used when you know the specific document types for which you are looking *and* you have many examples of those document types.

One simple but ubiquitous example of document classification is spam detection: an email is either an unwanted advertisement (spam) or it is not. Document classification techniques such as naïve Bayes [235] touch essentially every email sent worldwide, making email usable even though most emails are spam.

Sentiment analysis Instead of being what a document is about, a label y could also reveal the speaker. A recent subfield of natural language processing is to use machine learning to reveal the internal state of speakers based on what they say about a subject [295]. For example, given an example of sentence x, can we determine whether the speaker is a Liberal or a Conservative? Is the speaker happy or sad?

Simple approaches use dictionaries and word counting methods [299], but more nuanced approaches make use of *domain*-specific information to make better predictions. One uses different approaches to praise a toaster than to praise an air conditioner [44]; liberals and conservatives each frame health care differently from how they frame energy policy [277].

Part-of-speech tagging When the examples x are individual words and the labels y represent the grammatical function of a word (e.g., whether a word is a noun, verb, or adjective), the task is called part-of-speech tagging. This level of analysis can be useful for discovering simple patterns in text: distinguishing between when “hit” is used as a noun (a Hollywood hit) and when “hit” is used as a verb (the car hit the guard rail).

Unlike document classification, the examples x are not independent: knowing whether the previous word was an adjective makes it far more likely that the next word will be a noun than a verb. Thus, the classification algorithms need to incorporate structure into the decisions. Two common algorithms for this problem are hidden Markov models [313] and conditional random fields [220].

7.4 Evaluation

Evaluation techniques are common in economics, policy analysis, and development. They allow researchers to justify their conclusions using statistical means of validation and assessment. Text, however, is less amenable to standard definitions of error: it is clear that predicting that revenue will be $110 when it is really $100 is far better than predicting $900; however, it is hard to say how far "potato harvest" is from "journalism" if you are attempting to automatically label documents. Documents are hard to transform into numbers without losing semantic meanings and context.

Content analysis, discourse analysis, and bibliometrics are all common tools used by social scientists in their text mining exercises [134, 358]. However, they are rarely presented with robust evaluation metrics, such as type I and type II error rates, when retrieving data for further analysis. For example, bibliometricians often rely on search strings derived from expert interviews and workshops. However, it is hard to certify that those search strings are optimal. For instance, in nanotechnology research, Porter et al. [305] developed a canonical search strategy for retrieving nano-related papers from major scientific databases. Nevertheless, others adopt their own search string modifications and claim similar validity [144, 374].

▸ Chapter 10 discusses how to measure and diagnose errors in big data.

Evaluating these methods depends on reference corpora. We discuss metrics that help you understand whether a collection of documents for a query is a good one or not or whether a labeling of a document collection is consistent with an existing set of labels.

Purity Suppose you are tasked with categorizing a collection of documents based on what they are about. Reasonable people may disagree: I might put "science and medicine" together, while another person may create separate categories for "energy," "scientific research," and "health care," none of which is a strict subset of my "science and medicine" category. Nevertheless, we still want to know whether two categorizations are consistent.

Let us first consider the case where the labels differ but all categories match (i.e., even though you call one category "taxes" and I call it "taxation," it has exactly the same constituent documents). This should be the best case; it should have the highest score possible. Let us say that this maximum score should be 1.

The opposite case is if we both simply assign labels randomly. There will still be some overlap in our labeling: we will agree sometimes, purely by chance. On average, if we both assign one label, selected from the same set of K labels, to each document, then we should expect to agree on about $\frac{1}{K}$ of the labels. This is a lower bound on performance.

The formalization of this measure is called *purity*: how much overlap there is between each of my labels and the "best" match from your labels. Box 7.2 shows how to calculate it.

Box 7.2: Purity calculation

We compute purity by assigning each cluster to the class that is most frequent in the cluster, and then measuring the accuracy of this assignment by counting correctly assigned documents and dividing by the number of all documents, N [248]. In formal terms,

$$\text{Purity}(\Omega, \mathbb{C}) = \frac{1}{N} \sum_{k} \max_{j} |w_k \cap c_j|,$$

where $\Omega = \{w_1, w_2, \ldots, w_k\}$ is the set of candidate clusters and $\mathbb{C} = \{c_1, c_2, \ldots, c_j\}$ is the gold set of classes.

Precision and recall Chapter 6 already touched on the importance of precision and recall for evaluating the results of information retrieval and machine learning models (Box 7.3 provides a reminder of the formulae). Here we look at a particular example of how these metrics can be computed when working with scientific documents.

We assume that a user has three sets of documents $D_a = \{d_{a1}, d_{a2}, \ldots, d_n\}$, $D_b = \{d_{b1}, d_{b2}, \ldots, d_k\}$, and $D_c = \{d_{c1}, d_{c2}, \ldots, d_i\}$. All three sets are clearly tagged with a disciplinary label: D_a are computer science documents, D_b are physics, and D_c are chemistry.

The user also has a different set of documents—Wikipedia pages on "Computer Science," "Chemistry," and "Physics." Knowing that all documents in D_a, D_b, and D_c have clear disciplinary assignments, let us map the given Wikipedia pages to all documents within those three sets. For example, the Wikipedia-based query on "Computer Science" should return all computer science documents and none in physics or chemistry. So, if the query based on the "Computer Science" Wikipedia page returns only 50% of all computer science documents, then 50% of the relevant documents are lost: the recall is 0.5.

On the other hand, if the same "Computer Science" query returns 50% of all computer science documents but also 20% of the

Box 7.3: Precision and recall

These two metrics are commonly used in information retrieval and computational linguistics [318]. Precision computes the type I errors—*false positives*—in a similar manner to the purity measure; it is formally defined as

$$\text{Precision} = \frac{|\{\text{relevant documents}\} \cap \{\text{retrieved documents}\}|}{|\{\text{retrieved documents}\}|}.$$

Recall accounts for type II errors—*false negatives*—and is defined as

$$\text{Recall} = \frac{|\{\text{relevant documents}\} \cap \{\text{retrieved documents}\}|}{|\{\text{relevant documents}\}|}.$$

physics documents and 50% of the chemistry documents, then all of the physics and chemistry documents returned are false positives. Assuming that all document sets are of equal size, so that $|D_a| = 10$, $|D_b| = 10$ and $|D_c| = 10$, then the precision is $\frac{5}{12} = 0.42$.

F score The *F score* takes precision and recall measures a step further and considers the general accuracy of the model. In formal terms, the F score is a weighted average of the precision and recall:

$$F_1 = 2 \cdot \frac{\text{Precision} \cdot \text{Recall}}{\text{Precision} + \text{Recall}}. \tag{7.1}$$

In terms of type I and type II errors:

$$F_\beta = \frac{(1+\beta^2) \cdot \text{true positive}}{(1+\beta^2) \cdot \text{true positive} + \beta^2 \cdot \text{false negative} + \text{false positive}},$$

where β is the balance between precision and recall. Thus, F_2 puts more emphasis on the recall measure and $F_{0.5}$ puts more emphasis on precision.

7.5 Text analysis tools

We are fortunate to have access to a set of powerful open source text analysis tools. We describe three here.

The Natural Language Toolkit The NLTK is a commonly used natural language toolkit that provides a large number of relevant solutions for text analysis. It is Python-based and can be easily integrated into data processing and analytical scripts by a simple `import nltk` (or similar for any one of its submodules).

The NLTK includes a set of tokenizers, stemmers, lemmatizers and other natural language processing tools typically applied in text analysis and machine learning. For example, a user can extract tokens from a document *doc* by running the command `tokens = nltk.word_tokenize(doc)`.

Useful text corpora are also present in the NLTK distribution. For example, the stop words list can be retrieved by running the command `stops=nltk.corpus.stopwords.words(language)`. These stop words are available for several languages within NTLK, including English, French, and Spanish.

Similarly, the Brown Corpus or WordNet can be called by running `from nltk.corpus import wordnet/brown`. After the corpora are loaded, their various properties can be explored and used in text analysis; for example, `dogsyn = wordnet.synsets('dog')` will return a list of WordNet synsets related to the word "dog."

Term frequency distribution and n-gram indexing are other techniques implemented in NLTK. For example, a user can compute frequency distribution of individual terms within a document *doc* by running a command in Python: `fdist=nltk.FreqDist(text)`. This command returns a dictionary of all tokens with associated frequency within *doc*.

N-gram indexing is implemented as a chain-linked collocations algorithm that takes into account the probability of any given two, three, or more words appearing together in the entire corpus. In general, n-grams can be discovered as easily as running `bigrams = nltk.bigrams(text)`. However, a more sophisticated approach is needed to discover statistically significant word collocations, as we show in Listing 7.4.

Bird et al. [39] provide a detailed description of NLTK tools and techniques. See also the official NLTK website [284].

Stanford CoreNLP While NLTK's emphasis is on simple reference implementations, Stanford's CoreNLP [249, 354] is focused on fast implementations of cutting-edge algorithms, particularly for syntactic analysis (e.g., determining the subject of a sentence).

```
def bigram_finder(texts):
  # NLTK bigrams from a corpus of documents separated by new line
  tokens_list = nltk.word_tokenize(re.sub("\n"," ",texts))
  bgm    = nltk.collocations.BigramAssocMeasures()
  finder = nltk.collocations.BigramCollocationFinder.from_words(
    tokens_list)
  scored = finder.score_ngrams( bgm.likelihood_ratio  )

  # Group bigrams by first word in bigram.
  prefix_keys = collections.defaultdict(list)
  for key, scores in scored:
      prefix_keys[key[0]].append((key[1], scores))

  # Sort keyed bigrams by strongest association.
  for key in prefix_keys:
      prefix_keys[key].sort(key = lambda x: -x[1])
```

Listing 7.4. Python code to find bigrams using NLTK

MALLET For probabilistic models of text, MALLET, the MAchine Learning for LanguagE Toolkit [255], often strikes the right balance between usefulness and usability. It is written to be fast and efficient but with enough documentation and easy enough interfaces to be used by novices. It offers fast, popular implementations of conditional random fields (for part-of-speech tagging), text classification, and topic modeling.

7.6 Summary

Much "big data" of interest to social scientists is text: tweets, Facebook posts, corporate emails, and the news of the day. However, the meaning of these documents is buried beneath the ambiguities and noisiness of the informal, inconsistent ways by which humans communicate with each other. Despite attempts to formalize the meaning of text data through asking users to tag people, apply metadata, or to create structured representations, these attempts to manually curate meaning are often incomplete, inconsistent, or both.

These aspects make text data difficult to work with, but also a rewarding object of study. Unlocking the meaning of a piece of text helps bring machines closer to human-level intelligence—as language is one of the most quintessentially human activities—and helps overloaded information professionals do their jobs more effectively: understand large corpora, find the right documents, or automate repetitive tasks. And as an added bonus, the better computers become at understanding natural language, the easier it is for information professionals to communicate their needs: one day using computers to grapple with big data may be as natural as sitting down to a conversation over coffee with a knowledgeable, trusted friend.

7.7 Resources

Text analysis is one of the more complex tasks in big data analysis. Because it is unstructured, text (and natural language overall) requires significant processing and cleaning before we can engage in interesting analysis and learning. In this chapter we have referenced several resources that can be helpful in mastering text mining techniques:

- The Natural Language Toolkit is one of the most popular Python-based tools for natural language processing. It has a variety of methods and examples that are easily accessible online [284]. The book by Bird et al. [39], available online, contains multiple examples and tips on how to use NLTK.

- The book *Pattern Recognition and Machine Learning* by Christopher Bishop [40] is a useful introduction to computational techniques, including probabilistic methods, text analysis, and machine learning. It has a number of tips and examples that are helpful to both learning and experienced researchers.

- A paper by Anna Huang [171] provides a brief overview of the key similarity measures for text document clustering discussed in this chapter, including their strengths and weaknesses in different contexts.

- Materials at the MALLET website [255] can be specialized for the unprepared reader but are helpful when looking for specific solutions with topic modeling and machine classification using this toolkit.

- David Blei, one of the authors of the latent Dirichlet allocation algorithm (topic modeling), maintains a helpful web page with introductory resources for those interested in topic modeling [41].

- We provide an example of how to run topic modeling using MALLET on textual data from the National Science Foundation and Norwegian Research Council award abstracts [49].
- Weka, developed at the University of Waikato in New Zealand, is a useful resource for running both complex text analysis and other machine learning tasks and evaluations [150, 384].

Chapter 8

Networks: The Basics

Jason Owen-Smith

Social scientists are typically interested in describing the activities of individuals and organizations (such as households and firms) in a variety of economic and social contexts. The frame within which data has been collected will typically have been generated from tax or other programmatic sources. The new types of data permit new units of analysis—particularly network analysis—largely enabled by advances in mathematical graph theory. This chapter provides an overview of how social scientists can use network theory to generate measurable representations of patterns of relationships connecting entities. As the author points out, the value of the new framework is not only in constructing different right-hand-side variables but also in studying an entirely new unit of analysis that lies somewhere between the largely atomistic actors that occupy the markets of neo-classical theory and the tightly managed hierarchies that are the traditional object of inquiry of sociologists and organizational theorists.

8.1 Introduction

This chapter provides a basic introduction to the analysis of large networks. The following introduces the basic logic of network analysis, then turn to a summary of data structures and essential measures before presenting a primer on network visualization and a more elaborated descriptive case comparison of the collaboration networks of two research-intensive universities. Both those collaboration networks and a grant co-employment network for a large public university also examined in this chapter are derived from data produced by the multi-university Committee on Institutional Cooperation (CIC)'s UMETRICS project [228]. The snippets of code that are provided are from the `igraph` package for network analysis as implemented in Python.

At their most basic, networks are measurable representations of patterns of relationships connecting entities in an abstract or actual space. What this means is that there are two fundamental questions to ask of any network presentation or measure, First, what are the nodes? Second, what are the ties? While the network methods sketched in this chapter are equally applicable to technical or biological networks (e.g., the hub-and-spoke structure of the worldwide air travel system, or the neuronal network of a nematode worm), I focus primarily on social networks; patterns of relationships among people or organizations that are created by and come to influence individual action. This chapter draws most of its examples from the world of science, technology, and innovation. Thus, this chapter focuses particularly on networks developed and maintained by the collaborations of scientists and by the contractual relationships organizations form in pursuit of innovation.

This substantive area is of great interest because a great deal of research in sociology, management, and related fields demonstrates that networks of just these sorts are essential to understanding the process of innovation and outcomes at both the individual and the organizational level.

In other words, networks offer not just another convenient set of right-hand-side variables, but an entirely new unit of analysis that lies somewhere between the largely atomistic actors that occupy the markets of neo-classical theory and the tightly managed hierarchies that are the traditional object of inquiry of sociologists and organizational theorists. As Walter W. Powell [308] puts it in a description of buyer supplier networks of small Italian firms: "when the entangling of obligation and reputation reaches a point where the actions of the parties are interdependent, but there is no common ownership or legal framework . . . such a transaction is neither a market exchange nor a hierarchical governance structure, but a separate, different mode of exchange."

Existing as they do between the uncoordinated actions of independent individuals and coordinated work of organizations, networks offer a unique level of analysis for the study of scientific and creative teams [410], collaborations [202], and clusters of entrepreneurial firms [294]. The following sections introduce you to this approach to studying innovation and discovery, focusing on examples drawn from high-technology industries and particularly from the scientific collaborations among grant-employed researchers at UMETRICS universities. I make particular use of a network that connects individual researchers to grants that paid their salaries in 2012 for a large public university. The grants network for university A includes information on 9,206 individuals who were employed on 3,389 research grants from federal science agencies in a single year.

Before turning to those more specific substantive topics, the chapter first introduces the most common structures for large network data, briefly introduce three key social "mechanisms of action" by which social networks are thought to have their effects, and then present a series of basic measures that can be used to quantify characteristics of entire networks and the relative position individual people or organizations hold in the differentiated social structure created by networks.

Taken together, these measures offer an excellent starting point for examining how global network structures create opportunities and challenges for the people in them, for comparing and explaining the productivity of collaborations and teams, and for making sense of the differences between organizations, industries, and markets that hinge on the pattern of relationships connecting their participants.

But what is a network? At its simplest, a network is a pattern of concrete, measurable relationships connecting entities engaged in some common activity. While this chapter focuses my attention on social networks, you could easily use the techniques described here to examine the structure of networks such as the World Wide Web, the national railway route map of the USA, the food web of an ecosystem, or the neuronal network of a particular species of animal. Networks can be found everywhere, but the primary example used here is a network connecting individual scientists through their shared work on particular federal grants. The web of partnerships that emerges from university scientists' decentralized efforts to build effective collaborations and teams generates a distinctive social infrastructure for cutting-edge science.

Understanding the productivity and effects of university research thus requires an effort to measure and characterize the networks on which it depends. As suggested, those networks influence outcomes in three ways: first, they distinguish among individuals; second, they differentiate among teams; and third, they help to distinguish among research-performing universities. Most research-intensive institutions have departments and programs that cover similar arrays of topics and areas of study. What distinguishes them from one another is not the topics they cover but the ways in which their distinctive collaboration networks lead them to have quite different scientific capabilities.

8.2 Network data

Networks are comprised of *nodes*, which represent things that can be connected to one another, and of ties that represent the relationships connecting nodes. When ties are undirected they are called *edges*. When they are directed (as when I lend money to you and you do or do not reciprocate) they are called *arcs*. Nodes, edges and arcs can, in principle, be anything: patents and citations, web pages and hypertext links, scientists and collaborations, teenagers and intimate relationships, nations and international trade agreements. The very flexibility of network approaches means that the first step toward doing a network analysis is to clearly define what counts as a node and what counts as a tie.

While this seems like an easy move, it often requires deep thought. For instance, an interest in innovation and discovery could take several forms. We could be interested in how universities differ in their capacity to respond to new requests for proposals (a macro question that would require the comparison of full networks across campuses). We could wonder what sorts of training arrangements lead to the best outcomes for graduate students (a more micro-level question that requires us to identify individual positions in larger networks). Or we could ask what team structure is likely to lead to more or less radical discoveries (a decidedly meso-level question that requires we identify substructures and measure their features).

Each of these is a network question that relies on the ways in which people are connected to one another. The first challenge of measurement is to identify the nodes (what is being connected) and the ties (the relationships that matter) in order to construct the relevant networks. The next is to collect and structure the data in a fashion that is sufficient for analysis. Finally, measurement and visualization decisions must be made.

8.2.1 Forms of network data

Network ties can be directed (flowing from one node to another) or undirected. In either case they can be binary (indicating the presence or absence of a tie) or valued (allowing for relationships of different types or strengths). Network data can be represented in matrices or as lists of edges and arcs. All these types of relationships can connect one type of node (what is commonly called *one-mode* network data) or multiple types of nodes (what is called *two-mode* or affiliation data). Varied data structures correspond to

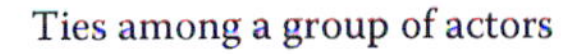

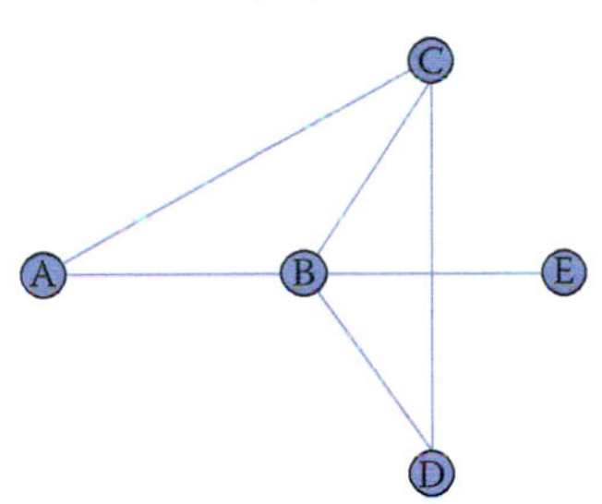

Undirected (edge), binary data

Example: Actors are firms, ties represent the presence of any strategic alliance

Represented as a symmetric square matrix

	A	B	C	D	E
A		1	1	0	0
B	1		1	1	1
C	1	1		1	0
D	0	1	1		0
E	0	1	0	0	

Represented as an edge list

A	B
A	C
B	C
B	D
B	E
C	D

Figure 8.1. Undirected, binary, one-mode network data

different classes of network data. The simplest form of network data represents instances where the same kinds of nodes are connected by undirected ties (edges) that are binary. An example of this type of data is a network where nodes are firms and ties indicate the presence of a strategic alliance connecting them [309]. This network would be represented as a square symmetric matrix or a list of edges connecting nodes. Figure 8.1 summarizes this simple data structure, highlighting the idea that network data of this form can be represented either as a matrix or as an edge list.

A much more complicated network would be one that is both directed and valued. One example might be a network of nations connected by flows of international trade. Goods and services flow from one nation to another and the value of those goods and services (or their volume) represents ties of different strengths. When networks connecting one class of nodes (in this case nations) are directed and valued, they can be represented as asymmetric valued matrices or lists of arcs with associated values. (See Figure 8.2 for an example.)

While many studies of small- to medium-sized social networks rely on one-mode data. Large-scale social network data of this type are relatively rare, but one-mode data of this sort are fairly common in relationships among other types of nodes such as hyperlinks

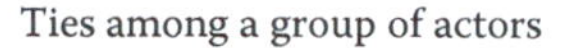

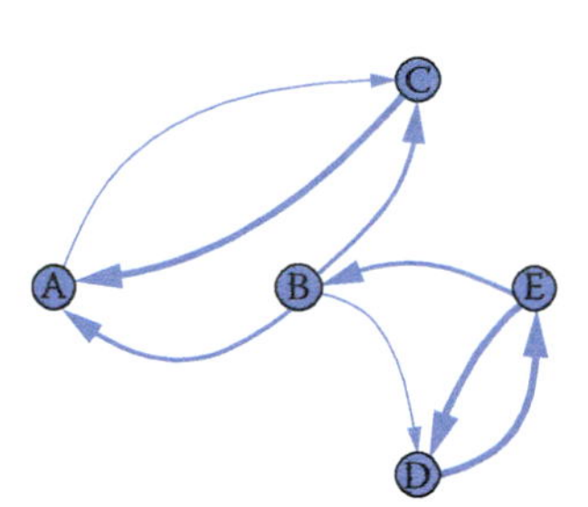

Directed (arc), valued data

Example: Nodes are faculty, ties represent the number of payments from one principal investigator's grant to another's

Represented as an asymmetric square matrix

	A	B	C	D	E
A		0	1	0	0
B	3		3	1	0
C	4	0		0	0
D	0	0	0		5
E	0	2	0	5	

Represented as an arc list

A C 1
B A 3
B C 3
B D 1
C A 4
D E 5
E B 2
E D 5

Figure 8.2. Directed, valued, one-mode network data

connecting web pages or citations connecting patents or publications. Nevertheless, much "big" social network analysis is conducted using two-mode data. The UMETRICS employee data set is a two-mode network that connects people (research employees) to the grants that pay their wages. These two types of nodes can be represented as a rectangular matrix that is either valued or binary. It is relatively rare to analyze untransformed two-mode network data. Instead, most analyses take advantage of the fact that such networks are *dual* [398]. In other words, a two-mode network connecting grants and people can be conceptualized (and analyzed) as two one-mode networks, or *projections*.*

★ Key insight: A two-mode network can be conceptualized and analyzed as two one-mode networks, or projections.

8.2.2 Inducing one-mode networks from two-mode data

The most important trick in large-scale social network analysis is that of inducing one-mode, or unipartite, networks (e.g., employee × employee relationships) from two-mode, or bipartite, data. But the ubiquity and potential value of two-mode data can come at a cost. Not all affiliations are equally likely to represent real, meaningful relationships. While it seems plausible to assume that two individuals paid by the same grant have interactions that reasonably pertain to the work funded by the grant, this need not be the case.

For example, consider the two-mode grant × person network for university A. I used SQL to create a representation of this network that is readable by a freeware network visualization program called Pajek [30]. In this format, a network is represented as two lists: a *vertex list* that lists the nodes in the graph and an *edge list* that lists the connections between those nodes. In our grant × person network, we have two types of nodes, people and grants, and one kind of edge, used to represent wage payments from grants to individuals.

I present a brief snippet of the resulting network file in what follows, showing first the initial 10 elements of the vertex list and then the initial 10 elements of the edge list, presented in two columns for compactness. (The complete file comprises information on 9,206 employees and 3,389 grants, for a total of 12,595 vertices and 15,255 edges. The employees come first in the vertex list, and so the 10 rows shown below all represent employees.) Each vertex is represented by a vertex number–label pair and each edge by a pair of vertices plus an optional value. Thus, the first entry in the edge list (1 10419) specifies that the vertex with identifier 1 (which happens to be the first element of the vertex list, which has value "00100679") is connected to the vertex with identifier 10419 by an edge with value 1, indicating that employee "00100679" is paid by the grant described by vertex 10419.

```
*Grant-Person-Network
*Vertices 12595 9206                 *Edges
    1   "00100679"                       1   10419
    2   "00107462"                       2   10422
    3   "00109569"                       3   9855
    4   "00145355"                       3   9873
    5   "00153190"                       4   9891
    6   "00163131"                       7   10432
    7   "00170348"                       7   12226
    8   "00172339"                       8   10419
    9   "00176582"                       9   11574
   10   "00203529"                      10   11196
```

The network excerpted above is two-mode because it represents relationships between two different classes of nodes, grants, and people. In order to use data of this form to address questions about patterns of collaboration on UMETRICS campuses, we must first transform it to represent collaborative relationships.

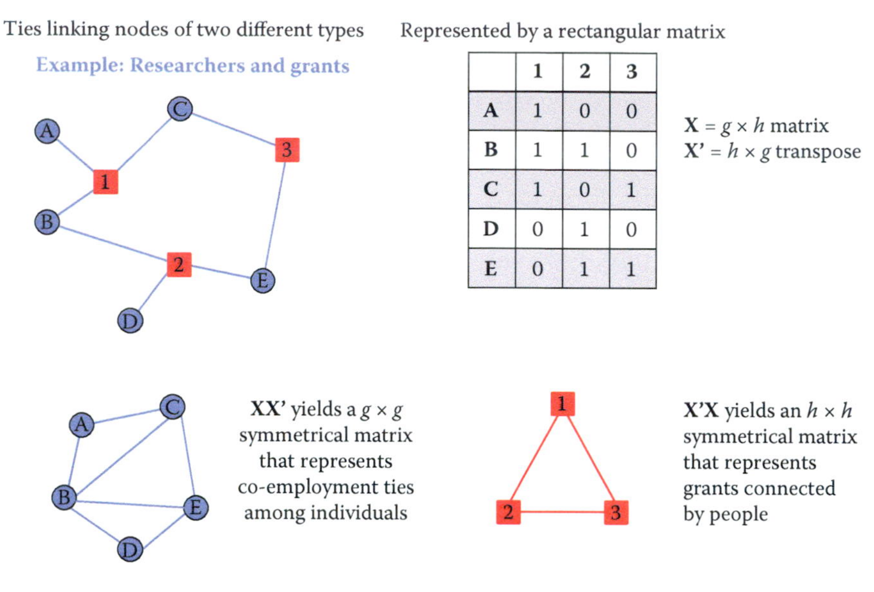

Figure 8.3. Two-mode affiliation data

A person-by-person projection of the original two-mode network assumes that ties exist between people when they are paid by the same grant. By the same token, a grant-by-grant projection of the original two-mode network assumes that ties exist between grants when they pay the same people. Transforming two-mode data into one-mode projections is a fairly simple matter. If $\mathbf{X}$ is a rectangular matrix, $p \times g$, then a one-mode projection, $p \times p$, can be obtained by multiplying $\mathbf{X}$ by its transpose $\mathbf{X}'$. Figure 8.3 summarizes this transformation.

In the following snippet of code, I use the `igraph` package in Python to read in a Pajek file and then transform the original two-mode network into two separate projections. Because my focus in this discussion is on relationships among people, I then move on to work exclusively with the employee-by-employee projection. However, every technique that I describe below can also be used with the grant-by-grant projection, which provides a different view of how federally funded research is put together by collaborative relationships on campus.

```
from igraph import *

# Read the graph
g = Graph.Read_Pajek("public_a_2m.net")
```

```
# Look at result
summary(g)


# IGRAPH U-WT 12595 15252 --
# + attr: color (v), id (v), shape (v), type (v), x (v), y (v), z
    (v), weight (e)
# ...
# ...

# Transform to get 1M projection
pr_g_proj1, pr_g_proj2= g.bipartite_projection()

# Look at results
summary(pr_g_proj1)

# IGRAPH U-WT 9206 65040 --
# + attr: color (v), id (v), shape (v), type (v), x (v), y (v), z
    (v), weight (e)

summary(pr_g_proj2)
# IGRAPH U-WT 3389 12510 --
# + attr: color (v), id (v), shape (v), type (v), x (v), y (v), z
    (v), weight (e)

# pr_g_proj1 is the employeeXemployee projection, n=9,206 nodes
# Rename to emp for use in future calculations

emp=pr_g_proj1
```

We now can work with the graph `emp`, which represents the collaborative network of federally funded research on this campus. Care must be taken when inducing one-mode network projections from two-mode network data because not all affiliations provide equally compelling evidence of actual social relationships. While assuming that people who are paid by the same research grants are collaborating on the same project seems plausible, it might be less realistic to assume that all students who take the same university classes have meaningful relationships. For the remainder of this chapter, the examples I discuss are based on UMETRICS employee data rendered as a one-mode person-by-person projection of the original two-mode person-by-grants data. In constructing these networks I assume that a tie exists between two university research employees when they are paid any wages from the same grant during the same year. Other time frames or thresholds might be used to define ties if appropriate for particular analyses.*

★ Key insight: Care must be taken when inducing one-mode network projections from two-mode network data because not all affiliations provide equally compelling evidence of actual social relationships.

8.3 Network measures

The power of networks lies in their unique flexibility and ability to address many phenomena at multiple levels of analysis. But harnessing that power requires the application of measures that take into account the overall structure of relationships represented in a given network. The key insight of structural analysis is that outcomes for any individual or group are a function of the complete pattern of connections among them. In other words, the explanatory power of networks is driven as much by the pathways that *indirectly* connect nodes as by the particular relationships that *directly* link members of a given dyad. Indirect ties create reachability in a network.*

★ Key insight: Structural analysis of outcomes for any individual or group are a function of the complete pattern of connections among them.

8.3.1 Reachability

Two nodes are said to be reachable when they are connected by an unbroken chain of relationships through other nodes. For instance, two people who have never met may nonetheless be able to reach each other through a common acquaintance who is positioned to broker an introduction [286] or the transfer of information and resources [60]. It is the reachability that networks create that makes them so important for understanding the work of science and innovation.

Consider Figure 8.4, which presents three schematic networks. In each, one focal node, ego, is colored orange. Each ego has four alters, but the fact that each has connections to four other nodes masks important differences in their structural positions. Those differences have to do with the number of other nodes they can reach through the network and the extent to which the other nodes in the network are connected to each other. The orange node (ego) in each network has four partners, but their positions are far from equivalent. Centrality measures on full network data can tease out the differences. The networks also vary in their gross characteristics. Those differences, too, are measurable.*

★ Key insight: Much of the power of networks (and their systemic features) is due to indirect ties that create reachability. Two nodes can reach each other if they are connected by an unbroken chain of relationships. These are often called indirect ties.

Networks in which more of the possible connections among nodes are realized are denser and more cohesive than networks in which fewer potential connections are realized. Consider the two smaller networks in Figure 8.4, each of which is comprised of five nodes. Just five ties connect those nodes in the network on the far right of the figure. One smaller subset of that network, the triangle connecting ego and two alters at the center of the image, represents

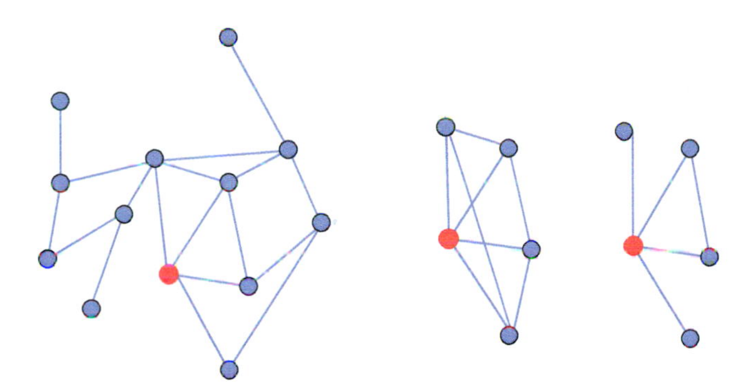

Figure 8.4. Reachability and indirect ties

a more cohesively connected subset of the networks. In contrast, eight of the nine ties that are possible connect the five nodes in the middle figure; no subset of those nodes is clearly more interconnected than any other. While these kinds of differences may seem trivial, they have implications for the orange nodes, and for the functioning of the networks as a whole. Structural differences between the positions of nodes, the presence and characteristics of cohesive "communities" within larger networks [133], and many important properties of entire structures can be quantified using different classes of network measures. Newman [272] provides the most recent and most comprehensive look at measures and algorithms for network research.

The most essential thing to be able to understand about larger scale networks is the pattern of indirect connections among nodes. What is most important about the structure of networks is not necessarily the ties that link particular pairs of nodes to one another. Instead, it is the chains of indirect connections that make networks function as a system and thus make them worthwhile as new levels of analysis for understanding social and other dynamics.

8.3.2 Whole-network measures

The basic terms needed to characterize whole networks are fairly simple. It is useful to know the size (in terms of nodes and ties) of each network you study. This is true both for the purposes of being able to generally gauge the size and connectivity of an entire network and because many of the measures that one might calculate using such networks should be standardized for analytic use. While the list of possible network measures is long, a few commonly used indices offer useful insights into the structure and implications of entire network structures.

Components and reachability As we have seen, a key feature of networks is reachability. The reachability of participants in a network is determined by their membership in what network theorists call *components*, subsets of larger networks where every member of a group is indirectly connected to every other. If you imagine a standard node and line drawing of a network, a component is a portion of the network where you can trace paths between every pair of nodes without ever having to lift your pen.

Most large networks have a single dominant component that typically includes anywhere from 50% to 90% of its participants as well as many smaller components and isolated nodes that are disconnected from the larger portion of the network. Because the path length centrality measures described below can only be computed on connected subsets of networks, it is typical to analyze the largest component of any given network. Thus any description of a network or any effort to compare networks should report the number of components and the percentage of nodes reachable through the largest component. In the code snippet below, I identify the weakly connected components of the employee network, `emp`.

```
# Add component membership
emp.vs["membership"] = emp.clusters(mode="weak").membership

# Add component size
emp.vs["csize"] = [emp.clusters(mode="weak").sizes()[i] for i in
    emp.clusters(mode="weak").membership]

# Identify the main component
# Get indices of max clusters
maxSize = max(emp.clusters(mode="weak").sizes())
emp.vs["largestcomp"] = [1 if maxSize == x else 0 for x in emp.vs[
    "csize"]]

# Add component membership

emp.vs["membership"] = emp.clusters(mode="weak").membership
```

The main component of a network is commonly analyzed and visualized because the graph-theoretic distance among unconnected nodes is infinite, which renders calculation of many common network measures impossible without strong assumptions about just how far apart unconnected nodes actually are. While some researchers replace infinite path lengths with a value that is one plus the longest path, called the network's diameter, observed in a given structure, it is also common to simply analyze the largest connected component of the network.

Path length One of the most robust and reliable descriptive statistics about an entire network is the average path length, l_G, among nodes. Networks with shorter average path lengths have structures that may make it easier for information or resources to flow among members in the network. Longer path lengths, by comparison, are associated with greater difficulty in the diffusion and transmission of information or resources. Let g be the number of nodes or vertices in a network. Then

$$l_G = \frac{1}{g(g-1)} \sum_{i \neq j} d(n_i, n_j).$$

As with other measures based on reachability, it is most common to report the average path length for the largest connected component of the network because the graph-theoretic distance between two unconnected nodes is infinite. In an electronic network such as the World Wide Web, a shorter path length means that any two pages can be reached through fewer hyperlink clicks.

The snippet of code below identifies the distribution of shortest path lengths among all pairs of nodes in a network and the average path length. I also include a line of code that calculates the network distance among all nodes and returns a matrix of those distances. That matrix (saved as `empdist`) can be used to calculate additional measures or to visualize the graph-theoretic proximities among nodes.

```
# Calculate distances and construct distance table

dfreq=emp.path_length_hist(directed=False)
print(dfreq)

# N = 12506433, mean +- sd: 5.0302 +- 1.7830
# Each * represents 51657 items
# [ 1,  2): * (65040)
# [ 2,  3): ********* (487402)
# [ 3,  4): *********************************** (1831349)
# [ 4,  5): ******************************************************
     **** (2996157)
# [ 5,  6): ****************************************************
     (2733204)
# [ 6,  7): ************************************** (1984295)
# [ 7,  8): ************************ (1267465)
# [ 8,  9): ************ (649638)
# [ 9, 10): ***** (286475)
```

```
# [10, 11): ** (125695)
# [11, 12): * (52702)
# [12, 13):  (18821)
# [13, 14):  (5944)
# [14, 15):  (1682)
# [15, 16):  (403)
# [16, 17):  (128)
# [17, 18):  (28)
# [18, 19):  (5)
print(dfreq.unconnected)
# 29864182

print(emp.average_path_length(directed=False))
#[1] 5.030207

empdist= emp.shortest_paths()
```

These measures provide a few key insights into the employee network we have been considering. First, the average pair of nodes that are connected by indirect paths are slightly more than five steps from one another. Second, however, many node pairs in this network (`$unconnected` = 29,864,182) are unconnected and thus unreachable to each other. Figure 8.5 presents a histogram of the distribution of path lengths in the network. It represents the numeric values returned by the `distance.table` command in the code snippet above. In this case the diameter of the network is 18 and five pairs of nodes are reachable at this distance, but the largest group of dyads is reachable (N = 2,996,157 dyads) at distance 4. In short, nearly 3 million pairs of nodes are collaborators of collaborators of collaborators of collaborators.

Degree distribution Another powerful way to describe and compare networks is to look at the distribution of centralities across nodes. While any of the centrality measures described above could be summarized in terms of their distribution, it is most common to plot the degree distribution of large networks. Degree distributions commonly have extremely long tails. The implication of this pattern is that most nodes have a small number of ties (typically one or two) and that a small percentage of nodes account for the lion's share of a network's connectivity and reachability. Degree distributions are typically so skewed that it is common practice to plot degree against the percentage of nodes with that degree score on a log–log scale.

High-degree nodes are often particularly important actors. In the UMETRICS networks that are employee × employee projections of employee × grant networks, for instance, the nodes with the highest degree seem likely to include high-profile faculty—the investi-

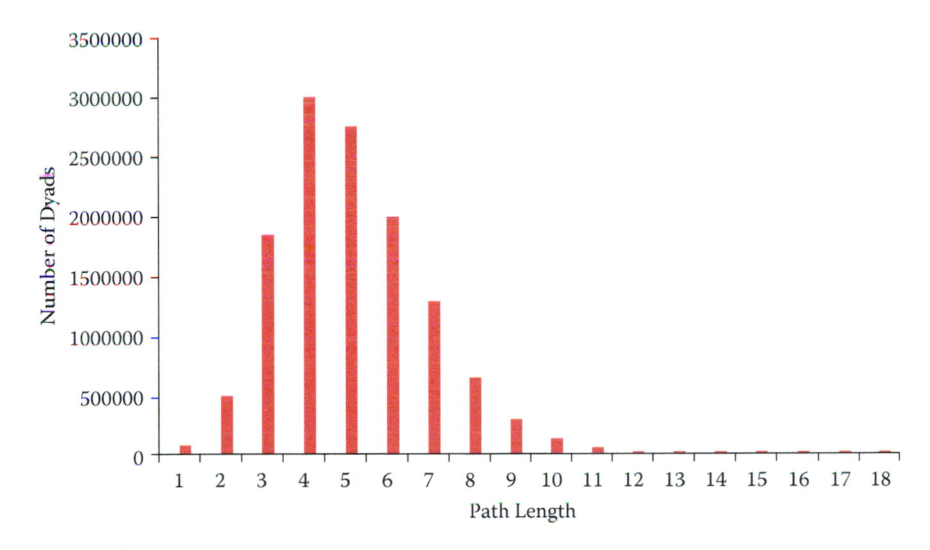

Figure 8.5. Histogram of path lengths for university A employee network

gators on larger institutional grants such as National Institutes of Health-funded Clinical and Translational Science Awards and National Science Foundation-funded Science and Technology Centers, and perhaps staff whose particular skills are in demand (and paid for) by multiple research teams. For instance, the head technician in a core microscopy facility or a laboratory manager who serves multiple groups might appear highly central in the degree distribution of a UMETRICS network.

Most importantly, the degree distribution is commonly taken to provide insight into the dynamics by which a network was created. Highly skewed degree distributions often represent scale-free networks [25, 271, 309], which grow in part through a process called *preferential attachment*, where new nodes entering the network are more likely to attach to already prominent participants. In the kinds of scientific collaboration networks that UMETRICS represents, a scale-free degree distribution might come about as faculty new to an institution attempt to enroll more established colleagues on grants as coinvestigators. In the comparison exercise outlined below, I plot degree distributions for the main components of two different university networks.

Clustering coefficient The third commonly used whole-network measure captures the extent to which a network is cohesive, with many nodes interconnected. In networks that are more cohesively clustered, there are fewer opportunities for individuals to play the kinds of brokering roles that we will discuss below in the context of betweenness centrality. Less cohesive networks, with lower levels of clustering, are potentially more conducive to brokerage and the kinds of innovation that accompany it.

However, the challenge of innovation and discovery is both the moment of invention, the "aha!" of a good new idea, and the often complicated, uncertain, and collaborative work that is required to turn an initial insight into a scientific finding. While less clustered, open networks are more likely to create opportunities for brokers to develop fresh ideas, more cohesive and clustered networks support the kinds of repeated interactions, trust, and integration that are necessary to do uncertain and difficult collaborative work.

While it is possible to generate a global measure of cohesiveness in networks, which is generically the number of closed triangles (groups of three nodes all connected to one another) as a proportion of the number of possible triads, it is more common to take a local measure of connectivity and average it across all nodes in a network. This local connectivity measure more closely approximates the notion of cohesion around nodes that is at the heart of studies of networks as means to coordinate difficult, risky work. The code snippet below calculates both the global clustering coefficient and a vector of node-specific clustering coefficients whose average represents the local measure for the employee × employee network projection of the university A UMETRICS data.

```
# Calculate clustering coefficients
emp.transitivity_undirected()
# 0.7241

local_clust=emp.transitivity_local_undirected(mode="zero")
# (isolates="zero" sets clustering to zero rather than undefined)

import pandas as pd
print(pd.Series(local_clust).describe())
# count    9206.000000
# mean        0.625161
# std         0.429687
# min         0.000000
# 25%         0.000000
# 50%         0.857143
# 75%         1.000000
# max         1.000000
#---------------------------------------------------#
```

Together, these summary statistics—number of nodes, average path length, distribution of path lengths, degree distribution, and the clustering coefficient—offer a robust set of measures to examine and compare whole networks. It is also possible to distinguish among the positions nodes hold in a particular network. Some of the most powerful centrality measures also rely on the idea of indirect ties.*

★ Key insight: Some of the most powerful centrality measures also rely on the idea of indirect ties.

Centrality measures This class of measures is the most common way to distinguish between the positions individual nodes hold in networks. There are many different measures of centrality that capture different aspects of network positions, but they fall into three general types. The most basic and intuitive measure of centrality, *degree centrality*, simply counts the number of ties that a node has. In a binary undirected network, this measure resolves into the number of unique alters each node is connected to. In mathematical terms it is the row or column sum of the adjacency matrix that characterizes a network. Degree centrality, $C_D(n_i)$, represents a clear measure of the prominence or visibility of a node. Let

$$C_D(n_i) = \sum_j x_{ij}.$$

The degree of a node is limited by the size of the network in which it is embedded. In a network of g nodes the maximum degree of any node is $g-1$. The two orange nodes in the small networks presented in Figure 8.4 have the maximum degree possible (4). In contrast, the orange node in the larger, 13-node network in that figure has the same number of alters but the possible number of partners is three times as large (12). For this reason it is problematic to compare raw degree centrality measures across networks of different sizes. Thus, it is common to normalize degree by the maximum value defined by $g-1$:

$$C'_D(n_i) = \frac{\sum_j x_{ij}}{g-1}.$$

While the normalized degree centrality of the two orange nodes of the smaller networks in Figure 8.4 is 1.0, the normalized value for the node in the large network of 13 nodes is 0.33. Despite the fact that the highlighted nodes in the two smaller networks have the same degree centrality, the pattern of indirect ties connecting their alters means they occupy meaningfully different positions. There are a number of degree-based centrality measures that take more of the structural information from a complete network into account by using a variety of methods to account not just for the number of partners a particular ego might have but also for the prominence of those partners. Two well-known examples are eigenvector centrality and page rank (see [272, Ch. 7.2 and 8.4]).

Consider two additional measures that capture aspects of centrality that have more to do with the indirect ties that increase reachability. Both make explicit use of the idea that reachability is the source of many of the important social and economic benefits of salutary network positions, but they do so with different substantive emphases. Both of these approaches rely on the idea of a network geodesic, the longest shortest path* connecting any pair of actors. Because these measures rely on reachability, they are only useful when applied to components. When nodes have no ties (degree 0) they are called *isolates*. The geodesic distances are infinite and thus path-based centrality measures cannot be calculated. This is a shortcoming of these measures, which can only be used on connected subsets of graphs where each node has at least one tie to another and all are indirectly connected.

★ A *shortest path* is a path that does not repeat any nodes or ties. Most pairs have several of those. The *geodesic* is the longest shortest path. So, if two people are directly connected (path length 1) and connected through shared ties to another person (path length 2), then their geodesic distance is two.

Closeness centrality, C_C, is based on the idea that networks position some individuals closer to or farther away from other participants. The primary idea is that shorter network paths between actors increase the likelihood of communication and with it the ability to coordinate complicated activities. Let $d(n_i, n_j)$ represent the number of network steps in the geodesic path connecting two nodes i and j. As d increases, the network distance between a pair of nodes grows. Thus a standard measure of closeness is the inverse of the sum of distances between any given node and all the others that are reachable in a network:

$$C_C(n_i) = \frac{1}{\sum_{j=1}^{g} d(n_i, n_j)}.$$

The maximum of closeness centrality occurs when a node is directly connected to every possible partner in the network. As with degree centrality, closeness depends on the number of nodes in a network. Thus, it is necessary to standardize the measure to allow comparisons across multiple networks:

$$C'_C(n_i) = \frac{g-1}{\sum_{j=1}^{g} d(n_i, n_j)}.$$

Like closeness centrality, betweenness centrality, C_B, relies on the concept of geodesic paths to capture nuanced differences between the positions of nodes in a connected network. Where closeness assumes that communication and the flow of information increase with proximity, betweenness captures the idea of brokerage that was made famous by Burt [59]. Here too the idea is that flows of information and resources pass between nodes that are not directly connected through indirect paths. The key to the idea of brokerage is that such paths pass through nodes that can interdict, or otherwise profit from their position "in between" unconnected alters. This idea has been particularly important in network studies of innovation [60, 293], where flows of information through strategic alliances among firms or social networks connecting individuals loom large in explanations of why some organizations or individuals are better able to develop creative ideas than others.

To calculate betweenness as originally specified, two strong assumptions are required [129]. First, one must assume that when people (or organizations) search for new information through their networks, they are capable of identifying the shortest path to what they seek. When multiple paths of equal length exist, we assume that each path is equally likely to be used. Newman [271] describes an alternative betweenness measure based on random paths through a network, rather than shortest paths, that relaxes these assumptions. For now, let g_{jk} equal the number of geodesic paths linking any two actors. Then $1/g_{jk}$ is the probability that any given path will be followed on a particular node's search for information or resources in a network. In order to calculate the betweenness score of a particular actor, i, it is then necessary to determine how many of the geodesic paths connecting j to k include i. That quantity is $g_{jk}(n_i)$. With these (unrealistic) assumptions in place, we calculate $C_B(n_i)$ as

$$C_B(n_i) = \sum_{j<k} g_{jk}^{(n_i)}/g_{jk}.$$

Here, too, the maximum value depends on the size of the network. $C_B(n_i) = 1$ if i sits on every geodesic path in the network. While this is only likely to occur in small, star-shaped networks, it is still common to standardize the measure. Instead of conceptualizing network size in terms of the number of nodes, however, this measure requires that we consider the number of possible pairs of actors (excluding ego) in a structure. When there are g nodes, that quantity is $(g-1)(g-2)/2$ and the standardized betweenness measure is

$$C_B'(n_i) = \frac{C_B(n_i)}{(g-1)(g-2)/2}.$$

Centrality measures of various sorts are the most commonly used means to examine network effects at the level of individual participants in a network. In the context of UMETRICS, such indices might be applied to examine the differential scientific or career success of graduate students as a function of their positions in the larger networks of their universities. In such an analysis, care must be taken to use the standardized measures as university collaboration networks can vary dramatically in size and structure. Describing and accounting for such variations and the possibility of analyses conducted at the level of entire networks or subsets of networks, such as teams and labs, requires a different set of measures. The code snippet presented below calculates each of these measures for the university A employee network we have been examining.

```
# Calculate centrality measures
emp.vs["degree"]=emp.degree()
emp.vs["close"]=emp.closeness(vertices=emp.vs)
emp.vs["btc"]=emp.betweenness(vertices=emp.vs, directed=False)
```

8.4 Comparing collaboration networks

Consider Figure 8.6, which presents visualizations of the main component of two university networks. Both of these representations are drawn from a single year (2012) of UMETRICS data. Nodes represent people, and ties reflect the fact that those individuals were paid with the same federal grant in the same year. The images are scaled so that the physical location of any node is a function of its position in the overall pattern of relationships in the network. The size and color of nodes represent their betweenness centrality. Larger, darker nodes are better positioned to play the role of brokers in the network. A complete review of the many approaches to network visualization and their dangers in the absence of descriptive statistics such as those presented above is beyond the scope of this chapter, but consider the guidelines presented in Chapter 9 on information visualization as well as useful discussions by Powell et al. [309] and Healy and Moody [163].

Consider the two images. University A is a major public institution with a significant medical school. University B, likewise, is a public institution but lacks a medical school. It is primarily known for strong engineering. The two networks manifest some interesting and suggestive differences. Note first that the network on the left (university A) appears much more tightly connected. There is a

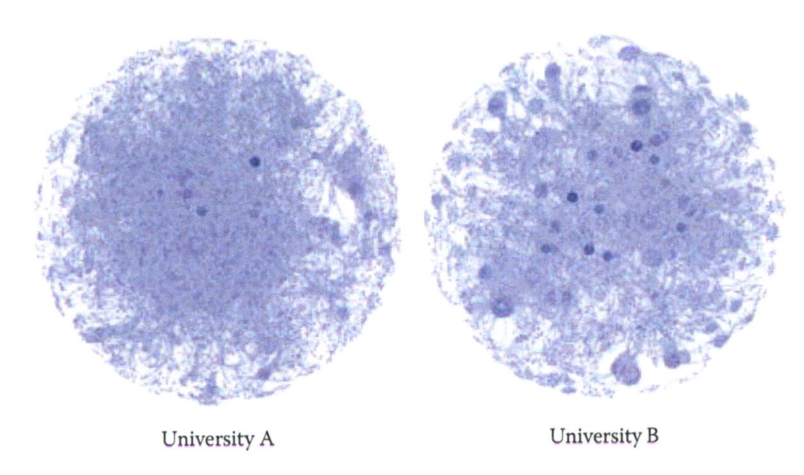

Figure 8.6. The main component of two university networks

dense center and there are fewer very large nodes whose positions bridge less well-connected clusters. Likewise, the network on the right (university B) seems at a glance to be characterized by a number of densely interconnected groups that are pulled together by ties through high-degree brokers. One part of this may have to do with the size and structure of university A's medical school, whose significant NIH funding dominates the network. In contrast, university B's engineering-dominated research portfolio seems to be arranged around clusters of researchers working on similar topic areas and lacks the dominant central core apparent in university B's image.

The implications of these kinds of university-level differences are just starting to be realized, and the UMETRICS data offer great possibilities for exactly this kind of study. These networks, in essence, represent the local social capacity to respond to new problems and to develop scientific findings. Two otherwise similar institutions might have quite different capabilities based on the structure and composition of their collaboration networks.

The intuitions suggested by Figure 8.6 can also be checked against some of the measures we have described. Figure 8.7, for instance, presents degree distributions for each of the two networks. Figure 8.8 presents the histogram of path lengths for each network.

It is evident from Figure 8.7 that they are quite different in character. University A's network follows a more classic skewed distribution of the sort that is often associated with the kinds of power-law degree distributions common to scale-free networks. In contrast, university B's distribution has some interesting features.

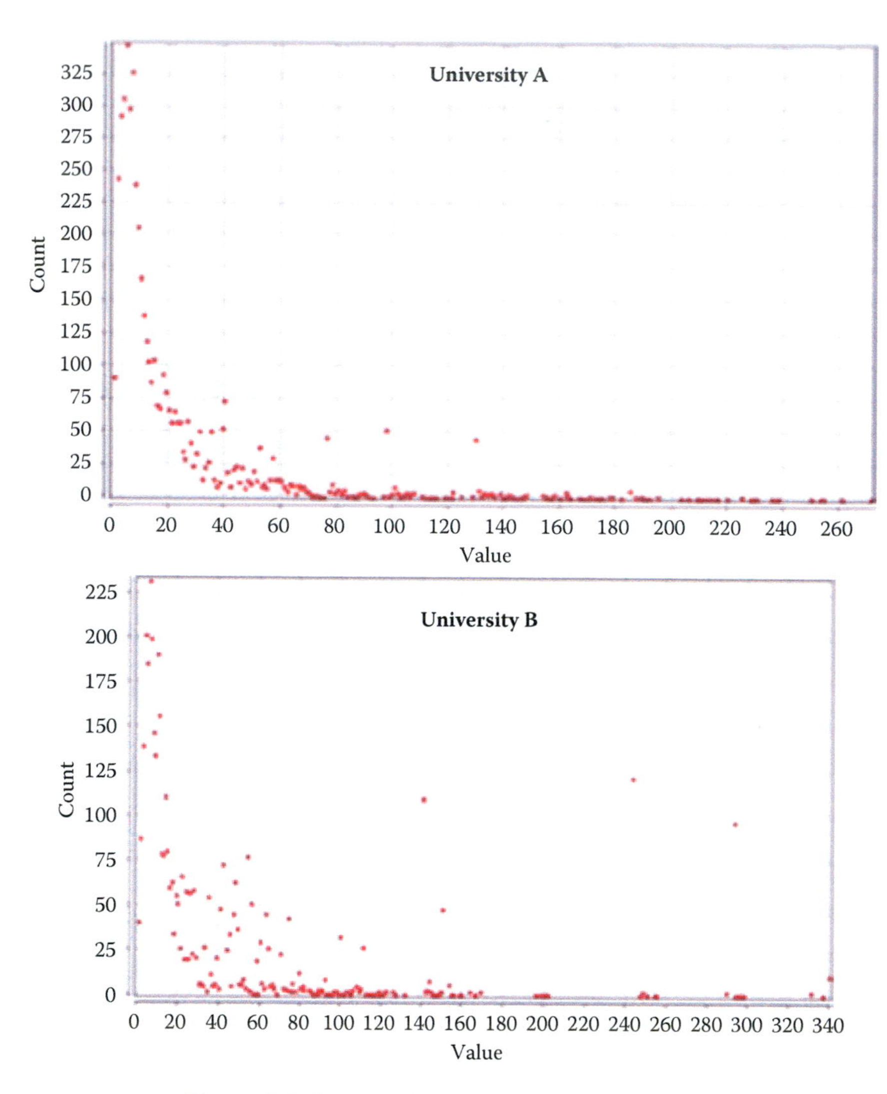

Figure 8.7. Degree distribution for two universities

First, the left-hand side of the distribution is more dispersed than it is for university A, suggesting that there are many nodes with moderate degree. These nodes may also have high betweenness centrality if their ties allow them to span different subgroups within the networks. Of course this might also reflect the fact that each cluster also has members that are more locally prominent. Finally, consider the few instances on the right-hand end of the distribution

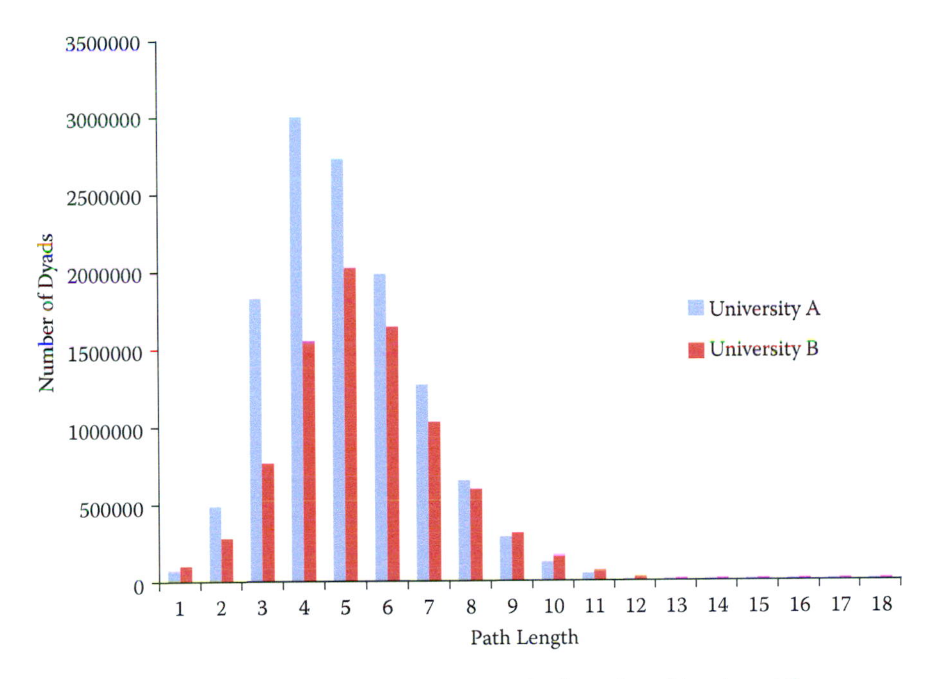

Figure 8.8. Distribution of path lengths for universities A and B

where there are relatively large numbers of people with surprisingly high degree. I suspect these are the result of large training grants or center grants that employ many people. A quirk of relying on one-mode projections of two-mode data is that every person associated with a particular grant is connected to every other. More work needs to be done to bear out these hypotheses, but for now it suffices to say that the degree distribution of the networks bears out the intuition we drew from the images that they are significantly different.

The path length histogram presented in Figure 8.8 suggests a similar pattern. While the average distance among any pair of connected nodes in both networks is fairly similar (see Table 8.1), university B has a larger number of unconnected nodes and university A has a greater concentration of more closely connected dyads. The preponderance of shorter paths in this network could also be a result of a few larger grants that connect many pairs of nodes at unit distance and thus shorten overall path lengths.

But how do the descriptive statistics shake out? Table 8.1 presents the basic descriptive statistics we have discussed for each

Table 8.1. Descriptive statistics for the main components of two university networks

	University A	University B
Nodes	4,999	4,144
Edges (total)	57,756	91,970
% nodes in main component	68.67%	67.34%
Diameter	18	18
Average degree	11.554	44.387
Clustering coefficient	0.855	0.913
Density	0.005	0.011
Average path length	5.034	5.463

network. University A's network includes 855 more nodes than university B's, a difference of about 20%. In contrast, there are far fewer edges connecting university A's research employees than connecting university B's, a difference that appears particularly starkly in the much higher density of university B's network. Part of the story can be found in the average degree of nodes in each network. As the degree distributions presented in Figure 8.6 suggested, the average researcher at university B is much more highly connected to others than is the case at university A. The difference is stark and quite likely has to do with the presence of larger grants that employ many individuals.

Both schools have a low average path length (around 5), suggesting that no member of the network is more than five acquaintances away from any other. Likewise, the diameter of both networks is 18, which means that on each campus the most distant pair of nodes is separated by just 18 steps. University A's slightly lower path length may be accounted for by the centralizing effect of its large medical school grant infrastructure. Finally, consider the clustering coefficient. This measure approaches 1 as it becomes more likely that two partners to a third node will themselves be connected. The likelihood that collaborators of collaborators will collaborate is high on both campuses, but substantially higher at university B.

8.5 Summary

This chapter has provided a brief overview of the basics of networks, using UMETRICS data as a source for examples. While network measures can produce new and exciting ways to characterize social dynamics, they are also important levels of analysis in their own right. There are essential concepts that need to be mastered, such as reachability, cohesion, brokerage, and reciprocity. Numbers are important, for a variety of reasons—they can be used to describe networks in terms of their composition and community structure. This chapter provides a classic example of how well social science meets data science. Social science is needed to identify the nodes (what is being connected) and the ties (the relationships that matter) in order to construct the relevant networks. Computer science is necessary to collect and structure the data in a fashion that is sufficient for analysis. The combination of data science and social science is key to making the right measurement and visualization decisions.

8.6 Resources

For more information about network analysis *in general*, the International Network for Social Network Analysis (http://www.insna.org/) is a large, interdisciplinary association dedicated to network analysis. It publishes a traditional academic journal, *Social Networks*, an online journal, *Journal of Social Structure*, and a short-format journal, *Connections*, all dedicated to social network analysis. Its several listservs offer vibrant international forums for discussion of network issues and questions. Finally, its annual meetings include numerous opportunities for intensive workshops and training for both beginning and advanced analysts.

A new journal, *Network Science* (http://journals.cambridge.org/action/displayJournal?jid=NWS), published by Cambridge University Press and edited by a team of interdisciplinary network scholars, is a good venue to follow for cutting-edge articles on computational network methods and for substantive findings from a wide range of social, natural, and information science applications.

There are some good software packages available. *Pajek* (http://mrvar.fdv.uni-lj.si/pajek/) is a freeware package for network analysis and visualization. It is routinely updated and has a vibrant user group. Pajek is exceptionally flexible for large networks and has a number of utilities that allow import of its relatively simple file types into other programs and packages for network analysis. *Gephi* (https://gephi.org/) is another freeware package that supports large-scale network visualization. Though I find it less flexible than Pajek, it offers strong support for compelling visualizations.

Network Workbench (http://nwb.cns.iu.edu/) is a freeware package that supports extensive analysis and visualization of networks.

This package also includes numerous shared data sets from many different fields that can used to test and hone your network analytic skills.

iGraph (http://igraph.org/redirect.html) is my preferred package for network analysis. Implementations are available in R, in Python, and in C libraries. The examples in this chapter were coded in iGraph for Python.

Nexus (http://nexus.igraph.org/api/dataset_info?format=html&limit=10&offset=20&operator=or&order=date) is a growing repository for network data sets that includes some classic data dating back to the origins of social science network research as well as more recent data from some of the best-known publications and authors in network science.

Part III

Inference and Ethics

Chapter 9

Information Visualization

M. Adil Yalçın and Catherine Plaisant

This chapter will show you how to explore data and communicate results so that data can be turned into interpretable, actionable information. There are many ways of presenting statistical information that convey content in a rigorous manner. The goal of this chapter is to present an introductory overview of effective visualization techniques for a range of data types and tasks, and to explore the foundations and challenges of information visualization.

9.1 Introduction

One of the most famous discoveries in science—that disease was transmitted through germs, rather than through pollution—resulted from insights derived from a visualization of the location of London cholera deaths near a water pump [349]. Information visualization in the twenty-first century can be used to generate similar insights: detecting financial fraud, understanding the spread of a contagious illness, spotting terrorist activity, or evaluating the economic health of a country. But the challenge is greater: many (10^2–10^7) items may be manipulated and visualized, often extracted or aggregated from yet larger data sets, or generated by algorithms for analytics.

Visualization tools can organize data in a meaningful way that lowers the cognitive and analytical effort required to make sense of the data and make data-driven decisions. Users can scan, recognize, understand, and recall visually structured representations more rapidly than they can process nonstructured representations. The science of visualization draws on multiple fields such as perceptual psychology, statistics, and graphic design to present information, and on advances in rapid processing and dynamic displays to design user interfaces that permit powerful interactive visual analysis.

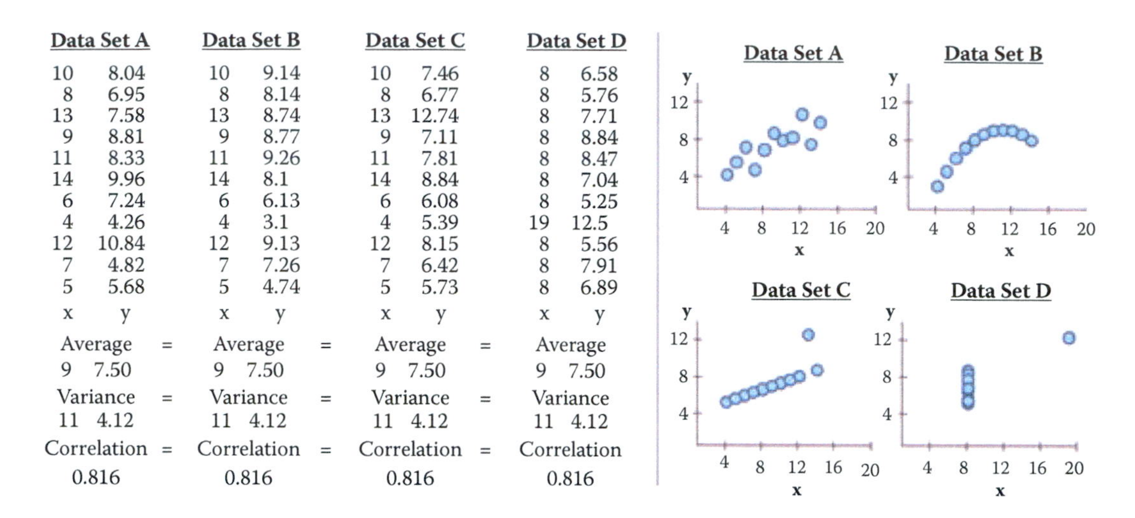

Data Set A		Data Set B		Data Set C		Data Set D	
10	8.04	10	9.14	10	7.46	8	6.58
8	6.95	8	8.14	8	6.77	8	5.76
13	7.58	13	8.74	13	12.74	8	7.71
9	8.81	9	8.77	9	7.11	8	8.84
11	8.33	11	9.26	11	7.81	8	8.47
14	9.96	14	8.1	14	8.84	8	7.04
6	7.24	6	6.13	6	6.08	8	5.25
4	4.26	4	3.1	4	5.39	19	12.5
12	10.84	12	9.13	12	8.15	8	5.56
7	4.82	7	7.26	7	6.42	8	7.91
5	5.68	5	4.74	5	5.73	8	6.89
x	y	x	y	x	y	x	y

Average = Average = Average = Average
9 7.50 9 7.50 9 7.50 9 7.50
Variance = Variance = Variance = Variance
11 4.12 11 4.12 11 4.12 11 4.12
Correlation = Correlation = Correlation = Correlation
0.816 0.816 0.816 0.816

Figure 9.1. Anscombe's quartet [16]

Figure 9.1, "Anscombe's quartet" [16], provides a classic example of the value of visualization compared to basic descriptive statistical analysis. The left-hand panel includes raw data of four small number-pair data sets (A, B, C, D), which have the same average, median, and standard deviation and have correlation across number pairs. The right-hand panel shows these data sets visualized with each point plotted on perpendicular axes (scatterplots), revealing dramatic differences between the data sets, trends, and outliers visually.

In broad terms, visualizations are used either to present results or for analysis and open-ended exploration. This chapter provides an overview of how modern information visualization, or visual data mining, can be used to in the context of big data.

9.2 Developing effective visualizations

The effectiveness of a visualization depends on both analysis needs and design goals. Sometimes, questions about the data are known in advance; in other cases, the goal may be to explore new data sets, generate insights, and answer questions that are unknown before starting the analysis. The design, development, and evaluation of a visualization is guided by understanding the background and goals of the target audience (see Box 9.1).

Box 9.1:

The development of an effective visualization is an iterative process that generally includes the following steps:

- Specify user needs, tasks, accessibility requirements and criteria for success.
- Prepare data (clean, transform).
- Design visual representations.
- Design interaction.
- Plan sharing of insights, provenance.
- Prototype/evaluate, including usability testing.
- Deploy (monitor usage, provide user support, manage revision process).

▶ See Chapters 2, 3, 4, and 5 for an overview of collecting, merging, storing, and processing data sets.

If the goal is to present results, there is a wide spectrum of users and a wide range of options. If the audience is broad, then *infographics* can be developed by graphic designers, as described in classic texts (see [119,380,381] or the examples compiled by Harrison et al. [157,206]). If, on the other hand, the audience comprises domain experts interested in monitoring the overview status of dynamic processes on a continuous basis, *dashboards* can be used. Examples include the monitoring of sales, or the number of tweets about people, or symptoms of the flu and how they compare to a baseline [120]. Dashboards can increase situational awareness so that problems can be noticed and solved early and better decisions can be made with up-to-date information.

Another goal of visualization is to enable *interactive exploratory analysis* for casual users as well as professional analysts. One casual example is BabyNameVoyager [156], which lets users type in a name and see a graph of its popularity over the past century. *

★ As the baby name is typed letter by letter, BabyNameVoyager visualizes the popularity of all the names starting with the letters entered so far, and it animates smoothly with each new letter input. For example, typing "Jo" shows all the names starting with "Jo". This view reveals that Joyce and Joan were popular girl names in the 1930s, and the use of "John" has declined since the 1960s.

Data analysis tools can enable analysis of structured and generic data sets. Figure 9.2 shows an exploratory browser of a selection of awards (grants) from the US Department of Agriculture, created using the web-based data exploration tool Keshif (http://www.keshif.me). The awards (records) are listed in the middle panel by latest start date first. Attributes of the awards, such as funding source, are shown on side panels, revealing the range of their values. The awards with status *new* are focused per the filtering selection, and

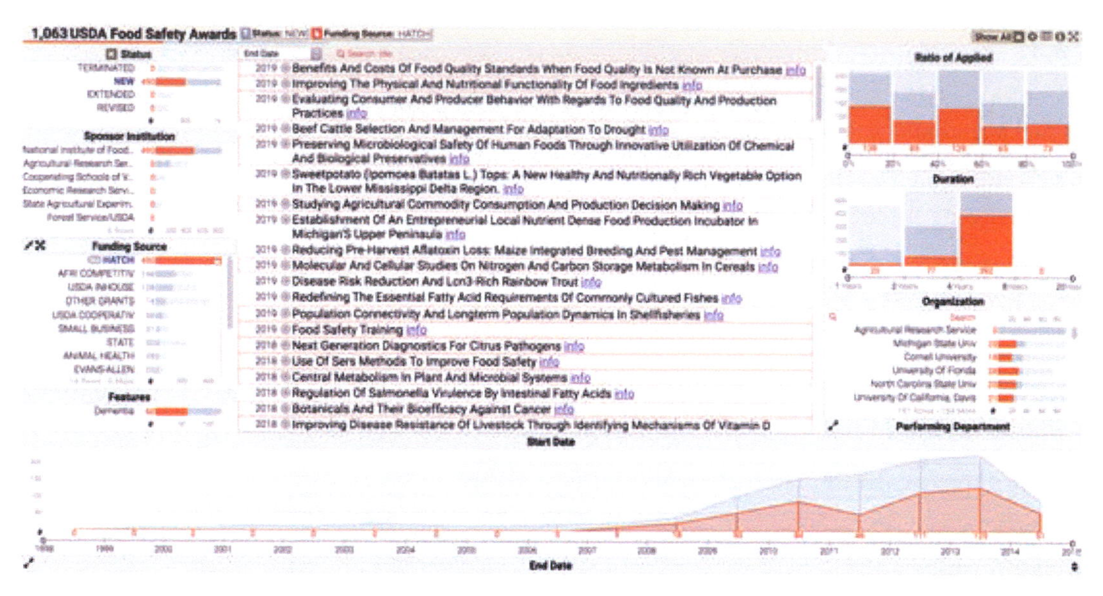

Figure 9.2. A data analysis browser of a selection of grants from the US Department of Agriculture was created by using the web-based tool Keshif

are visualized using darker gray colors. Lighter gray colors provide visual cues to the distribution of all awards. The orange selection highlights distributions of awards with formula funding, known as *Hatch funding*. The analyst can explore trends in new or old awards, funding resources, and other attributes in this rich award portfolio.

Commercial tools such as Spotfire and Tableau, among many other tools (see Section 9.6), allow users to create visualizations by offering chart types and visual design environments to analyze their data, and to combine them in potent dashboards rapidly shared with colleagues. For example, Figure 9.3 shows the charting interface of Tableau on a transaction data set. The left-hand panel shows the list of attributes associated with vendor transactions for a given university. The visualization (center) is constructed by placing the month of spending in chart columns, and the sum of payment amount on the chart row, with data encoded using line mark type. Agencies are broken down by color mapping. The agency list, to the right, allows filtering the agencies, which can be used to simplify the chart view. A peak in the line chart is annotated with an explanation of the spike. On the rightmost side, the Show Me panel suggests the applicable chart types potentially appropriate for the selected attributes. This chart can be combined with other charts focusing on other aspects in interactive dashboards. Figure 9.4 shows

Figure 9.3. Charting interface of Tableau

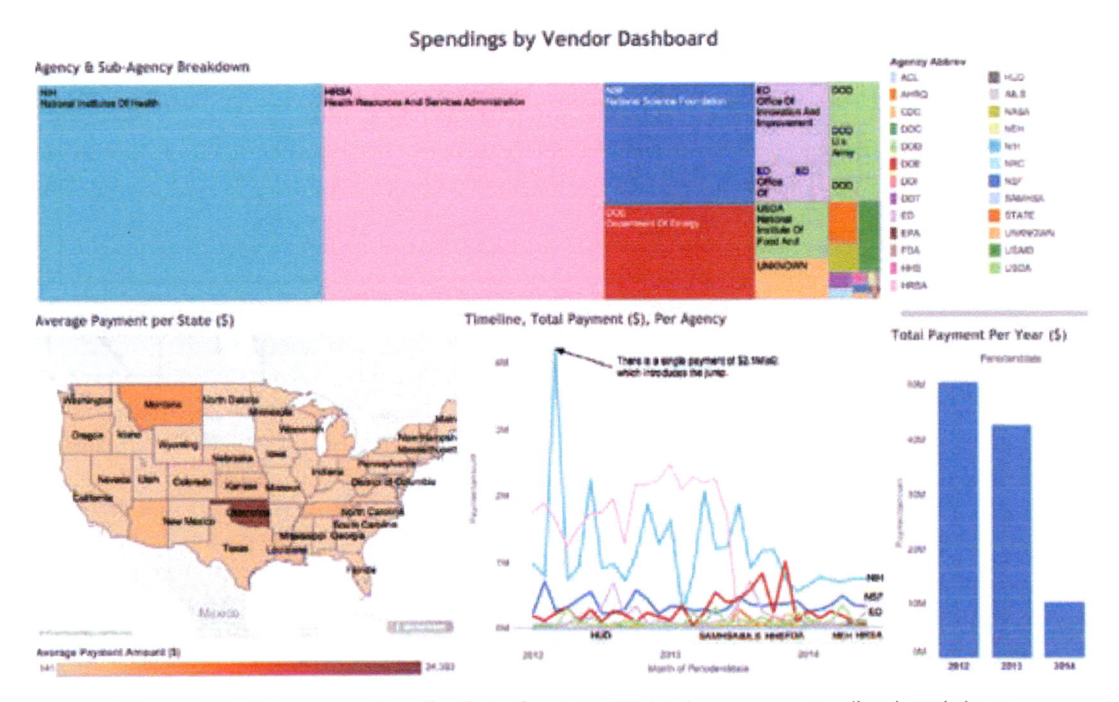

Figure 9.4. A treemap visualization of agency and sub-agency spending breakdown

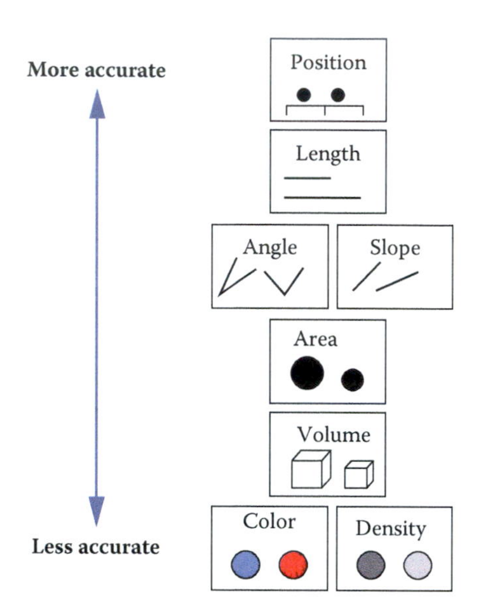

Figure 9.5. Visual elements described by MacKinlay [245]

a treemap [192] for agency and sub-agency spending breakdown, combined with a map showing average spending per state. Oklahoma state stands out with few but large expenditures. Mousing-over Oklahoma reveals details of these expenditures. An additional histogram provides an overview of spending change across three years.

Creating effective visualizations requires careful consideration of many components. Data values may be encoded using one or more visual elements, like position, length, color, angle, area, and texture (Figure 9.5; see also [79, 380]). Each of these can be organized in a multitude of ways, discussed in more detail by Munzner [264]. In addition to visual data encoding, units for axes, labels, and legends need to be provided as well as explanations of the mappings when the design is unconventional. (The website "A World of Terror" provides some compelling examples [301].) Annotations or comments can be used to guide viewer attention and to describe related insights. Providing attribution and data source, where applicable, is an ethical practice that also enables validating data, and promotes reuse to explore new perspectives.

The following is a short list of guidelines: provide immediate feedback upon interaction with the visualization; generate tightly coupled views (i.e., so that selection in one view updates the others); and use a high "data to ink ratio" [380]. Use color carefully and ensure that the visualization is truthful (e.g., watch for perceptual biases or distortion). Avoid use of three-dimensional representations or embellishments, since comparing 3D volumes is perceptually challenging and occlusion is a problem. Labels and legends should be meaningful, novel layouts should be carefully explained, and online visualizations should adapt to different screen sizes. For extended and in-depth discussions, see various textbooks [119, 208, 264, 380, 381, 395].

We provide a summary of the basic tasks that users typically perform during visual analysis of data in the next section.

9.3 A data-by-tasks taxonomy

We give an overview of visualization approaches for six common data types: multivariate, spatial, temporal, hierarchical, network, and text [344]. For each data type listed in this section, we discuss its distinctive properties, the common analytical questions, and examples. Real-life data sets often include multiple data types coming from multiple sources. Even a single data source can include a variety of data types. For example, a single data table of countries (as rows) can have a list of attributes with varying types: the growth rate in the last 10 years (one observation per year, time series data), their current population (single numerical data), the amount of trade with other countries (networked/linked data), and the top 10 exported products (if grouped by industry, hierarchical data). Furthermore, we provide an overview of common tasks for visual data analysis in Box 9.2, which can be applied across different data types based on goals and types of visualizations.

Interactive visualization design is also closely coupled with the targeted devices. Conventionally, visualizations have been designed for mouse and keyboard interaction on desktop computers. However, a wider range of device forms, such as mobile devices with small displays and touch interaction, is becoming common. Creating visualizations for new forms requires special care, though basic design principles such as “less is more” still apply.

9.3.1 Multivariate data

In common tabular data, each record (row) has a list of attributes (columns), whose value is mostly categorical or numerical. The analysis of multivariate data with basic categorical and interval types aims to understand patterns within and across data attributes.

Box 9.2: A task categorization for visual data analysis

Select/Query

- Filter to focus on a subset of the data
- Retrieve details of item
- Brush linked selections across multiple charts
- Compare across multiple selections

Navigate

- Scroll along a dimension (1D)
- Pan along two dimensions (2D)
- Zoom along the third dimension (3D)

Derive

- Aggregate item groups and generate characteristics
- Cluster item groups by algorithmic techniques
- Rank items to define ordering

Organize

- Select chart type and data encodings to organize data
- Layout multiple components or panels in the interface

Understand

- Observe distributions
- Compare items and distributions
- Relate items and patterns

Communicate

- Annotate findings
- Share results
- Trace action histories

Given a larger number of attributes, one of the challenges in data exploration and analytics is to select the attributes and relations to focus on. Expertise in the data domain can be helpful for targeting relevant attributes.

Multivariate data can be presented in multiple forms of charts depending on the data and relations being explored. One-dimensional (1D) charts present data on a single axis only. An example is a *box-plot*, which shows quartile ranges for numerical data. So-called 1.5D charts list the range of possible values on one axis, and describe a measurement of data on the other. *Bar charts* are a ubiquitous example, in order to show, for example, a numeric grade per student, or grade average for aggregated student groups by gender. Records can also be grouped over numerical ranges, and bars can show the number of items in each grouping, which generates a *histogram* chart. Two-dimensional charts plot data along two attributes, such as *scatterplots*. Matrix (grid) charts can also be used to show relations between two attributes. *Heatmaps* visualize each matrix cell using color to represent its value. *Correlation matrices* show the relation between attribute pairs.

To show relations of more than two attributes (3D+), one option is to use additional visual encodings in a single chart, for example, by adding point size/shape as a data variable in scatterplots. Another option is to use alternative visual designs that can encode multiple relations within a single chart. For example, a *parallel coordinate plot* [180] has multiple parallel axes, each one representing an attribute; each record is shown as connected lines passing through the record's values on each attribute. Charts can also show part-of-whole relations using appropriate mappings based on subdividing the chart space, such as stacked charts or pie charts.

Finally, another approach to analyzing multidimensional data is to use clustering algorithms to identify similar items. Clusters are typically represented as a tree structure (see Section 9.3.4). For example, *k*-means clustering starts by users specifying how many clusters to create; the algorithm then places every item into the most appropriate cluster. Surprising relationships and interesting outliers may be identified by these techniques on mechanical analysis algorithms. However, such results may require more effort to interpret.

9.3.2 Spatial data

Spatial data convey a physical context, commonly in a 2D space, such as geographical maps or floor plans. Several of the most famous examples of information visualization include maps, from the 1861 representation of Napoleon's ill-fated Russian campaign by Minard (popularized by Tufte [380] and Kraak [214]) to the interactive HomeFinder application that introduced the concept of dynamic queries [7]. The tasks include finding adjacent items, regions containing certain items or with specific characteristics, and paths between items—and performing the basic tasks listed in Box 9.1.

The primary form of visualizing spatial data is *maps*. In *choropleth maps*, color encoding is used to add represent one data attribute. *Cartograms* aim to encode the attribute value with the size of regions by distorting the underlying physical space. *Tile grid maps* reduce each spatial area to a uniform size and shape (e.g., a square) so that the color-coded data are easier to observe and compare, and they arrange the tiles to approximate the neighbor relations between physical locations [92,355]. Grid maps also make selection of smaller areas (such as small cities or states) easier. *Contour (isopleth) maps* connect areas with similar measurements and color each one separately. *Network maps* aim to show network connectivity between locations, such as flights to/from many regions of the world. Spatial data can be also presented with a nonspatial emphasis (e.g., as a hierarchy of continents, countries, and cities).

Maps are commonly combined with other visualizations. For example, in Figure 9.6, the US Cancer Atlas combines a map showing patterns across states on one attribute, with a sortable table providing additional statistical information and a scatterplot that allows users to explore correlations between attributes.

9.3.3 Temporal data

Time is the unique dimension in our physical world that steadily flows forward. While we cannot control time, we frequently record it as a point or interval. Time has multiple levels of representation (year, month, day, hour, minute, and so on) with irregularities (leap year, different days per month, etc.). As we measure time based on cyclic events in nature (day/night), our representations are also commonly cyclic. For example, January follows December (first month follows last). This cyclic nature can be captured by circular visual encodings, such as the the conventional clock with hour, minute, and second hands.

Time series data (Figures 9.7 and 9.8) describe values measured at regular intervals, such as stock market or weather data. The focus of analysis is to understand temporal trends and anomalies, querying for specific patterns, or prediction. To show multiple time-

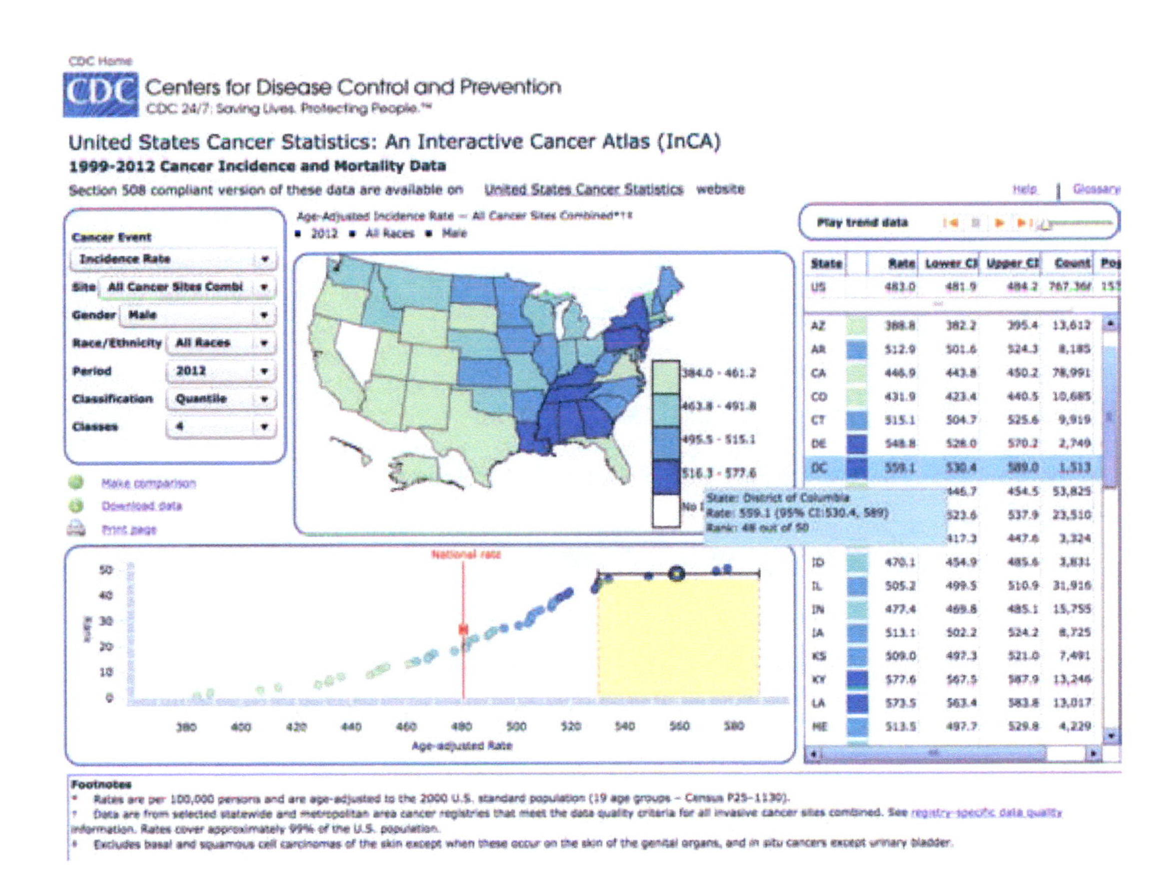

Figure 9.6. The US Cancer Atlas [66]. Interface based on [244]

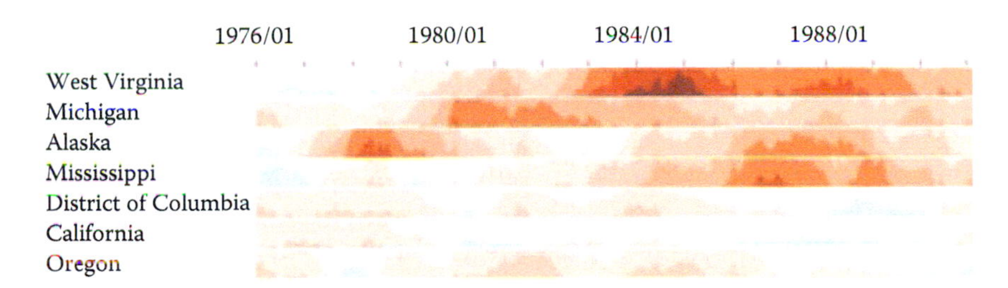

Figure 9.7. Horizon graphs used to display time series

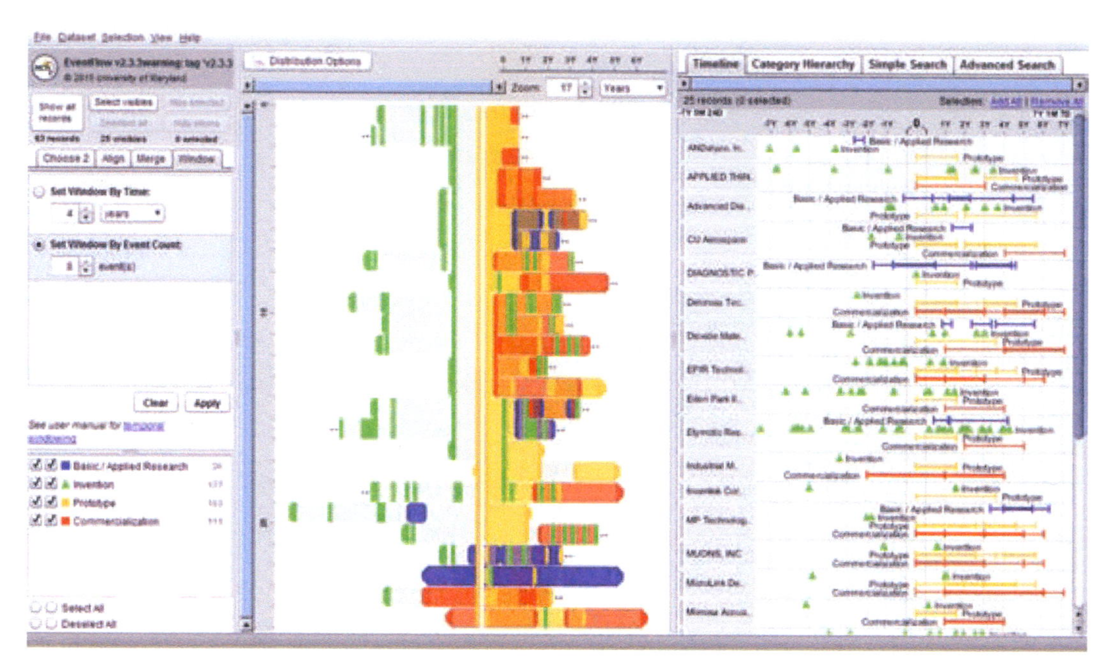

Figure 9.8. EventFlow (www.cs.umd.edu/hcil/eventflow) is used to visualize sequences of innovation activities by Illinois companies. Created with EventFlow; data sources include NIH, NSF, USPTO, SBIR. Image created by C. Scott Dempwolf, used with permission

series trends across different data categories in a very compact chart area, each trend can be shown with small height using a multi-layered color approach, creating horizon graphs. While perceptually effective after learning to read its encoding, this chart design may not be appropriate for audiences who may lack such training or familiarity.

Another form of temporal analysis is understanding sequences of events. The study of human activity often includes analyzing event sequences. For example, students' records include events such as attending orientation, getting a grade in a class, going on internship, and graduation. In the analysis of event sequences, finding the most common patterns, spotting rare ones, searching for specific sequences, or understanding what leads to particular types of events is important (e.g., what events lead to a student dropping out, precede a medical error, or a company filing bankruptcy). Figure 9.8 shows EventFlow used to visualize sequences of innovation activities by Illinois companies. Activity types include research, invention, prototyping, and commercialization. The timeline (right panel) shows the sequence of activities for each company. The overview panel (center) summarizes all the records aligned by the first prototyping activity of the company. In most of the sequences shown here, the company's first prototype is preceded by two or more patents with a lag of about a year.

9.3.4 Hierarchical data

Data are often organized in a hierarchical fashion. Each item appears in one grouping (e.g., like a file in a folder), and groups can be grouped to form larger groups (e.g., a folder within a folder), up to the root (e.g., a hard disk). Items, and the relations between items and their grouping, can have their own attributes. For example, the National Science Foundation is organized into directorates and divisions, each with a budget and a number of grant recipients.

Analysis may focus on the structure of the relations, by questions such as "how deep is the tree?", "how many items does this branch have?", or "what are the characteristics of one branch compared to another?" In such cases, the most appropriate representation is usually a node-link diagram [62, 303]. In Figure 9.9,

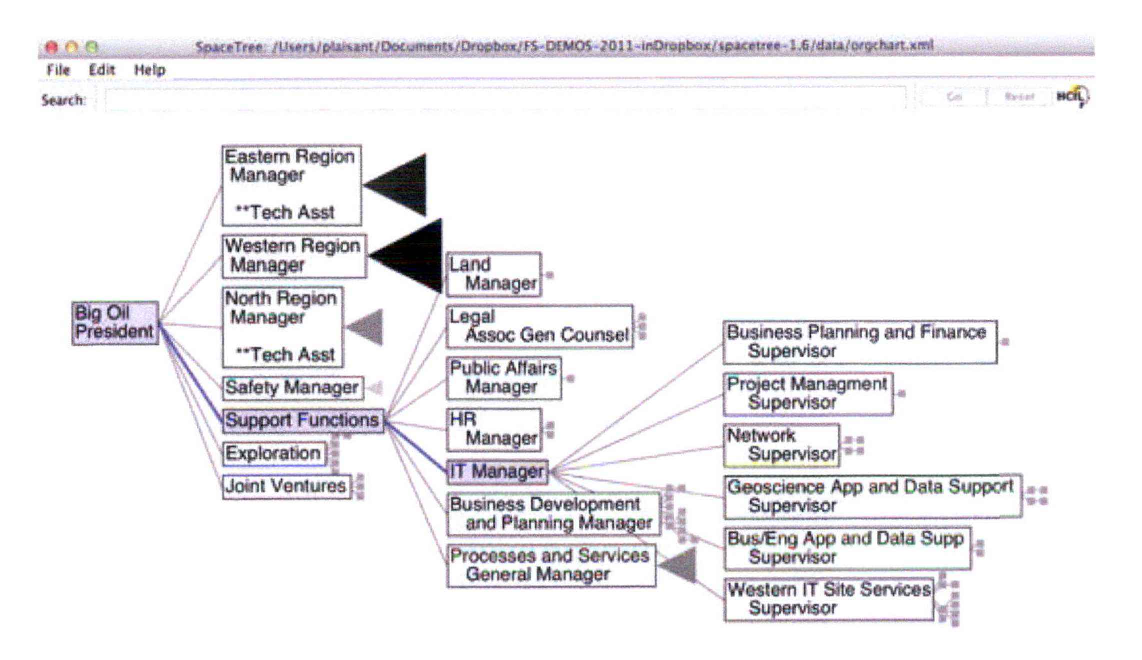

Figure 9.9. SpaceTree (www.cs.umd.edu/hcil/spacetree/)

Figure 9.10. The Finviz treemap helps users monitor the stock market (www.finviz.com)

Spacetree is used to browse a company organizational chart. Since not all the nodes of the tree fit on the screen, we see an iconic representation of the branches that cannot be displayed, indicating the size of each branch. As the tree branches are opened or closed, the layout is updated with smooth multiple-step animations to help users remain oriented.

When the structure is less important but the attribute values of the leaf nodes are of primary interest, treemaps, a space-filling approach, are preferable as they can show arbitrary-sized trees in a fixed rectangular space and map one attribute to the size of each rectangle and another to color. For example, Figure 9.10 shows the Finviz treemap that helps users monitor the stock market. Each stock is shown as a rectangle. The size of the rectangle represents market capitalization, and color indicates whether the stock is going up or down. Treemaps are effective for situation awareness: we can see that today is a fairly bad day as most stocks are red (i.e., down). Stocks are organized in a hierarchy of industries, allowing users to see that "healthcare technology" is not doing as poorly as most other industries. Users can also zoom on healthcare to focus on that industry.

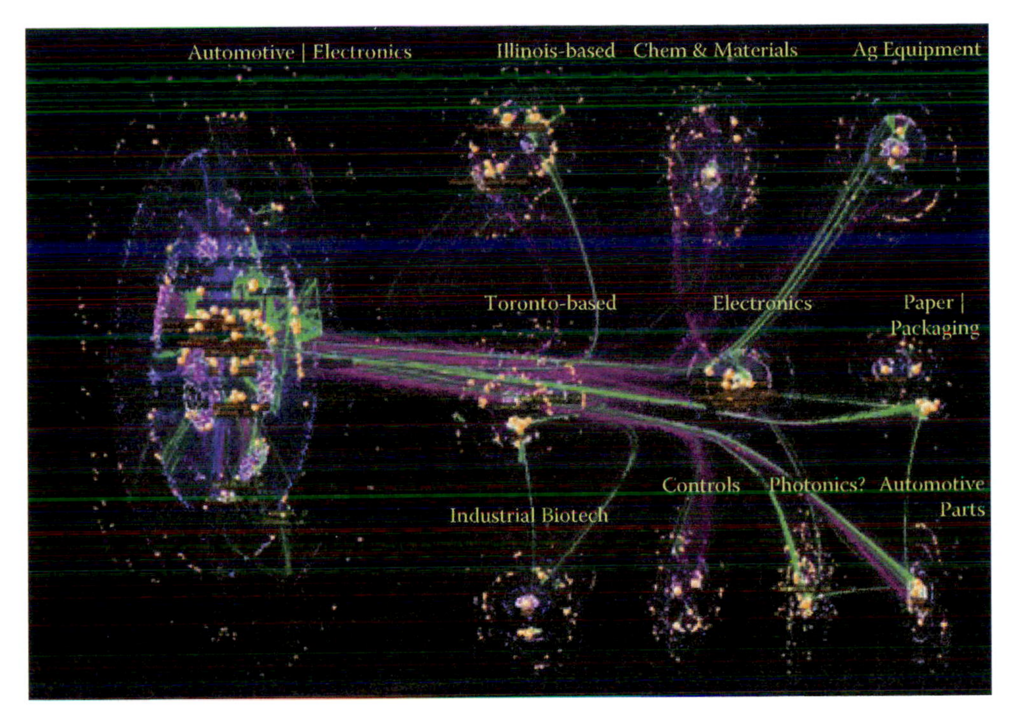

Figure 9.11. NodeXL showing innovation networks of the Great Lakes manufacturing region. Created with NodeXL. Data source: USPTO. Image created by C. Scott Dempwolf, used with permission

9.3.5 Network data

▶ See Chapter 8.

Network data encode relationships between items: for example, social connection patterns (friendships, follows and reposts, etc.), travel patterns (such as trips between metro stations), and communication patterns (such as emails). The network overviews attempt to reveal the structure of the network, show clusters of related items (e.g., groups of tightly connected people), and allow the path between items to be traced. Analysis can also focus on attributes of the items and the links in between, such as age of people in communication or the average duration of communications.

Node-link diagrams are the most common representation of network structures and overviews (Figures 9.11 and 9.12), and may use linear (arc), circular, or force-directed layouts for positioning the nodes (items). Matrices or grid layouts are also a valuable way to represent networks [164]. Hybrid solutions have been proposed,

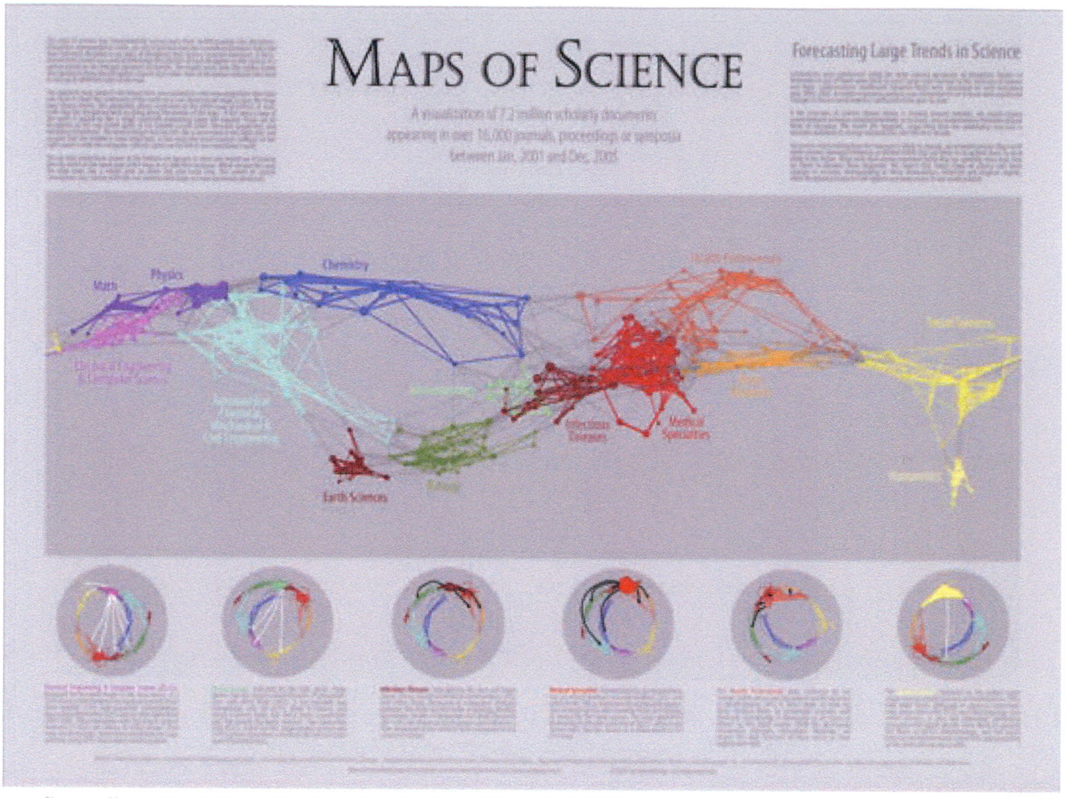

Figure 9.12. An example from "Maps of Science: Forecasting Large Trends in Science," 2007, The Regents of the University of California, all rights reserved [45]

with powerful ordering algorithms to reveal clusters [153]. A major challenge in network data exploration is in dealing with larger networks where nodes and edges inevitably overlap by virtue of the underlying network structure, and where aggregation and filtering may be needed before effective overviews can be presented to users.

Figure 9.11 shows the networks of inventors (white) and companies (orange) and their patenting connections (purple lines) in the network visualization NodeXL. Each company and inventor is also connected to a location node (blue = USA; yellow = Canada). Green lines are weak ties based on patenting in the same class and subclass, and they represent potential economic development leads.

The largest of the technology clusters are shown using the *group-in-a-box* layout option, which makes the clusters more visible. Note the increasing level of structure moving from the cluster in the lower right to the main cluster in the upper left. NodeXL is designed for interactive network exploration; many controls (not shown in the figure) allow users to zoom on areas of interest or change options. Figure 9.12 shows an example of network visualization on science as a topic used for data presentation in a book and a traveling exhibit. Designed for print media, it includes a clear title and annotations and shows a series of topic clusters at the bottom with a summary of the insights gathered by analysts.

9.3.6 Text data

▶ See Chapter 7 for text analysis approaches.

Text is usually preprocessed (for word/paragraph counts, sentiment analysis, categorization, etc.) to generate metadata about text segments, which are then visualized. Simple visualizations like tag clouds display statistics about word usage in a text collection, or can be used to compare two collections or text segments. While visually appealing, they can easily be misinterpreted and are often replaced by word indexes sorted by some count of interest. Specialized visual text analysis tools combine multiple visualizations of data extracted from the text collections, such as matrices to see relations, network diagrams, or parallel coordinates to see entity relationships (e.g., between what, who, where, and when). Timelines can be mapped to the linear dimension of text. Figure 9.13 shows an example using Jigsaw [357] for the exploration of car reviews. Entities have been extracted automatically (in this case make, model, features, etc.), and a cluster analysis has been performed, visualized in the bottom right. A separate view (rightmost) allows analysts to review links between entities. Another view allows traversing word sequences as a tree. Reading original documents is critical, so all the visualization elements are linked to the corresponding text.

9.4 Challenges

While information visualization is a powerful tool, there are many obstacles to its effective use. We note here four areas of particular concern: scalability, evaluation, visual impairment, and visual literacy.

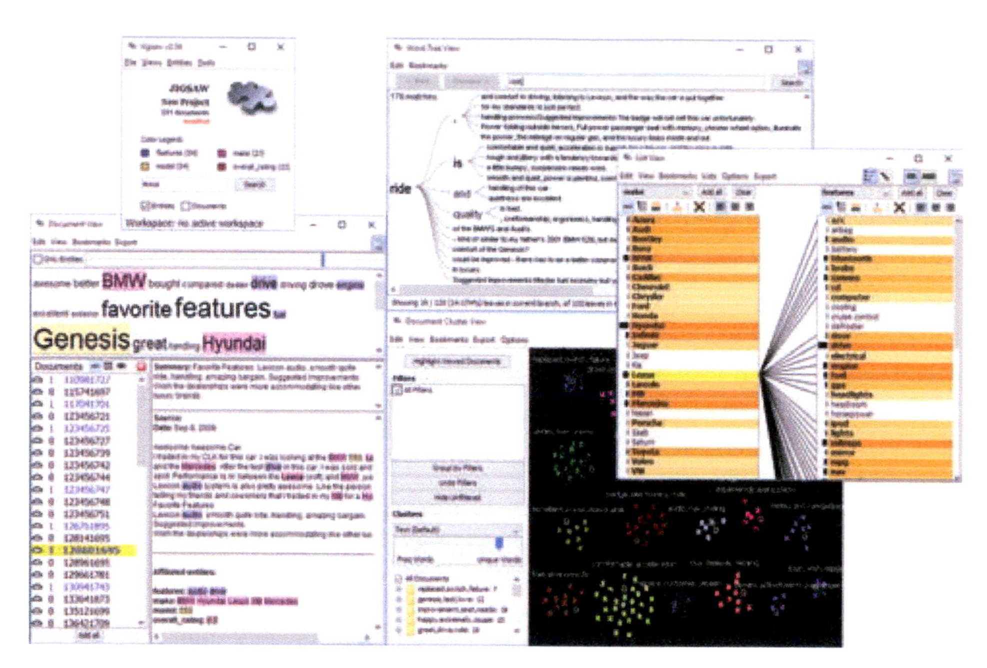

Figure 9.13. Jigsaw used to explore a collection of car reviews

9.4.1 Scalability

Most visualizations handle relatively small data sets (between a thousand and a hundred thousand, sometimes up to millions, depending on the technique) but scaling visualizations from millions to billions of records does require careful coordination of analytic algorithms to filter data or perform rapid aggregation, effective visual summary designs, and rapid refreshing of displays [343]. The visual information seeking mantra, "Overview first, zoom and filter, then details on demand," remains useful with data at scale. To accommodate a billion records, aggregate markers (which may represent thousands of records) and density plots are useful [101]. In some cases the large volume of data can be aggregated meaningfully into a small number of pixels. One example is Google Maps and its visualization of road conditions. A quick glance at the map allows drivers to use a highly aggregated summary of the speed of a large number of vehicles and only a few red pixels are enough to decide when to get on the road.

While millions of graphic elements may be represented on large screens [116], perception issues need to be taken into consideration [411]. Extraction and filtering may be necessary before even attempting to visualize individual records [408]. Preserving interactive rates in querying big data sources is a challenge, with a variety of methods proposed, such as approximations [123] and compact caching of aggregated query results [239]. Progressive loading and processing will help users review the results as they appear and steer the lengthy data processing [115, 135]. Systems are starting to emerge, and strategies to cope with volume and variety of patterns are being described [344].

9.4.2 Evaluation

Human-centric evaluation of visualization techniques can generate qualitative and quantitative assessments of their potential quality, with early studies focusing on the effectiveness of basic visual variables [245]. To this day, user studies remain the workhorse of evaluation. In laboratory settings, experiments can demonstrate faster task completion, reduced error rates, or increased user satisfaction. These studies are helpful for comparing visual and interaction designs. For example, studies are reporting on the effects of latency on interaction and understanding [241], and often reveal that different visualizations perform better for different tasks [303, 321]. Evaluations may also aim to measure and study the amount and value of the insights revealed by the use of exploratory visualization tools [325]. Diagnostic usability evaluation remains a cornerstone of user-centered design. Usability studies can be conducted at various stages of the development process to verify that users are able to complete benchmark tasks with adequate speed and accuracy. Comparisons with the technology previously used by target users may also be possible to verify improvements. Metrics need to address the learnability and utility of the system, in addition to performance and user satisfaction [223]. Usage data logging, user interviews, and surveys can also help identification of potential improvements in visualization and interaction design.

9.4.3 Visual impairment

Color impairment is a common condition that needs to be taken into consideration [291]. For example, red and green are appealing for their intuitive mapping to positive or negative outcomes (also depending on cultural associations); however, users with red-green color blindness, one of the most common forms, would not be able to differentiate such scales clearly. To assess and assist visual design under different color deficiencies, color simulation tools can be used (see additional resources). The impact of color impairment can be mitigated by careful selection of limited color schemes, using double encoding when appropriate (i.e., using symbols that vary by both shape and color), and allowing users to change or customize color palettes. To accommodate users with low vision, adjustable size and zoom settings can be useful. Users with severe visual impairments may require alternative accessibility-first interface and interaction designs.

9.4.4 Visual literacy

While the number of people using visualization continues to grow, not everyone is able to accurately interpret graphs and charts. When designing a visualization for a population of users who are expected to make sense of the data without training, it is important to adequately estimate the level of visual literacy of those users. Even simple scatterplots can be overwhelming for some users. Recent work has proposed new methods for assessing visual literacy [47], but user testing with representative users in the early stages of design and development will remain necessary to verify that adequate designs are being used. Training is likely to be needed to help analysts get started when using more visual analytics tools. Recorded video demonstrations and online support for question answering are helpful to bring users from novice to expert levels.

9.5 Summary

The use of information visualization is spreading widely, with a growing number of commercial products and additions to statistical packages now available. Careful user testing should be conducted to verify that visual data presentations go beyond the desire for eye-candy in visualization, and to implement designs that have demonstrated benefits for realistic tasks. Visualization is becoming increasingly used by the general public and attention should be given to the goal of universal usability so the widest range of users can access and benefit from new approaches to data presentation and interactive analysis.

9.6 Resources

We have referred to various textbooks throughout this chapter. Tufte's books remain the classics, as inspiring to read as they are instructive [380, 381]. We also recommend Few's books on information visualization [119] and information dashboard design [120]. See also the book's website for further readings.

Given the wide variety of goals, tasks, and use cases of visualization, many different data visualization tools have been developed that address different needs and appeal to different skill levels. In this chapter we can only point to a few examples to get started. To generate a wide range of visualizations and dashboards, and to quickly share them online, Tableau and Tableau Public provide a flexible visualization design platform. If a custom design is required and programmers are available, d3 is the de facto low-level library of choice for many web-based visualizations, with its native integration to web standards and flexible ways to convert and manipulate data into visual objects as a JavaScript library. There exist other JavaScript web libraries that offer chart templates (such as Highcharts), or web services that can be used to create a range of charts from given (small) data sets, such as Raw or DataWrapper. To clean, transform, merge, and restructure data sources so that they can be visualized appropriately, tools like Trifacta and Alteryx can be used to create pipelines for data wrangling. For statistical analysis and batch-processing data, programming environments such as R or libraries for languages such as Python (for example, the Python Plotly library) can be used.

An extended list of tools and books is available at http://www.keshif.me/demo/VisTools.

Chapter 10

Errors and Inference

Paul P. Biemer

This chapter deals with inference and the errors associated with big data. Social scientists know only too well the cost associated with bad data—we highlighted both the classic *Literary Digest* example and the more recent Google Flu Trends problems in Chapter 1. Although the consequences are well understood, the new types of data are so large and complex that their properties often cannot be studied in traditional ways. In addition, the data generating function is such that the data are often selective, incomplete, and erroneous. Without proper data hygiene, the errors can quickly compound. This chapter provides, for the first time, a systematic way to think about the error framework in a big data setting.

10.1 Introduction

The massive amounts of high-dimensional and unstructured data in big data bring both new opportunities and new challenges to the data analyst. Many of the problems with big data are well known (see, for example, the AAPOR report by Japec et al. [188]). The volume, variety, and velocity of big data can overwhelm traditional methods of data analysis. Add to this complexity the fact that, as it is generated, big data is often selective, incomplete, and erroneous. As it is processed, new errors can be introduced in downstream operations.

Big data is typically aggregated from disparate sources at various points in time and integrated to form data sets. These processes involve linking records together, transforming them to form new attributes (or variables), documenting the actions taken (although sometimes inadequately), and interpreting the newly created features of the data. These activities may introduce new errors into the data set: errors that may be either *variable* (i.e., errors that create random noise resulting in poor reliability) or *systematic* (i.e., errors that tend to be directional, thus exacerbating biases). Using big data in statistically valid ways is increasingly challenging in this environment; however, it is important for data analysts to be aware of the error risks and the potential effects of big data error on inferences and decision-making. The massiveness, high dimensionality, and accelerating pace of big data, combined with the risks of variable and systematic data errors, requires new, robust approaches to data analysis.

The core issue confronting big data veracity is that such data may not be generated from instruments and methods designed to produce valid and reliable data for scientific analysis and discovery. Rather, this is data that are being repurposed for uses not originally intended. Big data has been referred to as "found" data or "data exhaust" because it is generated for purposes that often do not align with those of the data analyst. In addition to inadvertent errors, there are also errors from mischief in big data; for example, automated systems have been written to generate bogus content in the social media that is indistinguishable from legitimate or authentic data. Regardless of the source, many big data generators have little or no regard for the quality of the data that are cast off from their processes. Big data analysts must be keenly aware of these limitations and should take the necessary steps to understand and hopefully mitigate the effects of hidden errors on their results.

10.2 The total error paradigm

We now provide a framework for describing, mitigating, and interpreting the errors in essentially any data set, be it structured or unstructured, massive or small, static or dynamic. This framework has been referred to as the total error framework or paradigm. We begin by reviewing the traditional paradigm, acknowledging its limitations for truly big data sets, and we suggest how this framework can be extended to encompass the new error structures often associated with big data.

10.2.1 The traditional model

Dealing with the risks that errors introduce in big data analysis can be facilitated through a better understanding of the sources and nature of those errors. Such knowledge is gained through in-depth understanding of the data generating mechanism, the data processing/transformation infrastructure, and the approaches used to create a specific data set or the estimates derived from it. For survey data, this knowledge is embodied in the well-known *total survey error* (TSE) framework that identifies all the major sources of error contributing to data validity and estimator accuracy [33, 35, 143]. The TSE framework attempts to describe the nature of the error sources and what they may suggest about how the errors could affect inference. The framework parses the total error into bias and variance components that, in turn, may be further subdivided into subcomponents that map the specific types of errors to unique components of the total mean squared error. It should be noted that, while our discussion on issues regarding inference has quantitative analyses in mind, some of the issues discussed here are also of interest to more qualitative uses of big data.

For surveys, the TSE framework provides useful insights regarding how the many steps in the data generating, reformatting, and file preparation processes affect estimation and inference, and may also suggest methods for either reducing the errors at their source or adjusting for their effects in the final data products to produce inferences of higher quality. The AAPOR Task Force Report on big data referenced above [188] describes the concept of data quality for big data and provides a total error framework for big data that we consider and further extend in this chapter. This approach was closely modeled after the TSE framework since, as we shall see, both share a number of common error sources. However, the big data total error (BDTE) framework, as it is called in the AAPOR report, necessarily includes additional error sources that are unique to big data and can create substantial biases and uncertainties in big data products. Like the TSE framework, the BDTE framework aids our understanding of the limitations of the data, leading to better-informed analyses and applications of the results. It may also inform a research agenda for reducing the effects of error on big data analytics.

In Figure 10.1, a canonical data file is represented as an array consisting of rows (records) and columns (variables), with their size denoted by N and p, respectively. Many data sets derived from big data can be represented in this way, at least conceptually. Later, we will consider data sets that do not conform to this rectangular structure and may even be unstructured.

Many administrative data sets have a simple tabular structure, as do survey sampling frames, population registers, and accounting spreadsheets. Let us assume that the data set is intended to

Record #	V_1	V_2	...	V_P
1				
2				
...				
N				

Figure 10.1. A typical rectangular data file format

represent some target population to which inference will be made in the subsequent analysis. Thus, the rows are typically aligned with units or elements of this target population, the columns represent characteristics, variables (or features) of the row elements, and the cells correspond to values of the column features for elements on the rows.

The total error for this data set may be expressed by the following heuristic formula:

$$\text{Total error} = \text{Row error} + \text{Column error} + \text{Cell error}.$$

Row error For the situations considered in this chapter, the row errors may be of three types:

- Omissions: Some rows are missing, which implies that elements in the target population are not represented on the file.
- Duplications: Some population elements occupy more than one row.
- Erroneous inclusions: Some rows contain elements or entities that are not part of the target population.

For survey sample data sets, omissions include members of the target population that are either inadvertently or deliberately absent from the frame, as well as nonsampled frame members. For big data, the selectivity of the capture mechanism is a common form of omissions. For example, a data set consisting of persons who conducted a Google search in the past week necessarily excludes persons not satisfying that criterion who may have quite different characteristics from those who do. Such exclusions can therefore be viewed as a source of selectivity bias if inference is to be made to the general population. For one, persons who do not have access to the Internet are excluded from the data set. These exclusions may be biasing in that persons with Internet access may have quite different demographic characteristics from persons who do not have Internet access. The selectivity of big data capture is similar to frame noncoverage in survey sampling and can bias inferences when researchers fail to consider it and compensate for it in their analyses.

Example: Google searches

As an example, in the United States, the word "Jewish" is included in 3.2 times more Google searches than "Mormon" [360]. This does not mean that the Jewish population is 3.2 times larger than the Mormon population. Another possible explanation is that Jewish people use the Internet in higher proportions or have more questions that require using the word "Jewish." Thus Google search data are more useful for relative comparisons than for estimating absolute levels.

A well-known formula in the survey literature provides a useful expression for the so-called *coverage bias* in the mean of some variable, V. Denote the mean by $\bar{V}$, and let $\bar{V}_T$ denote the (possibly hypothetical because it may not be observable) mean of the target population of N_T elements, including the $N_T - N$ elements that are missing from the observed data set. Then the bias due to this *noncoverage* is $B_{NC} = \bar{V} - \bar{V}_T = (1 - N/N_T)(\bar{V}_C - \bar{V}_{NC})$, where $\bar{V}_C$ is the mean of the *covered* elements (i.e., the elements in the observed data set) and $\bar{V}_{NC}$ is the mean of the $N_T - N$ *noncovered* elements. Thus we see that, to the extent that the difference between the covered and noncovered elements is large or the fraction of missing elements $(1 - N/N_T)$ is large, the bias in the descriptive statistic will also be large. As in survey research, often we can only speculate about the sizes of these two components of bias. Nevertheless, speculation is useful for understanding and interpreting the results of data analysis and cautioning ourselves regarding the risks of false inference.

We can also expect that big data sets, such as a data set containing Google searches during the previous week, could have the same person represented many times. People who conducted many searches during the data capture period would be disproportionately represented relative to those who conducted fewer searchers. If the rows of the data set correspond to tweets in a Twitter feed, duplication can arise when the same tweet is retweeted or when some persons are quite active in tweeting while others lurk and tweet much less frequently. Whether such duplications should be regarded as "errors" depends upon the goals of the analysis.

For example, if inference is to be made to a population of persons, persons who tweet multiple times on a topic would be overrepresented. If inference is to be made to the population of tweets, including retweets, then such duplication does not bias inference.

When it is a problem, it still may not be possible to identify duplications in the data. Failing to account for them could generate duplication biases in the analysis. If these unwanted duplications can be identified, they can either be removed from the data file (i.e., deduplication). Alternatively, if a certain number of rows, say d, correspond to the same population unit, those row values can be weighted by $1/d$ to correct the estimates for the duplications.

Erroneous inclusions can also create biases. For example, Google searches or tweets may not be generated by a person but rather by a computer either maliciously or as part of an information-gathering or publicity-generating routine. Likewise, some rows may not satisfy the criteria for inclusion in an analysis—for example, an analysis by age or gender includes some row elements not satisfying the criteria. If the criteria can be applied accurately, the rows violating the criteria can be excluded prior to analysis. However, with big data, some out-of-scope elements may still be included as a result of missing or erroneous information, and these inclusions will bias inference.

Column error The most common type of column error in survey data analysis is caused by inaccurate or erroneous labeling of the column data—an example of metadata error. In the TSE framework, this is referred to as a *specification* error. For example, a business register may include a column labeled "number of employees," defined as the number of persons in the company who received a payroll check in the month preceding. Instead the column contains the number of persons on the payroll whether or not they received a check in the prior month, thus including, for example, persons on leave without pay.

For big data analysis, such errors would seem to be quite common because of the complexities involved in producing a data set. For example, data generated from a source, such as an individual tweet, may undergo a number of transformations before it is included in the analysis data set. This transformative process can be quite complex, involving parsing phrases, identifying words, and classifying them as to subject matter and then perhaps further classifying them as either positive or negative expressions about some phenomenon like the economy or a political figure. There is considerable risk of the resulting variables being either inaccurately defined or misinterpreted by the data analyst.

Example: Specification error with Twitter data

As an example, consider a Twitter data set where the rows correspond to tweets and one of the columns supposedly contains an indicator of whether the tweet contained one of the following key words: marijuana, pot, cannabis, weed, hemp, ganja, or THC. Instead, the indicator actually corresponds to whether the tweet contained a shorter list of words; say, either marijuana or pot. The mislabeled column is an example of specification error which could be a biasing factor in an analysis. For example, estimates of marijuana use based upon the indicator could be underestimates.

Cell errors Finally, cell errors can be of three types: content error, specification error, or missing data. A content error occurs when the value in a cell satisfies the column definition but still deviates from the true value, whether or not the true value is known. For example, the value satisfies the definition of "number of employees" but is outdated because it does not agree with the current number of employees. Errors in sensitive data such as drug use, prior arrests, and sexual misconduct may be deliberate. Thus, content errors may be the result of the measurement process, a transcription error, a data processing error (e.g., keying, coding, editing), an imputation error, or some other cause.

Specification error is just as described for column error but applied to a cell. For example, the column is correctly defined and labeled; however, a few companies provided values that, although otherwise highly accurate, were nevertheless inconsistent with the required definition. Missing data, as the name implies, are just empty cells that should be filled. As described in Kreuter and Peng [215], data sets derived from big data are notoriously affected by all three types of cell error, particularly missing or incomplete data, perhaps because that is the most obvious deficiency.

The traditional TSE framework is quite general in that it can be applied to essentially any data set that conform to the format in Figure 10.1. However, in most practical situations it is quite limited because it makes no attempt to describe how the processes that generated the data may have contributed to what could be construed as data errors. In some cases, these processes constitute a "black box," and the best approach is to attempt to evaluate the quality of the end product. For survey data, the TSE framework provides a fairly complete description of the error-generating processes for survey data and survey frames [33]. In addition, there has been some effort to describe these processes for population registers and administrative data [391]. But at this writing, little effort has been devoted to enumerating the error sources and the error generating processes for big data.

As previously noted, missing data can take two forms: missing information in a cell of a data matrix (referred to as *item missingness*) or missing rows (referred to as *unit missingness*), with the former being readily observable whereas the latter can be completely hidden from the analyst. Much is known from the survey research literature about how both types of missingness affect data analysis (see, for example, Little and Rubin [240, 320]). Rubin [320] introduced the term *missing completely at random (MCAR)* to describe data where the data that are available (say, the rows of a data set) can be considered as a simple random sample of the inferential population (i.e., the population to which inferences from the data analysis will be made). Since the data set represents the population, MCAR data provide results that are generalizable to this population.

A second possibility also exists for the reasons why data are missing. For example, students who have high absenteeism may be missing because they were ill on the day of the test. They may otherwise be average performers on the test so, in this case, it has little to do with how they would score. Thus, the values are missing for reasons related to another variable, health, that may be available in the data set and completely observed. Students with poor health tend to be missing test scores, regardless of those student's performance on the test. Rubin [320] uses the term *missing at random (MAR)* to describe data that are missing for reasons related to completely observed variables in the data set. It is possible to compensate for this type of missingness in statistical inferences by modeling the missing data mechanism.

However, most often, missing data may be related to factors that are not represented in the data set and, thus, the missing data mechanism cannot be adequately modeled. For example, there may be a tendency for test scores to be missing from school administrative data files for students who are poor academic performers. Rubin calls this form of missingness *nonignorable*. With nonignorable missing data, the reasons for the missing observations depend on the values that are missing. When we suspect a nonignorable missing data mechanism, we need to use procedures much more complex than will be described here. Little and Rubin [240] and Schafer [326] discuss methods that can be used for nonignorable missing data. Ruling out a nonignorable response mechanism can simplify the analysis considerably.

In practice, it is quite difficult to obtain empirical evidence about whether or not the data are MCAR or MAR. Understanding the data generation process is invaluable for specifying models that appropriately represent the missing data mechanism and that will then be successful in compensating for missing data in an analysis. (Schafer and Graham [327] provide a more thorough discussion of this issue.)

One strategy for ensuring that the missing data mechanism can be successfully modeled is to have available on the data set many variables that may be causally related to missing data. For example, features such as personal income are subject to high item missingness, and often the missingness is related to income. However, less sensitive, surrogate variables such as years of education or type of employment may be less subject to missingness. The statistical relationship between income and other income-related variables increases the chance that information lost in missing variables is supplemented by other completely observed variables. Model-based methods use the multivariate relationship between variables to handle the missing data. Thus, the more informative the data set, the more measures we have on important constructs, the more successfully we can compensate for missing data using model-based approaches.

10.2.2 Extending the framework to big data

The processes involved in generating big data are as varied as big data itself. This diversity substantially complicates the goal of providing a detailed error framework that is applicable to all big data. Nevertheless, some progress can be made by considering the steps that are typically involved in creating a data set from big data, which often includes three stages: (a) data generation (see Chapter 2), (b) extract, transform and load (ETL), and (c) analysis (see, for example, Chapter 7). As noted there, this mapping of the process is oversimplified for some applications. For example, data that flow continuously from their sources may not be directed through an ETL process at all. Rather it may be gathered in real time and processed on an ad hoc basis. Likewise, the ETL processes may

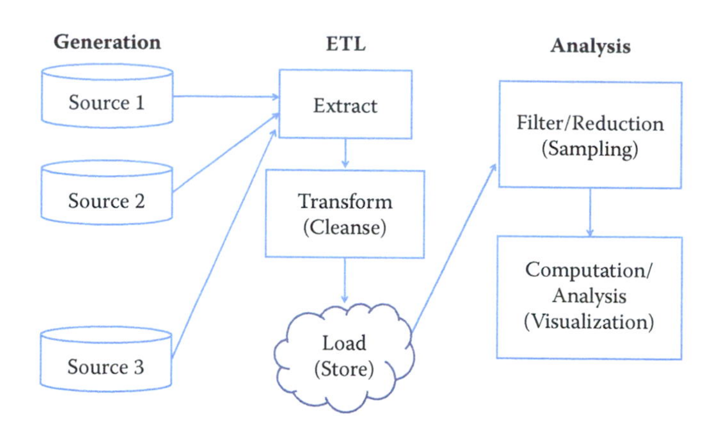

Figure 10.2. Big data process map

be recursive in that, as errors are identified, the processes may be altered and refined. In addition, the transform stage may generate new data (e.g., proxy variables). The analysis stage may also involve language transformations such as translation and character set transformations. Finally, there may downstream processes that lead to additional iterations of the stages in the process map. Thus, the model outline here is but an initial attempt to capture the tremendous complexities that may be involved in the big data life cycle.

Figure 10.2 graphically depicts the flow of data along the major steps in the process. One might imagine many more arrows among the boxes within the ETL and analysis stages that represent the recursive nature of those stages. The severity of the errors that arise from these processes will depend on the specific data sources and analytic goals involved. Nevertheless, we can still consider how each stage might create errors in a more generic fashion.

For example, data generation error is somewhat analogous to errors arising in survey data collection. Like surveys, the data-generating process for big data is subject to erroneous, incomplete, and missing data. In addition, the data-generating sources may be selective in that the data collected may not represent a well-defined population or one that is representative of a target population of inference in an analysis. Thus, data generation errors include low signal-to-noise ratio, lost signals, incomplete or missing values, systematic errors, selective elements, and metadata that are lacking, absent, or erroneous.

ETL processes may be quite similar to various data processing stages for surveys. These may include creating or enhancing metadata, record matching, variable coding, editing, data munging (or scrubbing), and data integration (i.e., linking and merging records and files across disparate systems). ETL errors include specification errors (including errors in metadata), matching errors, coding errors, editing errors, data munging errors, and data integration errors.

Finally, the analysis of big data introduces risks for a number of errors that will be described in more detail in the next section. These risks may be due to noise accumulation, coincidental correlations, and incidental endogeneity. As we shall see, these errors arise as a consequence of big data volume and variety even when the big data itself is infallible. Erroneous data compound these problems and can lead to other issues that are described in some detail below. For example, big data can be subject to sampling errors when the data are filtered, sampled, or otherwise reduced to form more manageable or representative data sets. So-called nonsampling errors can arise when the analysis involves further transforming the data and weighting the data elements, as well as can be errors due to modeling and estimation. The latter errors are similar to modeling and estimation errors in surveys and may include inadequate or erroneous adjustments for representativeness, improper or erroneous weighting, computation and algorithmic errors, model misspecification, and so on.

10.3 Illustrations of errors in big data

▶ See the discussion in Section 1.3.

A well-known example of the risks of big data error is provided by the Google Flu Trends series that uses Google searches on flu symptoms, remedies, and other related key words to provide near-real-time estimates of flu activity in the USA and 24 other countries. Compared to CDC data, the Google Flu Trends provided remarkably accurate indicators of flu incidence in the USA between 2009 and 2011. However, for the 2012–2013 flu seasons, Google Flu Trends predicted more than double the proportion of doctor visits for flu-like symptoms compared to the CDC [61]. Lazer et al. [230] cite two causes of this error: big data hubris and algorithm dynamics.

Hubris occurs when the big data researcher believes that the volume of the data compensates for any of its deficiencies, thus obviating the need for traditional, scientific analytic approaches. As Lazer et al. [230] note, big data hubris fails to recognize that "quantity of data does not mean that one can ignore foundational issues of measurement and construct validity and reliability."

Algorithm dynamics refers to properties of algorithms that allow them to adapt and "learn" as the processes generating the data change over time. Although explanations vary, the fact remains that Google Flu Trends was too high and by considerable margins for 100 out of 108 weeks starting in July 2012. Lazer et al. [230] also blame "blue team dynamics," which arises when the data generating engine is modified in such a way that the formerly highly predictive search terms eventually failed to work. For example, when a Google user searched on "fever" or "cough," Google's other programs started recommending searches for flu symptoms and treatments—the very search terms the algorithm used to predict flu. Thus, flu-related searches artificially spiked as a result of these changes to the algorithm and the impact these changes had on user behavior. In survey research, this is similar to the measurement biases induced by interviewers who suggest to respondents who are coughing that they might have flu, then ask the same respondents if they think they might have flu.

Algorithm dynamic issues are not limited to Google. Platforms such as Twitter and Facebook are also frequently modified to improve the user experience. A key lesson provided by Google Flu Trends is that successful analyses using big data today may fail to produce good results tomorrow. All these platforms change their methodologies more or less frequently, with ambiguous results for any kind of long-term study unless highly nuanced methods are routinely used. Recommendation engines often exacerbate effects in a certain direction, but these effects are hard to tease out. Furthermore, other sources of error may affect Google Flu Trends to an unknown extent. For example, selectivity may be an important issue because the demographics of people with Internet access are quite different from the demographic characteristics related to flu incidence [375]. Thus, the "at risk" population for influenza and the implied population based on Google searches do not correspond. This illustrates just one type of representativeness issue that often plagues big data analysis. In general it is an issue that algorithms are not (publicly) measured for accuracy, since they are often proprietary. Google Flu Trends is special in that it publicly failed. From what we have seen, most models fail privately and often without anyone at all noticing.

In the next section, we consider the impact of errors on some forms of analysis that are common in the big data literature. Due to space limitations, our focus is limited to the effects of content errors on data analysis. However, there are numerous resources available for studying and mitigating the effects of missing data on analysis;

in particular, the books by Little and Rubin [240], Schafer [326], and Allison [10] are quite relevant to big data analytics.

10.4 Errors in big data analytics

In this section, we consider some common types of errors in big data analytics. The next section considers analytic errors caused by the so-called "three Vs" of big data, namely, volume, velocity, and variety, under the assumption of perfect "veracity." If we relax the assumption of perfect veracity, the errors that afflict big data as a result of the three Vs can be exacerbated in largely unpredictable ways. Section 10.4.2 considers three common types of analysis when data lack veracity: classification, correlation, and regression. While the scope of our review is necessarily limited, hopefully the discussion will lay the foundation for further study regarding the effects of errors on big data analytics.

10.4.1 Errors resulting from volume, velocity, and variety, assuming perfect veracity

Data deficiencies represent only one set of challenges for the big data analyst. Other challenges arise solely as a result of the massive size, rapid generation, and vast dimensionality of the data. As a consequence of these so-called three Vs of big data, Fan et al. [111] identify three issues—noise accumulation, spurious correlations, and incidental endogeneity—which will be briefly discussed in this section. These issues should concern big data analysts even if the data could be regarded as infallible. Content errors, missing data, and other data deficiencies will only exacerbate these problems.

Example: Noise accumulation

To illustrate noise accumulation, Fan et al. [111] consider the following scenario. Suppose an analyst is interested in classifying individuals into two categories, C_1 and C_2, based upon the values of 1,000 variables in a big data set. Suppose further that, unknown to the researcher, the mean value for persons in C_1 is 0 on all 1,000 variables while persons in C_2 have a mean of 3 on the first 10 variables and 0 on all other variables. Since we are assuming the data are error-free, a classification rule based upon the first $m \leq 10$ variables performs quite well, with little classification error. However, as more and more variables are included in the rule, classification error increases because the uninformative variables (i.e., the 990 variables having

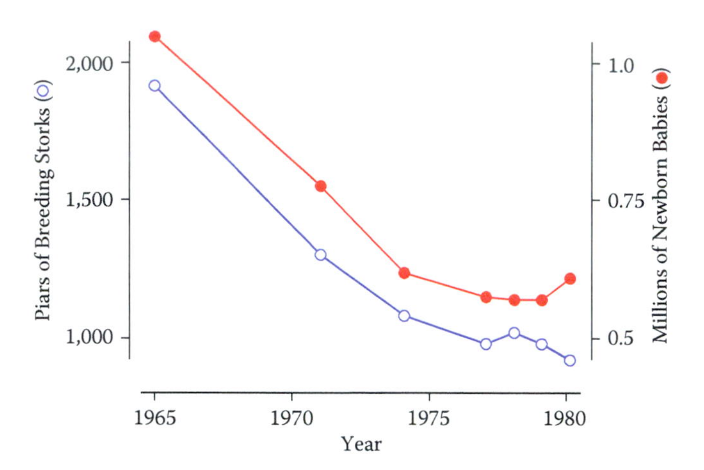

Figure 10.3. An illustration of coincidental correlation between two variables: stork die-off linked to human birth decline [346]

no discriminating power) eventually overwhelm the informative signals (i.e., the first 10 variables). In the Fan et al. [111] example, when $m > 200$, the accumulated noise exceeds the signal embedded in the first 10 variables and the classification rule becomes equivalent to a coin-flip classification rule.

High dimensionality can also introduce coincidental (or *spurious*) correlations in that many unrelated variables may be highly correlated simply by chance, resulting in false discoveries and erroneous inferences. The phenomenon depicted in Figure 10.3 is an illustration of this. Many more examples can be found on a website and in a book devoted to the topic [388, 389]. Fan et al. [111] explain this phenomenon using simulated populations and relatively small sample sizes. They illustrate how, with 800 independent (i.e., uncorrelated) variables, the analyst has a 50% chance of observing an absolute correlation that exceeds 0.4. Their results suggest that there are considerable risks of false inference associated with a purely empirical approach to predictive analytics using high-dimensional data.

Finally, turning to incidental endogeneity, a key assumption in regression analysis is that the model covariates are uncorrelated with the residual error; endogeneity refers to a violation of this assumption. For high-dimensional models, this can occur purely by chance—a phenomenon Fan and Liao [113] call *incidental endogeneity*. Incidental endogeneity leads to the modeling of spurious variation in the outcome variables resulting in errors in the model selection process and biases in the model predictions. The risks of incidental endogeneity increase as the number of variables in the model selection process grows large. Thus it is a particularly important concern for big data analytics.

Fan et al. [111] as well as a number of other authors [114, 361] (see, for example, Hall and Miller [151]; Fan and Liao, [112]) suggest robust statistical methods aimed at mitigating the risks of noise accumulation, spurious correlations, and incidental endogeneity. However, as previously noted, these issues and others are further compounded when data errors are present in a data set. Biemer and Trewin [37] show that data errors will bias the results of traditional data analysis and inflate the variance of estimates in ways that are difficult to evaluate or mitigate in the analysis process.

10.4.2 Errors resulting from lack of veracity

▶ The volume, velocity, and variety of big data make it challenging to analyze because of noise accumulation, spurious correlations, and incidental endogeneity. But, the three Vs are also the scourge of veracity—i.e., errors in big data.

The previous section examined some of the issues big data analysts face as either N or p in Figure 10.1 becomes extremely large. When row, column, and cell errors are added into the mix, these *volumatic* problems can be further exacerbated. For example, noise accumulation can be expected to accelerate when random noise (i.e., content errors) afflicts the data. Spurious correlations that give rise to both incidental endogeneity and coincidental correlations can render correlation analysis meaningless if the error levels in big data are not mitigated. In this section, we consider some of the issues that arise in classification, correlation, and regression analysis as a result of content errors that may be either variable or systematic. The current literature on big data error acknowledges that the data may be noisy, that is, subject to variable errors. However, there appears to be little recognition of the problems associated with systematic or correlated errors, particularly those introduced when data are combined from disparate sources that may be subject to source-specific errors.

There are various important findings in this section. First, for rare classes, even small levels of error can impart considerable biases in classification analysis. Second, variable errors will attenuate correlations and regression slope coefficients; however, these effects can be mitigated by forming meaningful aggregates of the data and substituting these aggregates for the individual units in these analyses. Third, unlike random noise, systematic errors can bias correlation and regression analysis is unpredictable ways, and these biases cannot be effectively mitigated by aggregating the data.

Finally, multilevel modeling can be an important mitigation strategy for dealing with systematic errors emanating from multiple data sources. These issues will be examined in some detail in the remainder of this section.

10.4.2.1 Variable and correlated error

Error models are essential for understanding the effects of error on data sets and the estimates that may be derived from them. They allow us to concisely and precisely communicate the nature of the errors that are being considered, the general conditions that give rise to them, how they affect the data, how they may affect the analysis of these data, and how their effects can be evaluated and mitigated. In the remainder of this chapter, we focus primarily on content errors and consider two types of error, variable errors and correlated errors, the latter a subcategory of systematic errors.

Variable errors are sometimes referred to as *random noise* or *uncorrelated* errors. For example, administrative databases often contain errors from a myriad of random causes, including mistakes in keying or other forms of data capture, errors on the part of the persons providing the data due to confusion about the information requested, difficulties in recalling information, the vagaries of the terms used to request the inputs, and other system deficiencies.

Correlated errors, on the other hand, carry a systematic effect that results in a nonzero covariance between the errors of two distinct units. For example, quite often, an analysis data set may combine multiple data sets from different sources and each source may impart errors that follow a somewhat different distribution. As we shall see, these differences in error distributions can induce correlated errors in the merged data set. It is also possible that correlated errors are induced from a single source as a result of different operators (e.g., computer programmers, data collection personnel, data editors, coders, data capture mechanisms) handling the data. Differences in the way these operators perform their tasks have the potential to alter the error distributions so that data elements handled by the same operator have errors that are correlated [35].

These concepts may be best expressed by a simple error model. Let y_{rc} denote the cell value for variable c on the rth unit in the data set, and let ε_{rc} denote the error associated with this value. Suppose it can be assumed that there is a true value underlying y_{rc}, which is denoted by μ_{rc}. Then we can write

$$y_{rc} = \mu_{rc} + \varepsilon_{rc}. \qquad (10.1)$$

At this point, ε_{rc} is not stochastic in nature because a statistical process for generating the data has not yet been assumed. Therefore, it is not clear what *correlated error* really means. To remedy this problem, we can consider the hypothetical situation where the processes generating the data set can be repeated under the same general conditions (i.e., at the same point in time with the same external and internal factors operating). Each time the processes are repeated, a different set of errors may be realized. Thus, it is assumed that although the true values, μ_{rc}, are fixed, the errors, ε_{rc}, can vary across the hypothetical, infinite repetitions of the data set generating process. Let $\mathrm{E}(\cdot)$ denote the expected value over all these hypothetical repetitions, and define the variance, $\mathrm{Var}(\cdot)$, and covariance, $\mathrm{Cov}(\cdot)$, analogously.

For the present, error correlations between variables are not considered, and thus the subscript, c, is dropped to simplify the notation. For the uncorrelated data model, we assume that $\mathrm{E}(y_r|r) = \mu_r$, $\mathrm{Var}(y_r|r) = \sigma_\varepsilon^2$, and $\mathrm{Cov}(y_r, y_s|r, s) = 0$, for $r \neq s$. For the correlated data model, the latter assumption is relaxed. To add a bit more structure to the model, suppose the data set is the product of combining data from multiple sources (or operators) denoted by $j = 1, 2, \ldots, J$, and let b_j denote the systematic effect of the jth source. Here we also assume that, with each hypothetical repetition of the data set generating process, these systematic effects can vary stochastically. (It is also possible to assume the systematic effects are fixed. See, for example, Biemer and Stokes [36] for more details on this model.) Thus, we assume that $\mathrm{E}(b_j) = 0$, $\mathrm{Var}(b_j) = \sigma_b^2$, and $\mathrm{Cov}(b_j, b_k) = 0$ for $j \neq k$.

Finally, for the rth unit within the jth source, let $\varepsilon_{rj} = b_j + e_{rj}$. Then it follows that

$$\mathrm{Cov}(\varepsilon_{rj}, \varepsilon_{sk}) = \begin{cases} \sigma_b^2 + \sigma_\varepsilon^2 & \text{for } r = s, j = k, \\ \sigma_\varepsilon^2 & \text{for } r = s, j \neq k, \\ 0 & \text{for } r \neq s, j \neq k. \end{cases}$$

The case where $\sigma_b^2 = 0$ corresponds to the uncorrelated error model (i.e., $b_j = 0$) and thus ε_{rj} is purely random noise.

Example: Speed sensor

Suppose that, due to calibration error, the jth speed sensor in a traffic pattern study underestimates the speed of vehicle traffic on a highway by an average of 4 miles per hour. Thus, the model for this sensor is that the speed for the rth vehicle recorded by this sensor (y_{rj}) is the vehicle's true speed (μ_{rj}) minus 4 mph (b_j) plus a random departure from −4 for the rth vehicle (ε_{rj}). Note that to the extent that b_j varies across sensors $j = 1, \ldots, J$ in the study, σ_b^2 will be large. Further, to the extent that ambient noise in the readings for jth sensor causes variation around the values $\mu_{rc} + b_j$, then σ_ε^2 will be large. Both sources of variation will reduce the reliability of the measurements. However, as shown in Section 10.4.2.4, the systematic error component is particularly problematic for many types of analysis.

10.4.2.2 Models for categorical data

For variables that are categorical, the model of the previous section is not appropriate because the assumptions it makes about the error structure do not hold. For example, consider the case of a binary (0/1) variable. Since both y_r and μ_r should be either 1 or 0, the error in equation (10.1) must assume the values of −1, 0, or 1. A more appropriate model is the misclassification model described by Biemer [34], which we summarize here.

Let ϕ_r denote the probability of a false positive error (i.e., $\phi_r = \Pr(y_r = 1|\mu_r = 0)$), and let ∂_r denote the probability of a false negative error (i.e., $\partial_r = \Pr(y_r = 0|\mu_r = 1)$). Thus, the probability that the value for row r is correct is $1 - \partial_r$ if the true value is 1, and $1 - \phi_r$ if the true value is 0.

As an example, suppose an analyst wishes to compute the proportion, $P = \sum_r y_r/N$, of the units in the file that are classified as 1, and let $\pi = \sum_r \mu_r/N$ denote the true proportion. Then under the assumption of uncorrelated error, Biemer [34] shows that

$$P = \pi(1 - \partial) + (1 - \pi)\phi,$$

where $\partial = \sum_r \partial_r/N$ and $\phi = \sum_r \phi_r/N$.

In the classification error literature, the sensitivity of a classifier is defined as $1 - \partial$, that is, the probability that a true positive is correctly classified. Correspondingly, $1 - \phi$ is referred to as the specificity of the classifier, that is, the probability that a true negative is correctly classified. Two other quantities that will be useful in our study of misclassification error are the positive predictive value (PPV) and negative predictive value (NPV) given by

$$\text{PPV} = \Pr(\mu_r = 1|y_r = 1), \quad \text{NPV} = \Pr(\mu_r = 0|y_r = 0).$$

The PPV (NPV) is the probability that a positive (negative) classification is correct.

10.4.2.3 Misclassification and rare classes

Fan et al. [111] and many others have stated that one of the strengths of big data is the ability to study rare population groups that seldom show up in large enough numbers in designed studies such as surveys and clinical trials. While this is true in theory, in practice content errors can quickly overwhelm such an analysis, rendering the data useless for this purpose. We illustrate this using the following contrived and somewhat amusing example. The results in this section are particularly relevant to the approaches considered in Chapter 6.

Example: Thinking about probabilities

Suppose, using big data and other resources, we construct a terrorist detector and boast that the detector is 99.9% accurate. In other words, both the probability of a false negative (i.e., classifying a terrorist as a nonterrorist, ∂) and the probability of a false positive (i.e., classifying a nonterrorist as a terrorist, ϕ) are 0.001. Assume that about 1 person in a million in the population is a terrorist, that is, $\pi = 0.000001$ (hopefully, somewhat of an overestimate). Your friend, Terry, steps into the machine and, to Terry's chagrin (and your surprise) the detector declares that he is a terrorist! What are the odds that the machine is right? The surprising answer is only about 1 in 1000. That is, 999 times out of 1,000 times the machine classifies a person as a terrorist, the machine will be wrong!

How could such an accurate machine be wrong so often in the terrorism example? Let us do the math.

The relevant probability is the PPV of the machine: given that the machine classifies an individual (Terry) as a terrorist, what is the probability the individual is truly a terrorist? Using the notation in Section 10.4.2.2 and Bayes' rule, we can derive the PPV as

$$\begin{aligned}
\Pr(\mu_r = 1|y_r = 1) &= \frac{\Pr(y_r = 1|\mu_r = 1)\Pr(\mu_r = 1)}{\Pr(y_r = 1)} \\
&= \frac{(1-\partial)\pi}{\pi(1-\partial) + (1-\pi)\phi} \\
&= \frac{0.999 \times 0.000001}{0.000001 \times 0.999 + 0.99999 \times 0.001} \\
&\approx 0.001.
\end{aligned}$$

This example calls into question whether security surveillance using emails, phone calls, etc. can ever be successful in finding

Table 10.1. Positive predictive value (%) for rare subgroups, high specificity, and perfect sensitivity

π_k	Specificity		
	99%	99.9%	99.99%
0.1	91.70	99.10	99.90
0.01	50.30	91.00	99.00
0.001	9.10	50.00	90.90
0.0001	1.00	9.10	50.00

rare threats such as terrorism since to achieve a reasonably high PPV (say, 90%) would require a sensitivity and specificity of at least $1 - 10^{-7}$, or less than 1 chance in 10 million of an error.

To generalize this approach, note that any population can be regarded as a *mixture* of subpopulations. Mathematically, this can be written as

$$f(y|\mathbf{x};\partial) = \pi_1 f(y|\mathbf{x};\partial_1) + \pi_2 f(y|\mathbf{x};\partial_2) + \ldots + \pi_K f(y|\mathbf{x};\partial_K),$$

where $f(y|\mathbf{x};\partial)$ denotes the population distribution of y given the vector of explanatory variables $\mathbf{x}$ and the parameter vector $\partial = (\partial_1, \partial_2, \ldots, \partial_K)$, π_k is the proportion of the population in the kth subgroup, and $f(y|\mathbf{x};\partial_k)$ is the distribution of y in the kth subgroup. A rare subgroup is one where π_k is quite small (say, less than 0.01).

Table 10.1 shows the PPV for a range of rare subgroup sizes when the sensitivity is perfect (i.e., no misclassification of true positives) and specificity is not perfect but still high. This table reveals the fallacy of identifying rare population subgroups using fallible classifiers unless the accuracy of the classifier is appropriately matched to the rarity of the subgroup. As an example, for a 0.1% subgroup, the specificity should be at least 99.99%, even with perfect sensitivity, to attain a 90% PPV.

10.4.2.4 Correlation analysis

In Section 10.4.1, we considered the problem of incidental correlation that occurs when an analyst correlates pairs of variables selected from big data stores containing thousands of variables. In this section, we discuss how errors in the data can exacerbate this problem or even lead to failure to recognize strong associations among the variables. We confine the discussion to the continuous variable model of Section 10.4.2.1 and begin with theoretical results that help explain what happens in correlation analysis when the data are subject to variable and systematic errors.

For any two variables in the data set, c and d, define the covariance between y_{rc} and y_{rd} as

$$\sigma_{y|cd} = \frac{\sum_r \mathrm{E}(y_{rc} - \bar{y}_c)(y_{rd} - \bar{y}_d)}{N},$$

where the expectation is with respect to the error distributions and the sum extends over all rows in the data set. Let

$$\sigma_{\mu|cd} = \frac{\sum_r (\mu_{rc} - \bar{\mu}_c)(\mu_{rd} - \bar{\mu}_d)}{N}$$

denote the *population* covariance. (The population is defined as the set of all units corresponding to the rows of the data set.) For any variable c, define the variance components

$$\sigma^2_{y|c} = \frac{\sum_r (y_{rc} - \bar{y}_c)^2}{N}, \quad \sigma^2_{\mu|c} = \frac{\sum_r (\mu_{rc} - \bar{\mu}_c)^2}{N},$$

and let

$$R_c = \frac{\sigma^2_{\mu|c}}{\sigma^2_{\mu|c} + \sigma^2_{b|c} + \sigma^2_{\varepsilon|c}}, \quad \rho_c = \frac{\sigma^2_{b|c}}{\sigma^2_{\mu|c} + \sigma^2_{b|c} + \sigma^2_{\varepsilon|c}},$$

with analogous definitions for d. The ratio R_c is known as the *reliability ratio*, and ρ_c will be referred to as the *intra-source correlation*. Note that the reliability ratio is the proportion of total variance that is due to the variation of true values in the data set. If there were no errors, either variable or systematic, then this ratio would be 1. To the extent that errors exist in the data, R_c will be less than 1.

Likewise, ρ_c is also a ratio of variance components that reflects the proportion of total variance that is due to systematic errors with biases that vary by data source. A value of ρ_c that exceeds 0 indicates the presence of systematic error variation in the data. As we shall see, even small values of ρ_c can cause big problems in correlation analysis.

Using the results in Biemer and Trewin [37], it can be shown that the correlation between y_{rc} and y_{rd}, defined as $\rho_{y|cd} = \sigma_{y|cd}/\sigma_{y|c}\sigma_{y|d}$, can be expressed as

$$\rho_{y|cd} = \sqrt{R_c R_d}\rho_{\mu|cd} + \sqrt{\rho_c \rho_d}. \tag{10.2}$$

Note that if there are no errors (i.e., when $\sigma^2_{b|c} = \sigma^2_{\varepsilon|c} = 0$), then $R_c = 1$, $\rho_c = 0$, and the correlation between y_c and y_d is just the population correlation.

Let us consider the implications of these results first without systematic errors (i.e., only variable errors) and then with the effects of systematic errors.

Variable errors only If the only errors are due to random noise, then the additive term on the right in equation (10.2) is 0 and $\rho_{y|cd} = \sqrt{R_c R_d}\rho_{\mu|cd}$, which says that the correlation is attenuated by the product of the root reliability ratios. For example, suppose $R_c = R_d = 0.8$, which is considered excellent reliability. Then the observed correlation in the data will be about 80% of the true correlation; that is, correlation is attenuated by random noise. Thus, $\sqrt{R_c R_d}$ will be referred to as the *attenuation factor* for the correlation between two variables.

Quite often in the analysis of big data, the correlations being explored are for aggregate measures, as in Figure 10.3. Therefore, suppose that, rather than being a single element, y_{rc} and y_{rd} are the means of n_{rc} and n_{rd} independent elements, respectively. For example, y_{rc} and y_{rd} may be the average rate of inflation and the average price of oil, respectively, for the rth year, for $r = 1, \ldots, N$ years. Aggregated data are less affected by variable errors because, as we sum up the values in a data set, the positive and negative values of the random noise components combine and cancel each other under our assumption that $\mathrm{E}(\varepsilon_{rc}) = 0$. In addition, the variance of the mean of the errors is of order $O(n_{rc}^{-1})$.

To simplify the result for the purposes of our discussion, suppose $n_{rc} = n_c$, that is, each aggregate is based upon the same sample size. It can be shown that equation (10.2) still applies if we replace R_c by its aggregated data counterpart denoted by $R_c^A = \sigma_{\mu|c}^2/(\sigma_{\mu|c}^2 + \sigma_{\varepsilon|c}^2/n_c)$. Note that R_c^A converges to 1 as n_c increases, which means that $\rho_{y|cd}$ will converge to $\rho_{\mu|cd}$. Figure 10.4 illustrates the speed at which this convergence occurs.

In this figure, we assume $n_c = n_d = n$ and vary n from 0 to 60. We set the reliability ratios for both variables to 0.5 (which is considered to be a "fair" reliability) and assume a population correlation of $\rho_{\mu|cd} = 0.5$. For n in the range [2, 10], the attenuation is pronounced. However, above 10 the correlation is quite close to the population value. Attenuation is negligible when $n > 30$. These results suggest that variable error can be mitigated by aggregating like elements that can be assumed to have independent errors.

Both variable and systematic errors If both systematic and variable errors contaminate the data, the additive term on the right in

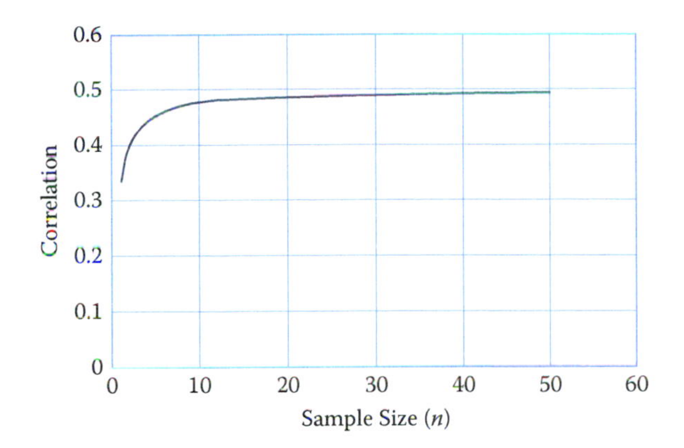

Figure 10.4. Correlation as a function of sample size for $\rho_{\mu|cd} = 0.5$ and $R_c = R_d = 0.5$

equation (10.2) is positive. For aggregate data, the reliability ratio takes the form

$$R_c^A = \frac{\sigma^2_{\mu|c}}{\sigma^2_{\mu|c} + \sigma^2_{b|c} + n_c^{-1}\sigma^2_{\varepsilon|c}},$$

which converges not to 1 as in the case of variable error only, but to $\sigma^2_{\mu|c}/(\sigma^2_{\mu|c} + \sigma^2_{b|c})$, which will be less than 1. Thus, some attenuation is possible regardless of the number of elements in the aggregate. In addition, the intra-source correlation takes the form

$$\rho_c^A = \frac{\sigma^2_{b|c}}{\sigma^2_{\mu|c} + \sigma^2_{b|c} + n_c^{-1}\sigma^2_{\varepsilon|c}},$$

which converges to $\rho_c^A = \sigma^2_{b|c}/(\sigma^2_{\mu|c} + \sigma^2_{b|c})$, which converges to $1 - R_c^A$. Thus, the systematic effects may still operate for correlation analysis without regard to the number of elements comprising the aggregates.

For example, consider the illustration in Figure 10.4 with $n_c = n_d = n$, reliability ratios (excluding systematic effects) set at 0.5 and population correlation at $\rho_{\mu|cd} = 0.5$. In this scenario, let $\rho_c = \rho_d = 0.25$. Figure 10.5 shows the correlation as a function of the sample size with systematic errors (dotted line) compared to the correlation without systematic errors (solid line). Correlation with systematic errors is both inflated and attenuated. However, at the assumed level of intra-source variation, the inflation factor overwhelms the

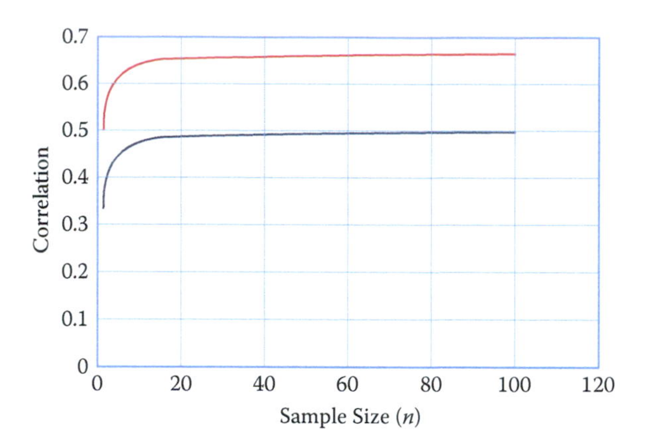

Figure 10.5. Correlation as a function of sample size for $\rho_{\mu|cd} = 0.5$, $R_c = R_d = 0.5$, and $\rho_c = \rho_d = 0.25$

attenuation factors and the result is a much inflated value of the correlation across all aggregate sizes.

To summarize these findings, correlation analysis is attenuated by variable errors, which can lead to null findings when conducting a correlation analysis and the failure to identify associations that exist in the data. Combined with systematic errors that may arise when data are extracted and combined from multiple sources, correlation analysis can be unpredictable because both attenuation and inflation of correlations can occur. Aggregating data mitigates the effects of variable error but may have little effect on systematic errors.

10.4.2.5 Regression analysis

The effects of variable errors on regression coefficients are well known [37, 82, 130]. The effects of systematic errors on regression have been less studied. We review some results for both types of errors in this section.

Consider the simple situation where we are interested in computing the population slope and intercept coefficients given by

$$b = \frac{\sum_r (y_r - \bar{y})(x_r - \bar{x})}{\sum_r (x_r - \bar{x})^2} \quad \text{and} \quad b_0 = \bar{y} - b\bar{x},$$

where, as before, the sum extends over all rows in the data set. When x is subject to variable errors, it can be shown that the observed regression coefficient will be attenuated from its error-free

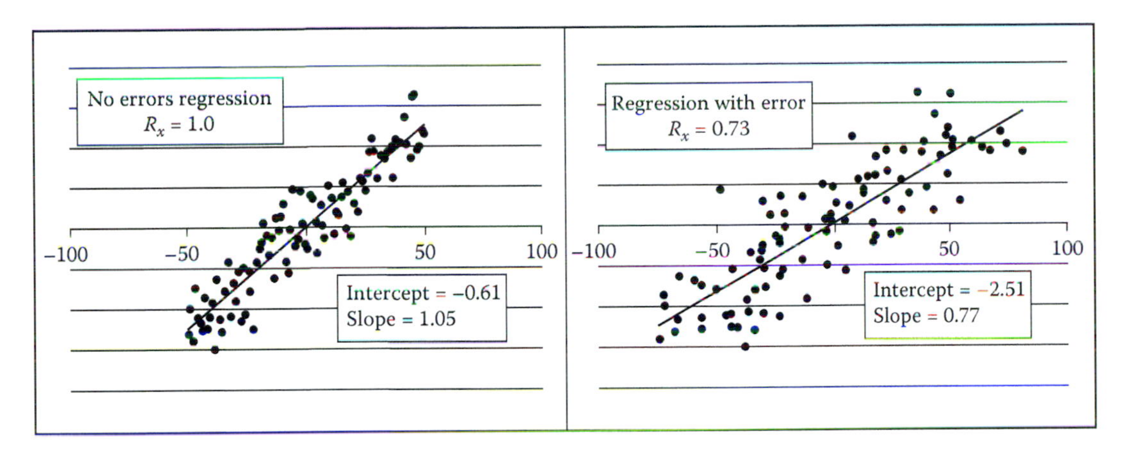

Figure 10.6. Regression of y on x with and without variable error. On the left is the population regression with no error in the x variable. On the right, variable error was added to the x-values with a reliability ratio of 0.73. Note its attenuated slope, which is very near the theoretical value of 0.77

counterpart. Let R_x denote the reliability ratio for x. Then

$$b = R_x B,$$

where $B = \sum_r (y_r - \bar{y})(\mu_{r|x} - \bar{\mu}_x) / \sum_r (\mu_{r|x} - \bar{\mu}_x)^2$ is the population slope coefficient, with $x_r = \mu_{r|x} + \varepsilon_{r|x}$, where $\varepsilon_{r|x}$ is the variable error with mean 0 and variance $\sigma^2_{\varepsilon|x}$. It can also be shown that $\text{Bias}(b_0) \approx B(1 - R_x)\bar{\mu}_x$.

As an illustration of these effects, consider the regressions displayed in Figure 10.6, which are based upon contrived data. The regression on the left is the population (true) regression with a slope of 1.05 and an intercept of −0.61. The regression on the left uses the same y- and x-values. The only difference is that normal error was added to the x-values, resulting in a reliability ratio of 0.73. As the theory predicted, the slope was attenuated toward 0 in direct proportion to the reliability, R_x. As random error is added to the x-values, reliability is reduced and the fitted slope will approach 0.

When the dependent variable, y, only is subject to variable error, the regression deteriorates, but the expected values of the slope and intercept coefficients are still equal to true to their population values. To see this, suppose $y_r = \mu_{y|r} + \varepsilon_{y|r}$, where $\mu_{r|y}$ denotes the error-free value of y_r and $\varepsilon_{r|y}$ is the associated variable error with variance $\sigma^2_{\varepsilon|y}$. The regression of y on x can now be rewritten as

$$\mu_{y|r} = b_0 + bx_r + e_r - \varepsilon_{r|y}, \tag{10.3}$$

where e_r is the usual regression residual error with mean 0 and variance σ_e^2, which is assumed to be uncorrelated with $\varepsilon_{r|y}$. Letting $e' = e_r - \varepsilon_{r|y}$, it follows that the regression in equation (10.3) is equivalent to the previously considered regression of y on x where y is not subject to error, but now the residual variance is increased by the additive term, that is, $\sigma_e'^2 = \sigma_{\varepsilon|y}^2 + \sigma_e^2$.

Chai [67] considers the case of systematic errors in the regression variables that may induce correlations both within and between variables in the regression. He shows that, in the presence of systematic errors in the independent variable, the bias in the slope coefficient may either attenuate the slope or increase its magnitude in ways that cannot be predicted without extensive knowledge of the error properties. Thus, like the results from correlation analysis, systematic errors greatly increase the complexity of the bias effects and their effects on inference can be quite severe.

One approach for dealing with systematic error at the source level in regression analysis is to model it using, for example, random effects [169]. In brief, a random effects model specifies $y_{ijk} = \beta_{0i}^* + \beta x_{ijk} + \varepsilon_{ijk}$, where $\varepsilon_{ijk}' = b_i + \varepsilon_{ijk}$ and $\mathrm{Var}(\varepsilon_{ijk}') = \sigma_b^2 + \sigma_{\varepsilon|j}^2$. The next section considers other mitigation strategies that attempt to eliminate the error rather than model it.

10.5 Some methods for mitigating, detecting, and compensating for errors

► Data errors further complicate analysis and exacerbate the analytical problems. There are essentially three solutions: prevention, remediation, and the choice of analysis methodology.

For survey data and other *designed* data collections, error mitigation begins at the data generation stage by incorporating design strategies that generate high-quality data that are at least adequate for the purposes of the data users. For example, missing data can be mitigated by repeated follow-up of nonrespondents, questionnaires can be perfected through pretesting and experimentation, interviewers can be trained in the effective methods for obtaining highly accurate responses, and computer-assisted interviewing instruments can be programmed to correct errors in the data as they are generated. For big data, the data generation process is often outside the purview of the data collectors, as noted in Section 10.1, and there is limited opportunity to address deficiencies in the data generation process. This is because big data is often a by-product of systems designed with little or no thought given to the potential secondary uses of the data. Instead, error mitigation must necessarily begin at the data processing stage, which is the focus of this chapter, particularly with regard to data editing and cleaning.

In survey collections, data processing may involve data entry, coding textual responses, creating new variables, editing the data, imputing missing cells, weighting the observations, and preparing the file, including the application of disclosure-limiting processes. For big data, many of these same operations could be needed to some extent, depending upon requirements of the data, its uses, and the prevailing statutory requirements. Of all the operations comprising data processing, the area that has the greatest potential to both alter data quality and consume vast resources is data editing. For example, Biemer and Lyberg [35] note that national statistical offices may spend as much as 40% or more of their production budgets on data editing for some surveys. Recent computer technologies have automated many formerly manual processes, which has greatly reduced these expenditures. Nevertheless, editing remains a key component of the quality improvement process for big data. Therefore, the remainder of this chapter will discuss the issue of editing big data.

► For example, Title 13 in the US code is explicit in terms of data statutory protections and—as you will see in the next chapter—can have substantial impact on the quality of subsequent inference.

Biemer and Lyberg [35] define data editing as a set of methodologies for identifying and correcting (or transforming) anomalies in the data. It often involves verifying that various relationships among related variables of the data set are plausible and, if they are not, attempting to make them so. Editing is typically a rule-based approach where rules can apply to a particular variable, a combination of variables, or an aggregate value that is the sum over all the rows or a subset of the rows in a data set.

In small data sets, data editing usually starts as a *bottom-up* process that applies various rules to the cell values—often referred to as *micro-editing*. The rules specify that the values for some variable or combinations of variables (e.g., a ratio or difference between two values) should be within some specified range. Editing may reveal impossible values (so-called *fatal* edits) as well as values that are simply highly suspect (leading to *query* edits). For example, a pregnant male would generate a fatal edit, while a property that sold for $500,000 in a low-income neighborhood might generate a query edit. Once the anomalous values are identified, the process attempts to deduce more accurate values based upon other variables in the data set or, perhaps, from an external data set.

The identification of anomalies can be handled quite efficiently and automatically using editing software. Recently, data mining and machine learning techniques have been applied to data editing with excellent results (see Chandola et al. [69] for a review). Tree-based methods such as classification and regression trees and random forests are particularly useful for creating editing rules for anomaly identification and resolution [302]. However, some human review may be necessary to resolve the most complex situations.

For big data, the identification of data anomalies could result in possibly billions of edit failures. Even if only a tiny proportion of these required some form of manual review for resolution, the task could still require the inspection of tens or hundreds of thousands of query edits, which would be infeasible for most applications. Thus, micro-editing must necessarily be a completely automated process unless it can be confined to a relatively small subset of the data. As an example, a representative (random) subset of the data set could be edited using manual editing for purposes of evaluating the error levels for the larger data set, or possibly to be used as a training data set, benchmark, or reference distribution for further processing, including recursive learning.

To complement fully automated micro-editing, big data editing usually involves *top-down* or *macro-editing* approaches. For such approaches, analysts and systems inspect aggregated data for conformance to some benchmark values or data distributions that are known from either training data or prior experience. When unexpected or suspicious aggregates are identified, the analyst can "drill down" into the data to discover and, if possible, remove the discrepancy by either altering the value at the source (usually a micro-data element) or delete the edit-failed value.

Note that an aggregate that is not flagged as suspicious in macro-editing passes the edit and is deemed correct. However, because serious offsetting errors can be masked by aggregation, there is no guarantee that the elements comprising the aggregate are indeed accurate. Recalling the discussion in Section 10.4.2.1 regarding random noise, we know that the data may be quite noisy while the aggregates may appear to be essentially error-free. However, macro-editing can be an effective control for systematic errors and egregious and some random errors that create outliers and other data anomalies.

Given the volume, velocity, and variety of big data, even macro-editing can be challenging when thousands of variables and billions of records are involved. In such cases, the *selective editing* strategies developed in the survey research literature can be helpful [90, 136]. Using selective macro-editing, query edits are selected based upon the importance of the variable for the analysis under study, the severity of the error, and the cost or level of effort involved in investigating the suspicious aggregate. Thus, only extreme errors, or less extreme errors on critical variables, would be investigated.

There are a variety of methods that may be effective in macro-editing. Some of these are based upon data mining [267], machine learning [78], cluster analysis [98, 162], and various data visualization tools such as treemaps [192, 342, 372] and tableplots [312, 371, 373]. The tableplot seems particularly well suited for the three Vs of big data and will be discussed in some detail.

Like other visualization techniques examined in Chapter 9, the tableplot has the ability to summarize a large multivariate data set in a single plot [246]. In editing big data, it can be used to detect outliers and unusual data patterns. Software for implementing this technique has been written in R and is available without cost from the Comprehensive R Archive Network (https://cran.r-project.org/). Figure 10.7 shows an example. The key idea is that micro-aggregates of two related variables should have similar data patterns. Inconsistent data patterns may signal errors in one of the aggregates that can be investigated and corrected in the editing process to improve data quality. The tableplot uses bar charts created for the micro-aggregates to identify these inconsistent data patterns.

Each column in the tableplot represents some variable in the data table, and each row is a "bin" containing a subset of the data. A statistic such as the mean or total is computed for the values in a bin and is displayed as a bar (for continuous variables) or as a stacked bar for categorical variables.

The sequence of steps typically involved in producing a tableplot is as follows:

1. Sort the records in the data set by the key variable.

2. Divide the sorted data set into B bins containing the same number of rows.

3. For continuous variables, compute the statistic to be compared across variables for each row bin, say T_b, for $b = 1, \dots, B$, for each continuous variable, V, ignoring missing values. The level of missingness for V may be represented by the color or brightness of the bar. For categorical variables with K categories, compute the proportion in the kth category, denoted by P_{bk}. Missing values are assigned to a new $(K + 1)$th category ("missing").

4. For continuous variables, plot the B values T_b as a bar chart. For categorical variables, plot the B proportions P_{bk} as a stacked bar chart.

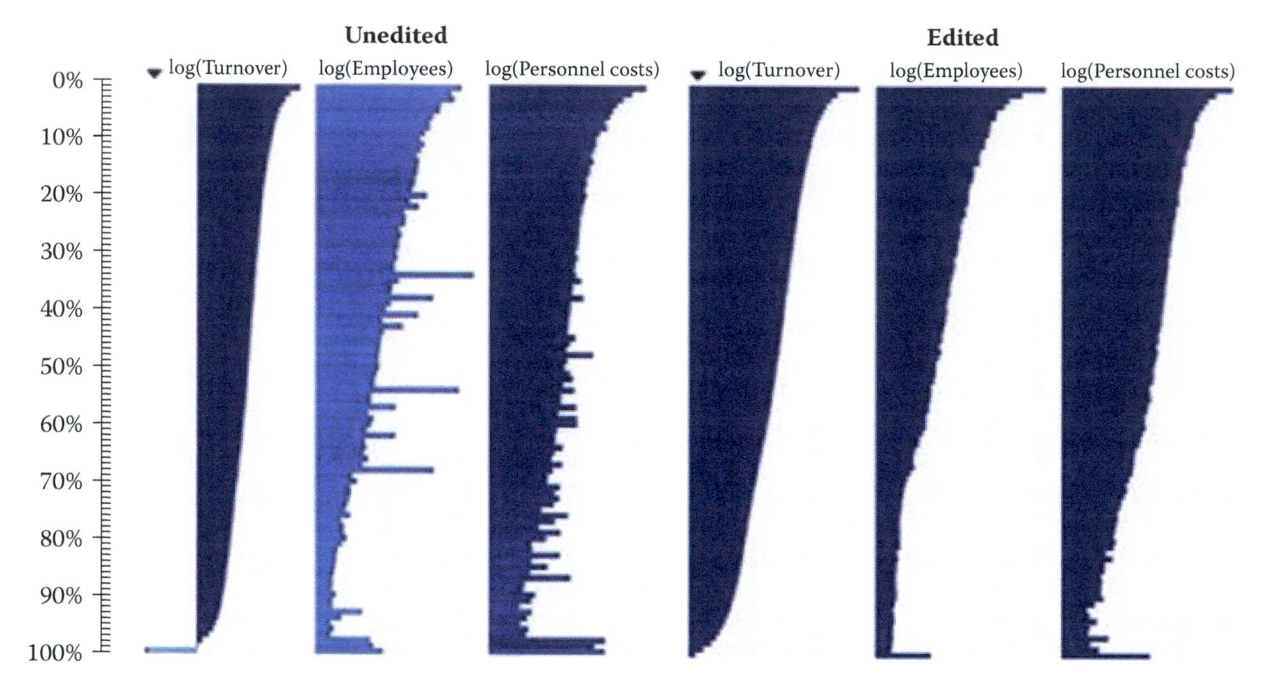

Figure 10.7. Comparison of tableplots for the Dutch Structural Business Statistics Survey for five variables before and after editing. Row bins with high missing and unknown numeric values are represented by lighter colored bars

Typically, T_b is the mean, but other statistics such as the median or range could be plotted if they aid in the outlier identification process. For highly skewed distributions, Tennekes and de Jonge [372] suggest transforming T_b by the log function to better capture the range of values in the data set. In that case, negative values can be plotted as $\log(-T_b)$ to the left of the origin and zero values can be plotted on the origin line. For categorical variables, each bar in the stack should be displayed using contrasting colors so that the divisions between categories are apparent.

Tableplots appear to be well suited for studying the distributions of variable values, the correlation between variables, and the occurrence and selectivity of missing values. Because they can help visualize massive, multivariate data sets, they seem particularly well suited for big data. Currently, the R implementation of tableplot is limited to 2 billion records.

The tableplot in Figure 10.7 is taken from Tennekes and de Jonge [372] for the annual Dutch Structural Business Statistics survey, a survey of approximately 58,000 business units annually. Topics covered in the questionnaire include turnover, number of employed persons, total purchases, and financial results. Figure 10.7 was created by sorting on the first column, viz., log(turnover), and dividing the 57,621 observed units into 100 bins, so that each row bin contains approximately 576 records. To aid the comparisons between unedited and edited data, the two tableplots are displayed side by side, with the unedited graph on the left and the edited graph on the right. All variables were transformed by the log function.

The unedited tableplot reveals that all four of the variables in the comparison with log(turnover) show some distortion by large values for some row bins. In particular, log(employees) has some fairly large nonconforming bins with considerable discrepancies. In addition, that variable suffers from a large number of missing values, as indicated by the brightness of the bar color. All in all, there are obvious data quality issues in the unprocessed data set for all four of these variables that should be dealt with in the subsequent processing steps.

The edited tableplot reveals the effect of the data checking and editing strategy used in the editing process. Notice the much darker color for the number of employees for the graph on the left compared to same graph on the right. In addition, the lack of data in the lowest part of the turnover column has been somewhat improved. The distributions for the graph on the right appear smoother and are less jagged.

10.6 Summary

This review of big data errors should dispel the misconception that the volume of the data can somehow compensate for other data deficiencies. To think otherwise is big data "hubris," according to Lazer et al. [230]. In fact, the errors in big data are at least as severe and impactful for data analysis as are the errors in designed data collections. For the latter, there is a vast literature that developed during the last century on mitigating and adjusting for the error effects on data analysis. But for the former, the corresponding literature is still in its infancy. While the survey literature can form a foundation for the treatment of errors in big data, its utility is also quite limited because, unlike for smaller data sets, it is often quite difficult to eliminate all but the most egregious errors from big data due to its size. Indeed, volume, velocity, and variety may be regarded as the three curses of big data veracity.

This chapter considers only a few types of data analysis that are common in big data analytics: classification, correlation, and regression analysis. We show that even when there is veracity, big data analytics can result in poor inference as a result of the three Vs. These issues include noise accumulation, spurious correlations, and incidental endogeneity. However, relaxing the assumption of veracity compounds these inferential issues and adds many new ones. Analytics can suffer in unpredictable ways.

One solution we propose is to clean up the data before analysis. Another option that was not discussed is the possibility of using analytical techniques that attempt to model errors and compensate for them in the analysis. Such techniques include the use of latent class analysis for classification error [34], multilevel modeling of systematic errors from multiple sources [169], and Bayesian statistics for partitioning massive data sets across multiple machines and then combining the results [176, 335].

The BDTE perspective can be helpful when we consider using a big data source. In the end, it is the responsibility of big data analysts to be aware of the many limitations of the data and to take the necessary steps to limit the effects of big data error on analytical results.

Finally, we note that the BDTE paradigm focuses on the accuracy dimension of Total Quality (writ large). Accuracy may also impinge on other quality dimensions such as timeliness, comparability, coherence, and relevance that we have not considered in this chapter. However, those other dimensions are equally important to some consumers of big data analytics. For example, timeliness often competes with accuracy because achieving acceptable levels of the latter often requires greater expenditures of resources and time. In fact, some consumers prefer analytic results that are less accurate for the sake of timeliness. Biemer and Lyberg [35] discuss these and other issues in some detail.

10.7 Resources

The American Association of Public Opinion Research has a number of resources on its website [1]. See, in particular, its report on big data [188].

The *Journal of Official Statistics* [194] is a standard resource with many relevant articles. There is also an annual international conference on the total survey error framework, supported by major survey organizations [378].

Chapter 11

Privacy and Confidentiality

Stefan Bender, Ron Jarmin, Frauke Kreuter, and Julia Lane

This chapter addresses the issue that sits at the core of any study of human beings—privacy and confidentiality. In a new field, like the one covered in this book, it is critical that many researchers have access to the data so that work can be replicated and built upon—that there be a scientific basis to data science. Yet the rules that social scientists have traditionally used for survey data, namely anonymity and informed consent, no longer apply when data are collected in the wild. This concluding chapter identifies the issues that must be addressed for responsible and ethical research to take place.

11.1 Introduction

Most big data applications in the social sciences involve data on units such as individuals, households, and different types of business, educational, and government organizations. Indeed, the example running throughout this book involves data on individuals (such as faculty and students) and organizations (such as universities and firms). In circumstances such as these, researchers must ensure that such data are used responsibly and ethically—that the subjects under study suffer no harm from their data being accessed, analyzed, and reported. A clear distinction must be made between analysis done for the public good and that done for private gain. In practical terms, this requires that the private interests of individual privacy and data confidentiality be balanced against the social benefits of research access and use.

Privacy "encompasses not only the famous 'right to be left alone,' or keeping one's personal matters and relationships secret, but also the ability to share information selectively but not publicly" [287].

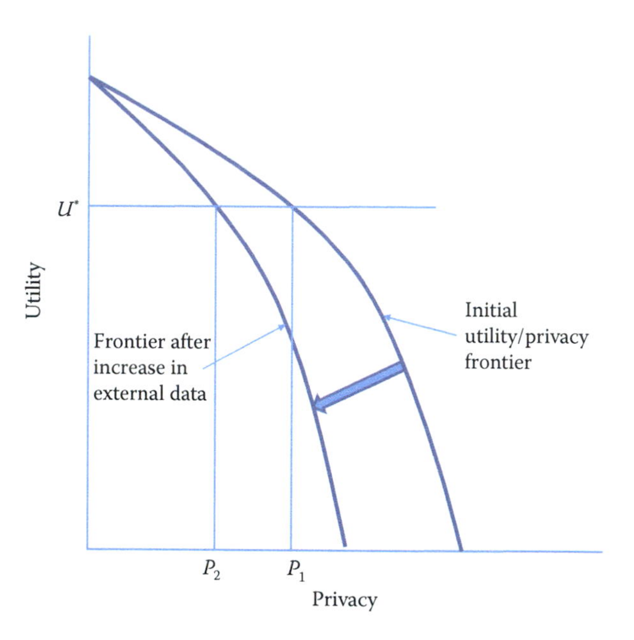

Figure 11.1. The privacy–utility tradeoff

Confidentiality is "preserving authorized restrictions on information access and disclosure, including means for protecting personal privacy and proprietary information" [254]. Doing so is not easy—the challenge to the research community is how to balance the *risk* of providing access with the associated utility [100]. To give a simple example, if means and percentages are presented for a *large* number of people, it will be difficult to infer an individual's value from such output, even if one knew that a certain individual or unit contributed to the formation of that mean or percentage. However, if those means and percentages are presented for subgroups or in multivariate tables with small cell sizes, the risk for disclosure increases [96]. As a result, the quality of data analysis is typically degraded with the production of public use data [100].

In general, the greater the access to data and their original values, the greater the risk of reidentification for individual units. We depict this tradeoff graphically in Figure 11.1. The concave curves in this hypothetical example depict the technological relationship between data utility and privacy for an organization such as a business firm or a statistical agency. At one extreme, all information is available to anybody about all units, and therefore high analytic utility is associated with the data that are not at all protected. At the other extreme, nobody has access to any data and no utility is achieved. Initially, assume the organization is on the outer frontier. Increased external data resources (those not used by the organization) increase the risk of reidentification. This is represented by an inward shift of the utility/privacy frontier in Figure 11.1. Before the increase in external data, the organization could achieve a level of data utility U^* and privacy P_1. The increase in externally available data now means that in order to maintain utility at U^*, privacy is reduced to P_2. This simple example represents the challenge to all organization that release statistical or analytical products obtained from underlying identifiable data in the era of big data. As more data become available externally, the more difficult it is to maintain privacy.

Before the big data era, national statistical agencies had the capacity and the mandate to make dissemination decisions: they assessed the risk, they understood the data user community and the associated utility from data releases. And they had the wherewithal to address the legal, technical, and statistical issues associated with protecting confidentiality [377].

But in a world of big data, many once-settled issues have new complications, and wholly new issues arise that need to be addressed, albeit under the same rubrics. The new types of data have much greater potential utility, often because it is possible to study small cells or the tails of a distribution in ways not possible with small data. In fact, in many social science applications, the tails of the distribution are often the most interesting and hardest-to-reach parts of the population being studied; consider health care costs for a small number of ill people [356], or economic activity such as rapid employment growth by a small number of firms [93].

Example: The importance of activity in the tails

Spending on health care services in the United States is highly concentrated among a small proportion of people with extremely high use. For the overall civilian population living in the community, the latest data indicate that more than 20% of all personal health care spending in 2009 ($275 billion) was on behalf of just 1% of the population [333].

It is important to understand where the risk of privacy breaches comes from. Let us assume for a moment that we conducted a traditional small-scale survey with 1,000 respondents. The survey contains information on political attitudes, spending and saving in a given year, and income, as well as background variables on income and education. If name and address are saved together with this data, and someone gets access to the data, obviously it is easy to identify individuals and gain access to information that is otherwise not public. If the personal identifiable information (name and address) are removed from this data file, the risk is much reduced. If someone has access to the survey data and sees all the individual values, it might be difficult to assess with certainty who among the 320 million inhabitants in the USA is associated with an individual data record. However, the risk is higher if one knows some of this information (say, income) for a person, and knows that this person is in the survey. With these two pieces of information, it is likely possible to uniquely identify the person in the survey data.

Big data increases the risk precisely for this reason. Much data is available for reidentification purposes [289]. Most obviously, the risk of reidentification is much greater because the new types of data have much richer detail and a much larger public community has access to ways to reidentify individuals. There are many famous examples of reidentification occurring even when obvious personal information, such as name and social security number, has been removed and the data provider thought that the data were consequently deidentified. In the 1990s, Massachusetts Group Insurance released "deidentified" data on the hospital visits of state employees; researcher Latanya Sweeney quickly reidentified the hospital records of the then Governor William Weld using nothing more than state voter records about residence and date of birth [366]. In 2006, the release of supposedly deidentified web search data by AOL resulted in two *New York Times* reports being able to reidentify a customer simply from her browsing habits [26]. And in 2012, statisticians at the department store, Target, used a young teenager's shopping patterns to determine that she was pregnant before her father did [166].

But there are also less obvious problems. What is the legal framework when the ownership of data is unclear? In the past, when data were more likely to be collected and used within the same entity—for example, within an agency that collects administrative data or within a university that collects data for research purposes—organization-specific procedures were (usually) in place and sufficient to regulate the usage of these data. In a world of big data, legal ownership is less clear.

Who has the legal authority to make decisions about permission, access, and dissemination and under what circumstances? The answer is often not clear. The challenge in the case of big data is that data sources are often combined, collected for one purpose, and used for another. Data providers often have a poor understanding of whether or how their data will be used.

Example: Knowledge is power

In a discussion of legal approaches to privacy in the context of big data, Strandburg [362] says: "'Big data' has great potential to benefit society. At the same time, its availability creates significant potential for mistaken, misguided or malevolent uses of personal information. The conundrum for the law is to provide space for big data to fulfill its potential for societal benefit, while protecting citizens adequately from related individual and social harms. Current privacy law evolved to address different concerns and must be adapted to confront big data's challenges."

It is critical to address privacy and confidentiality issues if the full public value of big data is to be realized. This chapter highlights why the challenges need to be met (i.e., why access to data is crucial), review the pre-big data past, point out challenges with this approach in the context of big data, briefly describe the current state of play from a legal, technical, and statistical perspective, and point to open questions that need to be addressed in the future.

11.2 Why is access important?

This book gives detailed examples of the potential of big data to provide insights into a variety of social science questions—particularly the relationship between investments in R&D and innovation. But that potential is only realized if researchers have access to the data [225]: not only to perform primary analyses but also to validate the data generation process (in particular, data linkage), replicate analyses, and build a knowledge infrastructure around complex data sets.

Validating the data generating process Research designs requiring a combination of data sources and/or analysis of the tails of populations challenge the traditional paradigm of conducting statistical analysis on deidentified or aggregated data. In order to combine data sets, someone in the chain that transforms raw data into research outputs needs access to link keys contained in the data sets to be combined. High-quality link keys uniquely identify the subjects under study and typically are derived from items such as individual names, birth dates, social security numbers, and business names, addresses, and tax ID numbers. From a privacy and confidentiality perspective, link keys are among most sensitive information in many data sets of interest to social scientists. This is why many organizations replace link keys containing personal identifiable information (PII)* with privacy-protecting identifiers [332]. Regardless, at some point in the process those must be generated out of the original information, thus access to the latter is important.

★ PII is "any information about an individual maintained by an agency, including (1) any information that can be used to distinguish or trace an individual's identity, such as name, social security number, date and place of birth, mother's maiden name, or biometric records; and (2) any other information that is linked or linkable to an individual, such as medical, educational, financial, and employment information" [254].

Replication John Ioannidis has claimed that most published research findings are false [184]; for example, the unsuccessful replication of genome-wide association studies, at less than 1%, is staggering [29]. Inadequate understanding of coverage, incentive, and quality issues, together with the lack of a comparison group, can result in biased analysis—famously in the case of using administrative records on crime to make inference about the role of death penalty policy in crime reduction [95, 234]. Similarly, overreliance on, say, Twitter data, in targeting resources after hurricanes can lead to the misallocation of resources towards young, Internet-savvy people with cell phones and away from elderly or impoverished neighborhoods [340], just as bad survey methodology led the *Literary Digest* to incorrectly call the 1936 election [353]. The first step to replication is data access; such access can enable other researchers to ascertain whether the assumptions of a particular statistical model are met, what relevant information is included or excluded, and whether valid inferences can be drawn from the data [215].

Building knowledge infrastructure Creating a community of practice around a data infrastructure can result in tremendous new insights, as the Sloan Digital Sky Survey and the Polymath project have shown [281]. In the social science arena, the Census Bureau has developed a productive ecosystem that is predicated on access to approved external experts to build, conduct research using, and improve key data assets such as the Longitudinal Business Database [190] and Longitudinal Employer Household Dynamics [4], which have yielded a host of new data products and critical policy-relevant insights on business dynamics [152] and labor market volatility [55], respectively. Without providing robust, but secure, access to confidential data, researchers at the Census Bureau would have been unable to undertake the innovations that made these new products and insights possible.

11.3 Providing access

The approaches to providing access have evolved over time. Statistical agencies often employ a range of approaches depending on the needs of heterogeneous data users [96, 126]. Dissemination of data to the public usually occurs in three steps: an evaluation of disclosure risks, followed by the application of an anonymization technique, and finally an evaluation of disclosure risks and analytical quality of the candidate data release(s). The two main approaches have been *statistical disclosure* control techniques to produce anonymized public use data sets, and controlled access through a *research data center*.

Statistical disclosure control techniques Statistical agencies have made data available in a number of ways: through tabular data, public use files, licensing agreements and, more recently, through synthetic data [317]. Hundepool et al. [174] define statistical disclosure control as follows:

> concepts and methods that ensure the confidentiality of micro and aggregated data that are to be published. It is methodology used to design statistical outputs in a way that someone with access to that output cannot relate a known individual (or other responding unit) to an element in the output.

Traditionally, confidentiality protection has been accomplished by releasing only *aggregated tabular data*. This practice works well in settings where the primary purpose is enumeration, such as census taking. However, tabular data are poorly suited to describing the underlying distributions and covariance across variables that are often the focus of applied social science research [100].

To provide researchers access to data that permitted analysis of the underlying variance–covariance structure of the data, some agencies have constructed public use micro-data samples. To product confidentiality in such *public use files*, a number of statistical disclosure control procedures are typically applied. These include stripping all identifying (e.g., PII) fields from the data, topcoding highly skewed variables (e.g., income), and swapping records [96, 413]. However, the mosaic effect—where disparate pieces of information can be combined to reidentify individuals—dramatically increases the risk of releasing public use files [88]. In addition, there is more and more evidence that the statistical disclosure procedure applied to produce them decreases their utility across many applications [58].

Some agencies provide access to confidential micro-data through *licensing* arrangements. A contract specifies the conditions of use and what safeguards must be in place. In some cases, the agency has the authority to conduct random inspections. However, this approach has led to a number of operational challenges, including version control, identifying and managing risky researcher behavior, and management costs [96].

More recently, *synthetic data* have been created whereby key features of the original data are preserved but the original data are replaced by results from estimations (synthetic data) so that no individual or business entity can be found in the released data [97]. Two examples of synthetic data sets are the SIPP Synthetic-Beta [5] of linked Survey of Income and Program Participation (SIPP) and Social Security Administration earnings data, and the Synthetic Longitudinal Business Database (SynLBD) [207]. Jarmin et al. [189] discuss how synthetic data sets lack utility in many research settings but are useful for generating flexible data sets underlying data tools and apps such as the Census Bureau's OnTheMap.

Research data centers The second approach is establishing research data centers. Here, qualified researchers gain access to micro-level data after they are sworn in to protect the confidentially of the data they access. Strong input and output controls are in place to ensure that published findings comply with the privacy and confidentiality regulations [161]. Some RDCs allow access through remote execution, where no direct access to the data is allowed, but it is not necessary to travel; others allow remote direct access.

11.4 The new challenges

While there are well-established policies and protocols surrounding access to and use of survey and administrative data, a major new challenge is the lack of clear guidelines governing the collection of data about human activity in a world in which all public, and some private, actions generate data that can be harvested [287, 289, 362]. The twin pillars on which so much of social science have rested—informed consent and anonymization—are virtually useless in a big data setting where multiple data sets can be and are linked together using individual identifiers by a variety of players beyond social scientists with formal training and whose work is overseen by institutional review boards. This rapid expansion in data and their use is very much driven by the increased utility of the linked information to businesses, policymakers, and ultimately the taxpayer. In addition, there are no obvious data stewards and custodians who can be entrusted with preserving the privacy and confidentiality with regard to both the source data collected from sensors, social media, and many other sources, and the related analyses [226].

It is clear that informed consent as historically construed is no longer feasible. As Nissenbaum [283] points out, notification is either comprehensive or comprehensible, but not both. While ideally human subjects are offered true freedom of choice based on a sound and sufficient understanding of what the choice entails, in reality the flow of data is so complex and the interest in the data usage so diverse that simplicity and clarity in the consent statement unavoidably result in losses of fidelity, as anyone who has accepted a Google Maps agreement is likely to understand [160]. In addition, informed consent requires a greater understanding of the breadth of type of privacy breaches, the nature of harm as diffused over time, and an improved valuation of privacy in the big data context. Consumers may value their own privacy in variously flawed ways. They may, for example, have incomplete information, or an overabundance of information rendering processing impossible, or use heuristics that establish and routinize deviations from rational decision-making [6].

It is also nearly impossible to anonymize data. Big data are often structured in such a way that essentially everyone in the file is unique, either because so many variables exist or because they are so frequent or geographically detailed, that they make it easy to reidentify individual patterns [266]. It is also no longer possible to rely on sampling or measurement error in external files as a buffer for data protection, since most data are not in the hands of statistical agencies.

There are no data stewards controlling access to individual data. Data are often so interconnected (think social media network data) that one person's action can disclose information about another person without that person even knowing that their data are being accessed. The group of students posting pictures about a beer party is an obvious example, but, in a research context, if the principal investigator grants access to the proposal, information could be divulged about colleagues and students. In other words, volunteered information of a minority of individuals can unlock the same information about many—a type of "tyranny of the minority" [28].

There are particular issues raised by the new potential to link information based on a variety of attributes that do not include PII. Barocas and Nissenbaum write as follows [27]:

> Rather than attempt to deanonymize medical records, for instance, an attacker (or commercial actor) might instead infer a rule that relates a string of more easily observable or accessible indicators to a specific medical condition, rendering large populations vulnerable to such inferences even in the absence of PII. Ironically, this is often the very thing about big data that generate the most excitement: the capacity to detect subtle correlations and draw actionable inferences. But it is this same feature that renders the traditional protections afforded by anonymity (again, more accurately, pseudonymity) much less effective.

In light of these challenges, Barocas and Nissenbaum continue

> the value of anonymity inheres not in namelessness, and not even in the extension of the previous value of namelessness to all uniquely identifying information, but instead to something we called "reachability," the possibility of knocking on your door, hauling you out of bed, calling your phone number, threatening you with sanction, holding you accountable—with or without access to identifying information.

It is clear that the concepts used in the larger discussion of privacy and big data require updating. How we understand and assess harms from privacy violations needs updating. And we must rethink established approaches to managing privacy in the big data context. The next section discusses the framework for doing so.

11.5 Legal and ethical framework

The Fourth Amendment to the US Constitution, which constrains the government's power to "search" the citizenry's "persons, houses, papers, and effects" is usually cited as the legal framework for privacy and confidentiality issues. In the US a "sectoral" approach to privacy regulation, for example, the Family Education Rights and

Privacy Act and the Health Insurance Portability and Accountability Act, is also used in situations where different economic areas have separate privacy laws [290]. In addition, current legal restrictions and guidance on data collection in the industrial setting include the Fair Information Practice Principles dating from 1973 and underlying the Fair Credit Reporting Act from 1970 and the Privacy Act from 1974 [362]. Federal agencies often have statutory oversight, such as Title 13 of the US Code for the Census Bureau, the Confidential Information Protection and Statistical Efficiency Act for federal statistical agencies, and Title 26 of the US Code for the Internal Revenue Service.

Yet the generation of big data often takes place in the open, or through commercial transactions with a business, and hence is not covered by these frameworks. There are major questions as to what is reasonably private and what constitutes unwarranted intrusion [362]. There is a lack of clarity on who owns the new types of data—whether it is the person who is the subject of the information, the person or organization who collects these data (the data custodian), the person who compiles, analyzes, or otherwise adds value to the information, the person who purchases an interest in the data, or society at large. The lack of clarity is exacerbated because some laws treat data as property and some treat it as information [65].

The ethics of the use of big data are also not clear, because analysis may result in being discriminately against unfairly, being limited in one's life choices, being trapped inside stereotypes, being unable to delineate personal boundaries, or being wrongly judged, embarrassed, or harassed. There is an entire research agenda to be pursued that examines the ways that big data may threaten interests and values, distinguishes the origins and nature of threats to individual and social integrity, and identifies different solutions [48]. The approach should be to describe what norms and expectations are likely to be violated if a person agrees to provide data, rather than to describe what will be done during the research.

What is clear is that most data are housed no longer in statistical agencies, with well-defined rules of conduct, but in businesses or administrative agencies. In addition, since digital data can be alive forever, ownership could be claimed by yet-to-be-born relatives whose personal privacy could be threatened by release of information about blood relations.

Traditional regulatory tools for managing privacy, notice, and consent have failed to provide a viable market mechanism allowing a form of self-regulation governing industry data collection. Going forward, a more nuanced assessment of tradeoffs in the big data context, moving away from individualized assessments of the costs of privacy violations, is needed [362]. Ohm advocates for a new conceptualization of legal policy regarding privacy in the big data context that uses five guiding principles for reform: first, that rules take into account the varying levels of inherent risk to individuals across different data sets; second, that traditional definitions of PII need to be rethought; third, that regulation has a role in creating and policing walls between data sets; fourth, that those analyzing big data must be reminded, with a frequency in proportion to the sensitivity of the data, that they are dealing with people; and finally, that the ethics of big data research must be an open topic for continual reassessment.

11.6 Summary

The excitement about how big data can change the social science research paradigm should be tempered by a recognition that existing ways of protecting confidentiality are no longer viable [203]. There is a great deal of research that can be used to inform the development of such a structure, but it has been siloed into disconnected research areas, such as statistics, cybersecurity, and cryptography, as well as a variety of different practical applications, including the successful development of remote access secure data enclaves. We must piece together the knowledge from these various fields to develop ways in which vast new sets of data on human beings can be collected, integrated, and analyzed while protecting them [227].

It is possible that the confidentiality risks of disseminating data may be so high that traditional access models will no longer hold; that the data access model of the future will be to take the analysis to the data rather than the data to the analyst or the analyst to the data. One potential approach is to create an integrated system including (a) unrestricted access to highly redacted data, most likely some version of synthetic data, followed by (b) means for approved researchers to access the confidential data via remote access solutions, combined with (c) verification servers that allows users to assess the quality of their inferences with the redacted data so as to be more efficient with their use (if necessary) of the remote data access. Such verification servers might be a web-accessible system based on a confidential database with an associated public microdata release, which helps to analyze the confidential database [203]. Such approaches are starting to be developed, both in the USA and in Europe [106, 193].

There is also some evidence that people do not require complete protection, and will gladly share even private information provided that certain social norms are met [300, 403]. There is a research agenda around identifying those norms as well; characterizing the interests and wishes of actors (the information senders and recipients or providers and users); the nature of the attributes (especially types of information about the providers, including how these might be transformed or linked); and identifying transmission principles (the constraints underlying the information flows).

However, it is likely that it is no longer possible for a lone social scientist to address these challenges. One-off access agreements to individuals are conducive to neither the production of high-quality science nor the high-quality protection of data [328]. The curation, protection, and dissemination of data on human subjects cannot be an artisan activity but should be seen as a major research infrastructure investment, like investments in the physical and life sciences [3, 38, 173]. In practice, this means that linkages become professionalized and replicable, research is fostered within research data centers that protect privacy in a systematic manner, knowledge is shared about the process of privacy protections disseminated in a professional fashion, and there is ongoing documentation about the value of evidence-based research. It is thus that the risk–utility tradeoff depicted in Figure 11.1 can be shifted in a manner that serves the public good.

11.7 Resources

The American Statistical Association's Privacy and Confidentiality website provides a useful source of information [12].

An overview of federal activities is provided by the Confidentiality and Data Access Committee of the Federal Committee on Statistics and Methodology [83].

The World Bank and International Household Survey Network provide a good overview of data dissemination "best practices" [183].

There is a *Journal of Privacy and Confidentiality* based at Carnegie Mellon University [195], and also a journal called *Transactions in Data Privacy* [370].

The United Nations Economic Commission on Europe hosts workshops and conferences and produces occasional reports [382].

Chapter 12

Workbooks

Jonathan Scott Morgan, Christina Jones, and Ahmad Emad

This final chapter provides an overview of the Python workbooks that accompany each chapter. These workbooks combine text explanation and code you can run, implemented in Jupyter notebooks, to explain techniques and approaches selected from each chapter and to provide thorough implementation details, enabling students and interested practitioners to quickly get up to speed on and start using the technologies covered in the book. We hope you have a lot of fun with them.

▸ See jupyter.org.

12.1 Introduction

We provide accompanying Juptyer IPython workbooks for most chapters in this book. These workbooks explain techniques and approaches selected from each chapter and provide thorough implementation details so that students and interested practitioners can quickly start using the technologies covered within.

The workbooks and related files are stored in the Big-Data-Workbooks GitHub repository, and so are freely available to be downloaded by anyone at any time and run on any appropriately configured computer. These workbooks are a live set of documents that could potentially change over time, so see the repository for the most recent set of information.

▸ github.com/BigDataSocialScience

These workbooks provide a thorough overview of the work needed to implement the selected technologies. They combine explanation, basic exercises, and substantial additional Python code to provide a conceptual understanding of each technology, give insight into how key parts of the process are implemented through exercises, and then lay out an end-to-end pattern for implementing each in your own work. The workbooks are implemented using IPython notebooks, interactive documents that mix formatted text and Python code samples that can be edited and run in real time in a Jupyter notebook server, allowing you to run and explore the code for each technology as you read about it.

12.2 Environment

The Big-Data-Workbooks GitHub repository provides two different types of workbooks, each needing a different Python setup to run. The first type of workbooks is intended to be downloaded and run locally by individual users. The second type is designed to be hosted, assigned, worked on, and graded on a single server, using `jupyterhub` (https://github.com/jupyter/jupyterhub) to host and run the notebooks and `nbgrader` (https://github.com/jupyter/nbgrader) to assign, collect, and grade.

The text, images, and Python code in the workbooks are the same between the two versions, as are the files and programs needed to complete each.

The differences in the workbooks themselves relate to the code cells within each notebook where users implement and test exercises. In the workbooks intended to be used locally, exercises are implemented in simple interactive code cells. In the `nbgrader` versions, these cells have additional metadata and contain the solutions for the exercises, making them a convenient answer key even if you are working on them locally.

12.2.1 Running workbooks locally

To run workbooks locally, you will need to install Python on your system, then install `ipython`, which includes a local Jupyter server you can use to run the workbooks. You will also need to install additional Python packages needed by the workbooks, and a few additional programs.

The easiest way to get this all working is to install the free Anaconda Python distribution provided by Continuum Analytics (https://www.continuum.io/downloads). Anaconda includes a Jupyter server and precompiled versions of many packages used in the workbooks. It includes multiple tools for installing and updating both Python and installed packages. It is separate from any OS-level version of Python, and is easy to completely uninstall.

Anaconda also works on Windows as it does on Mac and Linux. Windows is a much different operating system from Apple's OS X and Unix/Linux, and Python has historically been much trickier to install, configure, and use on Windows. Packages are harder to compile and install, the environment can be more difficult to set up, etc. Anaconda makes Python easier to work with on any OS, and on Windows, in a single run of the Anaconda installer, it integrates Python and common Python utilities like `pip` into Windows well enough that it approximates the ease and experience of using Python within OS X or Unix/Linux (no small feat).

You can also create your Python environment manually, installing Python, package managers, and Python packages separately. Packages like `numpy` and `pandas` can be difficult to get working, however, particularly on Windows, and Anaconda simplifies this setup considerably regardless of your OS.

12.2.2 Central workbook server

Setting up a server to host workbooks managed by `nbgrader` is more involved. Some of the workbooks consume multiple gigabytes of memory per user and substantial processing power. A hosted implementation where all users work on a single server requires substantial hardware, relatively complex configuration, and ongoing server maintenance. Detailed instructions are included in the Big-Data-Workbooks GitHub repository. It is not rocket science, but it is complicated, and you will likely need an IT professional to help you set up, maintain, and troubleshoot. Since all student work will be centralized in this one location, you will also want a robust, multi-destination backup plan.

For more information on installing and running the workbooks that accompany this book, see the Big-Data-Workbooks GitHub repository.

12.3 Workbook details

Most chapters have an associated workbook, each in its own directory in the Big-Data-Workbooks GitHub repository. Below is a list of the workbooks, along with a short summary of the topics that each covers.

12.3.1 Social Media and APIs

The Social Media and APIs workbook introduces you to the use of Internet-based web service APIs for retrieving data from online data stores. Examples include retrieving information on articles from Crossref (provider of Digital Object Identifiers used as unique IDs for publications) and using the PLOS Search and ALM APIs to retrieve information on how articles are shared and referenced in social media, focusing on Twitter. In this workbook, you will learn how to:

- Set up user API keys,
- Connect to Internet-based data stores using APIs,
- Collect DOIs and Article-Level Metrics data from web APIs,
- Conduct basic analysis of publication data.

12.3.2 Database basics

In the Database workbook you will learn the practical benefits that stem from using a database management system. You will implement basic SQL commands to query grants, patents, and vendor data, and thus learn how to interact with data stored in a relational database. You will also be introduced to using Python to execute and interact with the results of SQL queries, so you can write programs that interact with data stored in a database. In this workbook, you will learn how to:

- Connect to a database through Python,
- Query the database by using SQL in Python,
- Begin to understand to the SQL query language,
- Close database connections.

12.3.3 Data Linkage

In the Data Linkage workbook you will use Python to clean input data, including using regular expressions, then learn and implement the basic concepts behind the probabilistic record linkage: using different types of string comparators to compare multiple pieces of information between two records to produce a score that indicates how likely it is that the records are data about the same underlying entity. In this workbook, you will learn how to:

- Parse a name string into first, middle, and last names using Python's `split` method and regular expressions,

- Use and evaluate the results of common computational string comparison algorithms including Levenshtein distance, Levenshtein-Damerau distance, and Jaro-Winkler distance,
- Understand the Fellegi-Sunter probabilistic record linkage method, with step-by-step implementation guide.

12.3.4 Machine Learning

In the Machine Learning workbook you will train a machine learning model to predict missing information, working through the process of cleaning and prepping data for training and testing a model, then training and testing a model to impute values for a missing categorical variable, predicting the academic department of a given grant's primary investigator based on other traits of the grant. In this workbook, you will learn how to:

- Read, clean, filter, and store data with Python's `pandas` data analysis package,
- Recognize the types of data cleaning and refining needed to make data more compatible with machine learning models,
- Clean and refine data,
- Manage memory when working with large data sets,
- Employ strategies for dividing data to properly train and test a machine learning model,
- Use the `scikit-learn` Python package to train, fit, and evaluate machine learning models.

12.3.5 Text Analysis

In the Text Analysis workbook, you will derive a list of topics from text documents using MALLET, a Java-based tool that analyzes clusters of words across a set of documents to derive common topics within the documents, defined by sets of key words that are consistently used together. In this workbook, you will learn how to:

- Clean and prepare data for automated text analysis,
- Set up data for use in MALLET,
- Derive a set of topics from a collection of text documents,

- Create a model that detects these topics in documents, and use this model to categorize documents.

12.3.6 Networks

In the Networks workbook you will create network data where the nodes are researchers who have been awarded grants, and ties are created between each researcher on a given grant. You will use Python to read the grant data and translate them into network data, then use the `networkx` Python library to calculate node- and graph-level network statistics and `igraph` to create and refine network visualizations. You will also be introduced to graph databases, an alternative way of storing and querying network data. In this workbook, you will learn how to:

- Develop strategies for detecting potential network data in relational data sets,
- Use Python to derive network data from a relational database,
- Store and query network data using a graph database like `neo4j`,
- Load network data into `networkx`, then use it to calculate node- and graph-level network statistics,
- Use `networkx` to export graph data into commonly shared formats (`graphml`, edge lists, different tabular formats, etc.),
- Load network data into the `igraph` Python package and then create graph visualizations.

12.3.7 Visualization

The Visualization workbook introduces you to Tableau, a data analysis and visualization software package that is easy to learn and use. Tableau allows you to connect to and integrate multiple data sources into complex visualizations without writing code. It allows you to dynamically shift between views of data to build anything from single visualizations to an interactive dashboard that contains multiple views of your data. In this workbook, you will learn how to:

- Connect Tableau to a relational database,
- Interact with Tableau's interface,

- Select, combine, and filter the tables and columns included in visualizations,
- Create bar charts, timeline graphs, and heat maps,
- Group and aggregate data,
- Create a dashboard that combines multiple views of your data.

12.4 Resources

We noted in Section 1.8 the importance of Python, MySQL, and Git/GitHub for the social scientist who intends to work with large data. See that section for pointers to useful online resources, and also this book's website, at https://github.com/BigDataSocialScience, where we have collected many useful web links, including the following.

For more on getting started with Anaconda, see Continuum's Anaconda documentation [13], Anaconda FAQ [14], and Anaconda quick start guide [15].

For more information on IPython and the Jupyter notebook server, see the IPython site [186], IPython documentation [185], Jupyter Project site [197], and Jupyter Project documentation [196].

For more information on using `jupyterhub` and `nbgrader` to host, distribute, and grade workbooks using a central server, see the `jupyterhub` GitHub repository [198], `jupyterhub` documentation [199], `nbgrader` GitHub repository [201], and `nbgrader` documentation [200].

Bibliography

[1] AAPOR. American association for public opinion research website. http://www.aapor.org. Accessed February 1, 2016.

[2] Daniel Abadi, Rakesh Agrawal, Anastasia Ailamaki, Magdalena Balazinska, Philip A Bernstein, Michael J Carey, Surajit Chaudhuri, Jeffrey Dean, An-Hai Doan, Michael J Franklin, et al. The Beckman Report on Database Research. *ACM SIGMOD Record*, 43(3):61–70, 2014. http://beckman.cs.wisc.edu/beckman-report2013.pdf.

[3] Kevork N. Abazajian, Jennifer K. Adelman-McCarthy, Marcel A. Agüeros, Sahar S. Allam, Carlos Allende Prieto, Deokkeun An, Kurt S. J. Anderson, Scott F. Anderson, James Annis, Neta A. Bahcall, et al. The seventh data release of the Sloan Digital Sky Survey. *Astrophysical Journal Supplement Series*, 182(2):543, 2009.

[4] John M. Abowd, John Haltiwanger, and Julia Lane. Integrated longitudinal employer-employee data for the United States. *American Economic Review*, 94(2):224–229, 2004.

[5] John M. Abowd, Martha Stinson, and Gary Benedetto. Final report to the Social Security Administration on the SIPP/SSA/IRS Public Use File Project. Technical report, Census Bureau, Longitudinal Employer-Household Dynamics Program, 2006.

[6] Alessandro Acquisti. The economics and behavioral economics of privacy. In Julia Lane, Victoria Stodden, Stefan Bender, and Helen Nissenbaum, editors, *Privacy, Big Data, and the Public Good: Frameworks for Engagement*, pages 98–112. Cambridge University Press, 2014.

[7] Christopher Ahlberg, Christopher Williamson, and Ben Shneiderman. Dynamic queries for information exploration: An implementation and evaluation. In *Proceedings of the SIGCHI Conference on Human Factors in Computing Systems*, pages 619–626. ACM, 1992.

[8] Suha Alawadhi, Armando Aldama-Nalda, Hafedh Chourabi, J. Ramon Gil-Garcia, Sofia Leung, Sehl Mellouli, Taewoo Nam, Theresa A. Pardo, Hans J. Scholl, and Shawn Walker. Building understanding of smart city initiatives. In *Electronic Government*, pages 40–53. Springer, 2012.

[9] Ed Albanese. Scaling social science with Hadoop. http://blog.cloudera.com/blog/2010/04/scaling-social-science-with-hadoop/. Accessed February 1, 2016.

[10] Paul D Allison. *Missing Data.* Sage Publications, 2001.

[11] Amazon. AWS public data sets. http://aws.amazon.com/datasets.

[12] American Statistical Association. ASA Privacy and Confidentiality Subcommittee. http://community.amstat.org/cpc/home.

[13] Continuum Analytics. Anaconda. http://docs.continuum.io/anaconda. Accessed February 1, 2016.

[14] Continuum Analytics. Anaconda FAQ. http://docs.continuum.io/anaconda/faq. Accessed February 1, 2016.

[15] Continuum Analytics. Anaconda quick start guide. https://www.continuum.io/sites/default/files/Anaconda-Quickstart.pdf. Accessed February 1, 2016.

[16] Francis J. Anscombe. Graphs in statistical analysis. *American Statistician*, 27(1):17–21, 1973.

[17] Dolan Antenucci, Michael Cafarella, Margaret Levenstein, Christopher Ré, and Matthew D. Shapiro. Using social media to measure labor market flows. Technical report, National Bureau of Economic Research, 2014.

[18] Apache Software Foundation. Apache ambari. http://ambari.apache.org. Accessed February 1, 2016.

[19] Apache Software Foundation. Apache Hadoop documentation site. https://hadoop.apache.org/docs/current/. Accessed February 1, 2016.

[20] Apache Software Foundation. Apache Spark documentation site. https://spark.apache.org/docs/current/. Accessed February 1, 2016.

[21] Michael Armbrust, Armando Fox, Rean Griffith, Anthony D. Joseph, Randy Katz, Andy Konwinski, Gunho Lee, David Patterson, Ariel Rabkin, Ion Stoica, et al. A view of cloud computing. *Communications of the ACM*, 53(4):50–58, 2010.

[22] Art Branch Inc. SQL Cheatsheet. http://www.sql-tutorial.net/SQL-Cheat-Sheet.pdf. Accessed December 1, 2015.

[23] Eric Baldeschwieler. Best practices for selecting Apache Hadoop hardware. *Hortonworks*, http://hortonworks.com/blog/best-practices-for-selecting-apache-hadoop-hardware/, September 1, 2011.

[24] Anita Bandrowski, Matthew Brush, Jeffery S. Grethe, Melissa A. Haendel, David N. Kennedy, Sean Hill, Patrick R. Hof, Maryann E Martone, Maaike Pols, Serena S. Tan, et al. The Resource Identification Initiative: A cultural shift in publishing. *Brain and Behavior*, 2015.

[25] Albert-László Barabási and Réka Albert. Emergence of scaling in random networks. *Science*, 286(5439):509–512, 1999.

[26] Michael Barbaro, Tom Zeller, and Saul Hansell. A face is exposed for AOL searcher no. 4417749. *New York Times*, August 9, 2006.

[27] Solon Barocas and Helen Nissenbaum. Big data's end run around procedural privacy protections. *Communications of the ACM*, 57(11):31–33, 2014.

[28] Solon Barocas and Helen Nissenbaum. The limits of anonymity and consent in the big data age. In Julia Lane, Victoria Stodden, Stefan Bender, and Helen Nissenbaum, editors, *Privacy, Big Data, and the Public Good: Frameworks for Engagement*. Cambridge University Press, 2014.

[29] Hilda Bastian. Bad research rising: The 7th Olympiad of research on biomedical publication. *Scientific American*, http://blogs.scientificamerican.com/absolutely-maybe/bad-research-rising-the-7th-olympiad-of-research-on-biomedical-publication/, 2013.

[30] Vladimir Batagelj and Andrej Mrvar. Pajek—program for large network analysis. *Connections*, 21(2):47–57, 1998.

[31] Alex Bell. Python for economists. http://cs.brown.edu/~ambell/pyseminar/pyseminar.html, 2012.

[32] Krithika Bhuvaneshwar, Dinanath Sulakhe, Robinder Gauba, Alex Rodriguez, Ravi Madduri, Utpal Dave, Lukasz Lacinski, Ian Foster, Yuriy Gusev, and Subha Madhavan. A case study for cloud based high throughput analysis of NGS data using the Globus Genomics system. *Computational and Structural Biotechnology Journal*, 13:64–74, 2015.

[33] Paul P. Biemer. Total survey error: Design, implementation, and evaluation. *Public Opinion Quarterly*, 74(5):817–848, 2010.

[34] Paul P. Biemer. *Latent Class Analysis of Survey Error*. John Wiley & Sons, 2011.

[35] Paul P. Biemer and Lars E. Lyberg. *Introduction to Survey Quality*. John Wiley & Sons, 2003.

[36] Paul P. Biemer and S. L. Stokes. Approaches to modeling measurement error. In Paul P. Biemer, R. Groves, L. Lyberg, N. Mathiowetz, and S. Sudman, editors, *Measurement Errors in Surveys*, pages 54–68. John Wiley, 1991.

[37] Paul P. Biemer and Dennis Trewin. A review of measurement error effects on the analysis of survey data. In L. Lyberg, P. Biemer, M. Collins, E. De Leeuw, C. Dippo, N. Schwarz, and D. Trewin, editors, *Survey Measurement and Process Quality*, pages 601–632. John Wiley & Sons, 1997.

[38] Ian Bird. Computing for the Large Hadron Collider. *Annual Review of Nuclear and Particle Science*, 61:99–118, 2011.

[39] Steven Bird, Ewan Klein, and Edward Loper. *Natural Language Processing with Python: Analyzing Text with the Natural Language Toolkit*. O'Reilly Media, 2009. Available online at http://www.nltk.org/book/.

[40] Christopher M. Bishop. *Pattern Recognition and Machine Learning*. Springer, 2006.

[41] David M. Blei. Topic modeling. http://www.cs.columbia.edu/~blei/topicmodeling.html. Accessed February 1, 2016.

[42] David M. Blei and John Lafferty. Topic models. In Ashok Srivastava and Mehran Sahami, editors, *Text Mining: Theory and Applications*. Taylor & Francis, 2009.

[43] David M. Blei, Andrew Ng, and Michael Jordan. Latent Dirichlet allocation. *Journal of Machine Learning Research*, 3:993–1022, 2003.

[44] John Blitzer, Mark Dredze, and Fernando Pereira. Biographies, Bollywood, boom-boxes and blenders: Domain adaptation for sentiment classification. In *Proceedings of the Association for Computational Linguistics*, 2007.

[45] Katy Börner. *Atlas of Science: Visualizing What We Know*. MIT Press, 2010.

[46] Philip E. Bourne and J. Lynn Fink. I am not a scientist, I am a number. *PLoS Computational Biology*, 4(12):e1000247, 2008.

[47] Jeremy Boy, Ronald Rensink, Enrico Bertini, Jean-Daniel Fekete, et al. A principled way of assessing visualization literacy. *IEEE Transactions on Visualization and Computer Graphics*, 20(12):1963–1972, 2014.

[48] Danah Boyd and Kate Crawford. Critical questions for big data: Provocations for a cultural, technological, and scholarly phenomenon. *Information, Communication & Society*, 15(5):662–679, 2012.

[49] Jordan Boyd-Graber. http://www.umiacs.umd.edu/~jbg/lda_demo. Accessed February 1, 2016.

[50] Leo Breiman. Random forests. *Machine Learning*, 45(1):5–32, 2001.

[51] Eric Brewer. CAP twelve years later: How the "rules" have changed. *Computer*, 45(2):23–29, 2012.

[52] C. Brody, T. de Hoop, M. Vojtkova, R. Warnock, M. Dunbar, P. Murthy, and S. L. Dworkin. Economic self-help group programs for women's empowerment: A systematic review. *Campbell Systematic Reviews*, 11(19), 2015.

[53] Jeen Broekstra, Arjohn Kampman, and Frank Van Harmelen. Sesame: A generic architecture for storing and querying RDF and RDF schema. In *The Semantic Web—ISWC 2002*, pages 54–68. Springer, 2002.

[54] Marc Bron, Bouke Huurnink, and Maarten de Rijke. Linking archives using document enrichment and term selection. In *Proceedings of the 15th International Conference on Theory and Practice of Digital Libraries: Research and Advanced Technology for Digital Libraries*, pages 360–371. Springer, 2011.

[55] Clair Brown, John Haltiwanger, and Julia Lane. *Economic Turbulence: Is a Volatile Economy Good for America?* University of Chicago Press, 2008.

[56] Erik Brynjolfsson, Lorin M. Hitt, and Heekyung Hellen Kim. Strength in numbers: How does data-driven decisionmaking affect firm performance? Available at SSRN 1819486, 2011.

[57] Bureau of Labor Statistics. The employment situation—November 2015. http://www.bls.gov/news.release/archives/empsit_12042015.pdf, December 4, 2015.

[58] Richard V. Burkhauser, Shuaizhang Feng, and Jeff Larrimore. Improving imputations of top incomes in the public-use current population survey by using both cell-means and variances. *Economics Letters*, 108(1):69–72, 2010.

[59] Ronald S. Burt. The social structure of competition. *Explorations in Economic Sociology*, 65:103, 1993.

[60] Ronald S. Burt. Structural holes and good ideas. *American Journal of Sociology*, 110(2):349–399, 2004.

[61] Declan Butler. When Google got flu wrong. *Nature*, 494(7436):155, 2013.

[62] Stuart K. Card and David Nation. Degree-of-interest trees: A component of an attention-reactive user interface. In *Proceedings of the Working Conference on Advanced Visual Interfaces*, pages 231–245. ACM, 2002.

[63] Jillian B. Carr and Jennifer L. Doleac. The geography, incidence, and underreporting of gun violence: New evidence using ShotSpotter data. Technical report, http://jenniferdoleac.com/wp-content/uploads/2015/03/Carr_Doleac_gunfire_underreporting.pdf, 2015.

[64] Charlie Catlett, Tanu Malik, Brett Goldstein, Jonathan Giuffrida, Yetong Shao, Alessandro Panella, Derek Eder, Eric van Zanten, Robert Mitchum, Severin Thaler, and Ian Foster. Plenario: An open data discovery and exploration platform for urban science. *Bulletin of the IEEE Computer Society Technical Committee on Data Engineering*, pages 27–42, 2014.

[65] Joe Cecil and Donna Eden. The legal foundations of confidentiality. In *Key Issues in Confidentiality Research: Results of an NSF workshop*. National Science Foundation, 2003.

[66] Centers for Disease Control and Prevention. United States cancer statistic: An interactive cancer atlas. http://nccd.cdc.gov/DCPC_INCA. Accessed February 1, 2016.

[67] John J. Chai. Correlated measurement errors and the least squares estimator of the regression coefficient. *Journal of the American Statistical Association*, 66(335):478–483, 1971.

[68] Lilyan Chan, Hannah F. Cross, Joseph K. She, Gabriel Cavalli, Hugo F. P. Martins, and Cameron Neylon. Covalent attachment of proteins to solid supports and surfaces via Sortase-mediated ligation. *PLoS One*, 2(11):e1164, 2007.

[69] Varun Chandola, Arindam Banerjee, and Vipin Kumar. Anomaly detection: A survey. *ACM Computing Surveys*, 41(3):15, 2009.

[70] O. Chapelle and S. S. Keerthi. Efficient algorithms for ranking with SVMs. *Information Retrieval*, 13(3):201–215, 2010.

[71] Kyle Chard, Jim Pruyne, Ben Blaiszik, Rachana Ananthakrishnan, Steven Tuecke, and Ian Foster. Globus data publication as a service: Lowering barriers to reproducible science. In *11th IEEE International Conference on eScience*, 2015.

[72] Kyle Chard, Steven Tuecke, and Ian Foster. Efficient and secure transfer, synchronization, and sharing of big data. *Cloud Computing, IEEE*, 1(3):46–55, 2014. See also https://www.globus.org.

[73] Nitesh V. Chawla. Data mining for imbalanced datasets: An overview. In Oded Maimon and Lior Rokach, editors, *The Data Mining and Knowledge Discovery Handbook*, pages 853–867. Springer, 2005.

[74] Raj Chetty. The transformative potential of administrative data for microeconometric research. http://conference.nber.org/confer/2012/SI2012/LS/ChettySlides.pdf. Accessed February 1, 2016, 2012.

[75] James O. Chipperfield and Raymond L. Chambers. Using the bootstrap to account for linkage errors when analysing probabilistically linked categorical data. *Journal of Official Statistics*, 31(3):397–414, 2015.

[76] Peter Christen. *Data Matching: Concepts and Techniques for Record Linkage, Entity Resolution, and Duplicate Detection*. Springer, 2012.

[77] Peter Christen. A survey of indexing techniques for scalable record linkage and deduplication. *IEEE Transactions on Knowledge and Data Engineering*, 24(9):1537–1555, 2012.

[78] Claire Clarke. Editing big data with machine learning methods. Paper presented at the Australian Bureau of Statistics Symposium, Canberra, 2014.

[79] William S. Cleveland and Robert McGill. Graphical perception: Theory, experimentation, and application to the development of graphical methods. *Journal of the American Statistical Association*, 79(387):531–554, 1984.

[80] C. Clifton, M. Kantarcioglu, A. Doan, G. Schadow, J. Vaidya, A.K. Elmagarmid, and D. Suciu. Privacy-preserving data integration and sharing. In G. Das, B. Liu, and P. S. Yu, editors, *9th ACM SIGMOD Workshop on Research Issues in Data Mining and Knowledge Discovery*, pages 19–26. ACM, June 2006.

[81] Cloudera. Cloudera Manager. https://www.cloudera.com/content/www/en-us/products/cloudera-manager.html. Accessed April 16, 2016.

[82] William G. Cochran. Errors of measurement in statistics. *Technometrics*, 10(4):637–666, 1968.

[83] Confidentiality and Data Access Committee. Federal Committee on Statistics and Methodology. http://fcsm.sites.usa.gov/committees/cdac/. Accessed April 16, 2016.

[84] Consumer Financial Protection Bureau. Home mortgage disclosure act data. http://www.consumerfinance.gov/hmda/learn-more. Accessed April 16, 2016.

[85] Paolo Corti, Thomas J. Kraft, Stephen Vincent Mather, and Bborie Park. *PostGIS Cookbook*. Packt Publishing, 2014.

[86] Koby Crammer and Yoram Singer. On the algorithmic implementation of multiclass kernel-based vector machines. *Journal of Machine Learning Research*, 2:265–292, 2002.

[87] Patricia J. Crossno, Douglas D. Cline, and Jeffrey N Jortner. A heterogeneous graphics procedure for visualization of massively parallel solutions. *ASME FED*, 156:65–65, 1993.

[88] John Czajka, Craig Schneider, Amang Sukasih, and Kevin Collins. Minimizing disclosure risk in HHS open data initiatives. Technical report, US Department of Health & Human Services, 2014.

[89] DataCite. DataCite homepage. https://www.datacite.org. Accessed February 1, 2016.

[90] Ton De Waal, Jeroen Pannekoek, and Sander Scholtus. *Handbook of Statistical Data Editing and Imputation*. John Wiley & Sons, 2011.

[91] Jeffrey Dean and Sanjay Ghemawat. MapReduce: Simplified data processing on large clusters. In *Proceedings of the 6th Conference on Symposium on Opearting Systems Design & Implementation—Volume 6*, OSDI'04. USENIX Association, 2004.

[92] Danny DeBelius. Let's tesselate: Hexagons for tile grid maps. *NPR Visuals Team Blog*, http://blog.apps.npr.org/2015/05/11/hex-tile-maps.html, May 11, 2015.

[93] Ryan A. Decker, John Haltiwanger, Ron S. Jarmin, and Javier Miranda. Where has all the skewness gone? The decline in high-growth (young) firms in the US. *European Economic Review*, to appear.

[94] David J. DeWitt and Michael Stonebraker. MapReduce: A major step backwards. http://www.dcs.bbk.ac.uk/~dell/teaching/cc/paper/dbc08/dewitt_mr_db.pdf, January 17, 2008.

[95] John J. Donohue III and Justin Wolfers. Uses and abuses of empirical evidence in the death penalty debate. Technical report, National Bureau of Economic Research, 2006.

[96] Pat Doyle, Julia I. Lane, Jules J. M. Theeuwes, and Laura V. Zayatz. *Confidentiality, Disclosure, and Data Access: Theory and Practical Applications for Statistical Agencies*. Elsevier Science, 2001.

[97] Jörg Drechsler. *Synthetic Datasets for Statistical Disclosure Control: Theory and Implementation*. Springer, 2011.

[98] Lian Duan, Lida Xu, Ying Liu, and Jun Lee. Cluster-based outlier detection. *Annals of Operations Research*, 168(1):151–168, 2009.

[99] Eva H. DuGoff, Megan Schuler, and Elizabeth A. Stuart. Generalizing observational study results: Applying propensity score methods to complex surveys. *Health Services Research*, 49(1):284–303, 2014.

[100] G. Duncan, M. Elliot, and J. J. Salazar-González. *Statistical Confidentiality: Principles and Practice*. Springer, 2011.

[101] Cody Dunne and Ben Shneiderman. Motif simplification: Improving network visualization readability with fan, connector, and clique glyphs. In *Proceedings of the SIGCHI Conference on Human Factors in Computing Systems*, pages 3247–3256. ACM, 2013.

[102] Ted Dunning. Accurate methods for the statistics of surprise and coincidence. *Computational Linguistics*, 19(1):61–74, 1993.

[103] Economic and Social Research Council. Administrative Data Research Network, 2016.

[104] Liran Einav and Jonathan D. Levin. The data revolution and economic analysis. Technical report, National Bureau of Economic Research, 2013.

[105] B. Elbel, J. Gyamfi, and R. Kersh. Child and adolescent fast-food choice and the influence of calorie labeling: A natural experiment. *International Journal of Obesity*, 35(4):493–500, 2011.

[106] Peter Elias. A European perspective on research and big data access. In Julia Lane, Victoria Stodden, Stefan Bender, and Helen Nissenbaum, editors, *Privacy, Big Data, and the Public Good: Frameworks for Engagement*, pages 98–112. Cambridge University Press, 2014.

[107] Joshua Elliott, David Kelly, James Chryssanthacopoulos, Michael Glotter, Kanika Jhunjhnuwala, Neil Best, Michael Wilde, and Ian Foster. The parallel system for integrating impact models and sectors (pSIMS). *Environmental Modelling & Software*, 62:509–516, 2014.

[108] Ahmed K. Elmagarmid, Panagiotis G. Ipeirotis, and Vassilios S. Verykios. Duplicate record detection: A survey. *IEEE Transactions on Knowledge and Data Engineering*, 19(1):1–16, 2007.

[109] David S. Evans. Tests of alternative theories of firm growth. *Journal of Political Economy*, 95:657–674, 1987.

[110] J. A. Evans and J. G. Foster. Metaknowledge. *Science*, 331(6018):721–725, 2011.

[111] Jianqing Fan, Fang Han, and Han Liu. Challenges of big data analysis. *National Science Review*, 1(2):293–314, 2014.

[112] Jianqing Fan and Yuan Liao. Endogeneity in ultrahigh dimension. Technical report, Princeton University, 2012.

[113] Jianqing Fan and Yuan Liao. Endogeneity in high dimensions. *Annals of Statistics*, 42(3):872, 2014.

[114] Jianqing Fan, Richard Samworth, and Yichao Wu. Ultrahigh dimensional feature selection: Beyond the linear model. *Journal of Machine Learning Research*, 10:2013–2038, 2009.

[115] Jean-Daniel Fekete. ProgressiVis: A toolkit for steerable progressive analytics and visualization. Paper presented at 1st Workshop on Data Systems for Interactive Analysis, Chicago, IL, October 26, 2015.

[116] Jean-Daniel Fekete and Catherine Plaisant. Interactive information visualization of a million items. In *IEEE Symposium on Information Visualization*, pages 117–124. IEEE, 2002.

[117] Ronen Feldman and James Sanger. *Text Mining Handbook: Advanced Approaches in Analyzing Unstructured Data*. Cambridge University Press, 2006.

[118] Ivan P. Fellegi and Alan B. Sunter. A theory for record linkage. *Journal of the American Statistical Association*, 64(328):1183–1210, 1969.

[119] Stephen Few. *Now You See It: Simple Visualization Techniques for Quantitative Analysis*. Analytics Press, 2009.

[120] Stephen Few. *Information Dashboard Design: Displaying Data for At-a-Glance Monitoring*. Analytics Press, 2013.

[121] Roy T. Fielding and Richard N. Taylor. Principled design of the modern Web architecture. *ACM Transactions on Internet Technology*, 2(2):115–150, 2002.

[122] Figshare. Figshare homepage. http://figshare.com. Accessed February 1, 2016.

[123] Danyel Fisher, Igor Popov, Steven Drucker, and m. c. schraefel. Trust me, I'm partially right: Incremental visualization lets analysts explore large datasets faster. In *Proceedings of the SIGCHI Conference on Human Factors in Computing Systems*, pages 1673–1682. ACM, 2012.

[124] Peter Flach. *Machine Learning: The Art and Science of Algorithms That Make Sense of Data*. Cambridge University Press, 2012.

[125] Blaz Fortuna, Marko Grobelnik, and Dunja Mladenic. OntoGen: Semi-automatic ontology editor. In *Proceedings of the 2007 Conference on Human Interface: Part II*, pages 309–318. Springer, 2007.

[126] Lucia Foster, Ron S. Jarmin, and T. Lynn Riggs. Resolving the tension between access and confidentiality: Past experience and future plans at the US Census Bureau. Technical Report 09-33, US Census Bureau Center for Economic Studies, 2009.

[127] Armando Fox, Steven D. Gribble, Yatin Chawathe, Eric A. Brewer, and Paul Gauthier. Cluster-based scalable network services. *ACM SIGOPS Operating Systems Review*, 31(5), 1997.

[128] W. N. Francis and H. Kucera. Brown corpus manual. Technical report, Department of Linguistics, Brown University, Providence, Rhode Island, US, 1979.

[129] Linton C. Freeman. Centrality in social networks conceptual clarification. *Social Networks*, 1(3):215–239, 1979.

[130] Wayne A. Fuller. Regression estimation in the presence of measurement error. In Paul P. Biemer, Robert M. Groves, Lars E. Lyberg, Nancy A. Mathiowetz, and Seymour Sudman, editors, *Measurement Errors in Surveys*, pages 617–635. John Wiley & Sons, 1991.

[131] S. Geman and D. Geman. Stochastic relaxation, Gibbs distributions, and the Bayesian restoration of images. In Glenn Shafer and Judea Pearl, editors, *Readings in Uncertain Reasoning*, pages 452–472. Morgan Kaufmann, 1990.

[132] Maria Girone. CERN database services for the LHC computing grid. In *Journal of Physics: Conference Series*, volume 119, page 052017. IOP Publishing, 2008.

[133] Michelle Girvan and Mark E. J. Newman. Community structure in social and biological networks. *Proceedings of the National Academy of Sciences*, 99(12):7821–7826, 2002.

[134] Wolfgang Glänzel. Bibliometric methods for detecting and analysing emerging research topics. *El Profesional de la Información*, 21(1):194–201, 2012.

[135] Michael Glueck, Azam Khan, and Daniel J. Wigdor. Dive in! Enabling progressive loading for real-time navigation of data visualizations. In *Proceedings of the SIGCHI Conference on Human Factors in Computing Systems*, pages 561–570. ACM, 2014.

[136] Leopold Granquist and John G. Kovar. Editing of survey data: How much is enough? In L. Lyberg, P. Biemer, M. Collins, E. De Leeuw, C. Dippo, N. Schwarz, and D. Trewin, editors, *Survey Measurement and Process Quality*, pages 415–435. John Wiley & Sons, 1997.

[137] Jim Gray. The transaction concept: Virtues and limitations. In *Proceedings of the Seventh International Conference on Very Large Data Bases*, volume 7, pages 144–154, 1981.

[138] Donald P. Green and Holger L. Kern. Modeling heterogeneous treatment effects in survey experiments with Bayesian additive regression trees. *Public Opinion Quarterly*, 76:491–511, 2012.

[139] Daniel Greenwood, Arkadiusz Stopczynski, Brian Sweatt, Thomas Hardjono, and Alex Pentland. The new deal on data: A framework for institutional controls. In Julia Lane, Victoria Stodden, Stefan Bender, and Helen Nissenbaum, editors, *Privacy, Big Data, and the Public Good: Frameworks for Engagement*, page 192. Cambridge University Press, 2014.

[140] Thomas L. Griffiths and Mark Steyvers. Finding scientific topics. *Proceedings of the National Academy of Sciences*, 101(Suppl. 1):5228–5235, 2004.

[141] Justin Grimmer and Brandon M. Stewart. Text as data: The promise and pitfalls of automatic content analysis methods for political texts. *Political Analysis*, 21(3):267–297, 2013.

[142] William Gropp, Ewing Lusk, and Anthony Skjellum. *Using MPI: Portable Parallel Programming with the Message-Passing Interface*. MIT Press, 2014.

[143] Robert M. Groves. *Survey Errors and Survey Costs*. John Wiley & Sons, 2004.

[144] Jiancheng Guan and Nan Ma. China's emerging presence in nanoscience and nanotechnology: A comparative bibliometric study of several nanoscience 'giants'. *Research Policy*, 36(6):880–886, 2007.

[145] Laurel L. Haak, Martin Fenner, Laura Paglione, Ed Pentz, and Howard Ratner. ORCID: A system to uniquely identify researchers. *Learned Publishing*, 25(4):259–264, 2012.

[146] Marc Haber, Daniel E. Platt, Maziar Ashrafian Bonab, Sonia C. Youhanna, David F. Soria-Hernanz, Begoña Martínez-Cruz, Bouchra Douaihy, Michella Ghassibe-Sabbagh, Hoshang Rafatpanah, Mohsen Ghanbari, et al.

Afghanistan's ethnic groups share a Y-chromosomal heritage structured by historical events. *PLoS One*, 7(3):e34288, 2012.

[147] Apache Hadoop. HDFS architecture. http://spark.apache.org/docs/latest/programming-guide.html#transformations.

[148] Jens Hainmueller and Chad Hazlett. Kernel regularized least squares: Reducing misspecification bias with a flexible and interpretable machine learning approach. *Political Analysis*, 22(2):143–168, 2014.

[149] Alon Halevy, Peter Norvig, and Fernando Pereira. The unreasonable effectiveness of data. *IEEE Intelligent Systems*, 24(2):8–12, 2009.

[150] Mark Hall, Eibe Frank, Geoffrey Holmes, Bernhard Pfahringer, Peter Reutemann, and Ian H Witten. The Weka data mining software: An update. *ACM SIGKDD Explorations Newsletter*, 11(1):10–18, 2009.

[151] P. Hall and H. Miller. Using generalized correlation to effect variable selection in very high dimensional problems. *Journal of Computational and Graphical Statistics*, 18:533–550, 2009.

[152] John Haltiwanger, Ron S. Jarmin, and Javier Miranda. Who creates jobs? Small versus large versus young. *Review of Economics and Statistics*, 95(2):347–361, 2013.

[153] Derek Hansen, Ben Shneiderman, and Marc A. Smith. *Analyzing Social Media Networks with NodeXL: Insights from a Connected World*. Morgan Kaufmann, 2010.

[154] Morris H. Hansen, William N. Hurwitz, and William G. Madow. *Sample Survey Methods and Theory*. John Wiley & Sons, 1993.

[155] Tim Harford. Big data: A big mistake? *Significance*, 11(5):14–19, 2014.

[156] Lane Harrison, Katharina Reinecke, and Remco Chang. Baby Name Voyager. http://www.babynamewizard.com/voyager/. Accessed February 1, 2016.

[157] Lane Harrison, Katharina Reinecke, and Remco Chang. Infographic aesthetics: Designing for the first impression. In *Proceedings of the 33rd Annual ACM Conference on Human Factors in Computing Systems*, pages 1187–1190. ACM, 2015.

[158] Trevor Hastie and Rob Tibshirani. Statistical learning course. https://lagunita.stanford.edu/courses/HumanitiesandScience/StatLearning/Winter2015/about. Accessed February 1, 2016.

[159] Trevor Hastie, Robert Tibshirani, and Jerome Friedman. *The Elements of Statistical Learning*. Springer, 2001.

[160] Erica Check Hayden. Researchers wrestle with a privacy problem. *Nature*, 525(7570):440, 2015.

[161] Erika Check Hayden. A broken contract. *Nature*, 486(7403):312–314, 2012.

[162] Zengyou He, Xiaofei Xu, and Shengchun Deng. Discovering cluster-based local outliers. *Pattern Recognition Letters*, 24(9):1641–1650, 2003.

[163] Kieran Healy and James Moody. Data visualization in sociology. *Annual Review of Sociology*, 40:105–128, 2014.

[164] Nathalie Henry and Jean-Daniel Fekete. MatrixExplorer: A dual-representation system to explore social networks. *IEEE Transactions on Visualization and Computer Graphics*, 12(5):677–684, 2006.

[165] Thomas N. Herzog, Fritz J. Scheuren, and William E. Winkler. *Data Quality and Record Linkage Techniques*. Springer, 2007.

[166] Kashmir Hill. How Target figured out a teen girl was pregnant before her father did. *Forbes*, http://www.forbes.com/sites/kashmirhill/2012/02/16/how-target-figured-out-a-teen-girl-was-pregnant-before-her-father-did/#7280148734c6, February 16, 2012.

[167] Thomas Hofmann. Probabilistic latent semantic analysis. In *Proceedings of Uncertainty in Artificial Intelligence*, 1999.

[168] Torsten Hothorn. Cran task view: Machine learning & statistical learning. https://cran.r-project.org/web/views/MachineLearning.html. Accessed February 1, 2016.

[169] Joop Hox. *Multilevel Analysis: Techniques and Applications*. Routledge, 2010.

[170] Yuening Hu, Ke Zhai, Vlad Eidelman, and Jordan Boyd-Graber. Polylingual tree-based topic models for translation domain adaptation. In *Proceedings of the 52nd Annual Meeting of the Association for Computational Linguistics*, 2014.

[171] Anna Huang. Similarity measures for text document clustering. Paper presented at New Zealand Computer Science Research Student Conference, Christchurch, New Zealand, April 14–18, 2008.

[172] Jian Huang, Seyda Ertekin, and C. Lee Giles. Efficient name disambiguation for large-scale databases. In *Knowledge Discovery in Databases: PKDD 2006*, pages 536–544. Springer, 2006.

[173] Human Microbiome Jumpstart Reference Strains Consortium, K. E. Nelson, G. M. Weinstock, et al. A catalog of reference genomes from the human microbiome. *Science*, 328(5981):994–999, 2010.

[174] Anco Hundepool, Josep Domingo-Ferrer, Luisa Franconi, Sarah Giessing, Rainer Lenz, Jane Longhurst, E. Schulte Nordholt, Giovanni Seri, and P. Wolf. Handbook on statistical disclosure control. Technical report, Network of Excellence in the European Statistical System in the Field of Statistical Disclosure Control, 2010.

[175] Kaye Husband Fealing, Julia Ingrid Lane, Jack Marburger, and Stephanie Shipp. *Science of Science Policy: The Handbook*. Stanford University Press, 2011.

[176] Joseph G. Ibrahim and Ming-Hui Chen. Power prior distributions for regression models. *Statistical Science*, 15(1):46–60, 2000.

[177] ICML. International conference on machine learning. http://icml.cc/. Accessed February 1, 2016.

[178] Kosuke Imai, Marc Ratkovic, et al. Estimating treatment effect heterogeneity in randomized program evaluation. *Annals of Applied Statistics*, 7(1):443–470, 2013.

[179] Guido W. Imbens and Donald B. Rubin. *Causal Inference in Statistics, Social, and Biomedical Sciences*. Cambridge University Press, 2015.

[180] Alfred Inselberg. *Parallel Coordinates*. Springer, 2009.

[181] Institute for Social Research. Panel study of income dynamics. http://psidonline.isr.umich.edu. Accessed February 1, 2016.

[182] Institute for Social Research. PSID file structure and merging PSID data files. Technical report. http://psidonline.isr.umich.edu/Guide/FileStructure.pdf. September 17, 2013.

[183] International Household Survey Network. Data Dissemination. http://www.ihsn.org/home/projects/dissemination. Accessed April 16, 2016.

[184] J. P. A. Ioannidis. Why most published research findings are false. *PLoS Medicine*, 2(8):e124, 2005.

[185] IPython development team. IPython documentation. http://ipython.readthedocs.org/. Accessed February 1, 2016.

[186] IPython development team. IPython website. http://ipython.org/. Accessed February 1, 2016.

[187] Gareth James, Daniela Witten, Trevor Hastie, and Robert Tibshirani. *An Introduction to Statistical Learning*. Springer, 2013.

[188] Lilli Japec, Frauke Kreuter, Marcus Berg, Paul Biemer, Paul Decker, Cliff Lampe, Julia Lane, Cathy O'Neil, and Abe Usher. Big data in survey research: AAPOR Task Force Report. *Public Opinion Quarterly*, 79(4):839–880, 2015.

[189] Ron S. Jarmin, Thomas A. Louis, and Javier Miranda. Expanding the role of synthetic data at the US Census Bureau. *Statistical Journal of the IAOS*, 30(2):117–121, 2014.

[190] Ron S. Jarmin and Javier Miranda. The longitudinal business database. Available at SSRN 2128793, 2002.

[191] Rachel Jewkes, Yandisa Sikweyiya, Robert Morrell, and Kristin Dunkle. The relationship between intimate partner violence, rape and HIV amongst South African men: A cross-sectional study. *PLoS One*, 6(9):e24256, 2011.

[192] Brian Johnson and Ben Shneiderman. Tree-maps: A space-filling approach to the visualization of hierarchical information structures. In *Proceedings of the IEEE Conference on Visualization*, pages 284–291. IEEE, 1991.

[193] Paul Jones and Peter Elias. Administrative data as a research resource: A selected audit. Technical report, ESRC National Centre for Research Methods, 2006.

[194] JOS. Journal of official statistics website. http://www.jos.nu. Accessed February 1, 2016.

[195] JPC. Journal of Privacy and Confidentiality. http://repository.cmu.edu/jpc/. Accessed April 16, 2016.

[196] Jupyter. Jupyter project documentation. http://jupyter.readthedocs.org/. Accessed February 1, 2016.

[197] Jupyter. Jupyter project website. http://jupyter.org/. Accessed February 1, 2016.

[198] Jupyter. jupyterhub GitHub repository. https://github.com/jupyter/jupyterhub/. Accessed February 1, 2016.

[199] Jupyter. jupyterhyb documentation. http://jupyterhub.readthedocs.org/. Accessed February 1, 2016.

[200] Jupyter. nbgrader documentation. http://nbgrader.readthedocs.org/. Accessed February 1, 2016.

[201] Jupyter. nbgrader GitHub repository. https://github.com/jupyter/nbgrader/. Accessed February 1, 2016.

[202] Felichism Kabo, Yongha Hwang, Margaret Levenstein, and Jason Owen-Smith. Shared paths to the lab: A sociospatial network analysis of collaboration. *Environment and Behavior*, 47(1):57–84, 2015.

[203] Alan Karr and Jerome P. Reiter. Analytical frameworks for data release: A statistical view. In Julia Lane, Victoria Stodden, Stefan Bender, and Helen Nissenbaum, editors, *Privacy, Big Data, and the Public Good: Frameworks for Engagement*. Cambridge University Press, 2014.

[204] KDD. ACM international conference on knowledge discovery and data mining (KDD). http://www.kdd.org. Accessed February 1, 2016.

[205] Sallie Ann Keller, Steven E. Koonin, and Stephanie Shipp. Big data and city living: What can it do for us? *Significance*, 9(4):4–7, 2012.

[206] Keshif. Infographics aesthetics dataset browser. http://keshif.me/demo/infographics_aesthetics. Accessed February 1, 2016.

[207] Satkartar K. Kinney, Jerome P. Reiter, Arnold P. Reznek, Javier Miranda, Ron S. Jarmin, and John M. Abowd. Towards unrestricted public use business microdata: The synthetic Longitudinal Business Database. *International Statistical Review*, 79(3):362–384, 2011.

[208] Andy Kirk. *Data Visualization: A Successful Design Process*. Packt Publishing, 2012.

[209] Tibor Kiss and Jan Strunk. Unsupervised multilingual sentence boundary detection. *Computational Linguistics*, 32(4):485–525, 2006.

[210] Jon Kleinberg, Jens Ludwig, Sendhil Mullainathan, and Ziad Obermeyer. Prediction policy problems. *American Economic Review*, 105(5):491–95, 2015.

[211] Ulrich Kohler and Frauke Kreuter. *Data Analysis Using Stata, 3rd Edition.* Stata Press, 2012.

[212] Lingpeng Kong, Nathan Schneider, Swabha Swayamdipta, Archna Bhatia, Chris Dyer, and Noah A. Smith. A dependency parser for tweets. In *Proceedings of the 2014 Conference on Empirical Methods in Natural Language Processing (EMNLP)*, pages 1001–1012. Association for Computational Linguistics, October 2014.

[213] Hanna Köpcke, Andreas Thor, and Erhard Rahm. Evaluation of entity resolution approaches on real-world match problems. *Proceedings of the VLDB Endowment*, 3(1-2):484–493, 2010.

[214] Menno-Jan Kraak. *Mapping Time: Illustrated by Minard's Map of Napoleon's Russian Campaign of 1812.* ESRI Press, 2014.

[215] Frauke Kreuter and Roger D. Peng. Extracting information from big data: Issues of measurement, inference, and linkage. In Julia Lane, Victoria Stodden, Stefan Bender, and Helen Nissenbaum, editors, *Privacy, Big Data, and the Public Good: Frameworks for Engagement*, pages 257–275. Cambridge University Press, 2014.

[216] H. W. Kuhn. The Hungarian method for the assignment problem. *Naval Research Logistics*, 52(1):7–21, 2005.

[217] Max Kuhn and Kjell Johnson. *Applied Predictive Modeling.* Springer Science & Business Media, 2013.

[218] Solomon Kullback and Richard A. Leibler. On information and sufficiency. *Annals of Mathematical Statistics*, 22(1):79–86, 1951.

[219] Mohit Kumar, Rayid Ghani, and Zhu-Song Mei. Data mining to predict and prevent errors in health insurance claims processing. In *Proceedings of the 16th ACM SIGKDD International Conference on Knowledge Discovery and Data Mining*, KDD '10, pages 65–74. ACM, 2010.

[220] John D. Lafferty, Andrew McCallum, and Fernando C. N. Pereira. Conditional random fields: Probabilistic models for segmenting and labeling sequence data. In *Proceedings of the Eighteenth International Conference on Machine Learning*, pages 282–289. Morgan Kaufmann, 2001.

[221] Partha Lahiri and Michael D Larsen. Regression analysis with linked data. *Journal of the American Statistical Association*, 100(469):222–230, 2005.

[222] Himabindu Lakkaraju, Everaldo Aguiar, Carl Shan, David Miller, Nasir Bhanpuri, Rayid Ghani, and Kecia L. Addison. A machine learning framework to identify students at risk of adverse academic outcomes. In *Proceedings of the 21th ACM SIGKDD International Conference on Knowledge Discovery and Data Mining*, KDD '15, pages 1909–1918. ACM, 2015.

[223] Heidi Lam, Enrico Bertini, Petra Isenberg, Catherine Plaisant, and Sheelagh Carpendale. Empirical studies in information visualization: Seven scenarios. *IEEE Transactions on Visualization and Computer Graphics*, 18(9):1520–1536, 2012.

[224] T. Landauer and S. Dumais. Solutions to Plato's problem: The latent semantic analysis theory of acquisition, induction and representation of knowledge. *Psychological Review*, 104(2):211–240, 1997.

[225] Julia Lane. Optimizing access to micro data. *Journal of Official Statistics*, 23:299–317, 2007.

[226] Julia Lane and Victoria Stodden. What? Me worry? what to do about privacy, big data, and statistical research. *AMSTAT News*, 438:14, 2013.

[227] Julia Lane, Victoria Stodden, Stefan Bender, and Helen Nissenbaum, editors. *Privacy, Big Data, and the Public Good: Frameworks for Engagement*. Cambridge University Press, 2014.

[228] Julia I. Lane, Jason Owen-Smith, Rebecca F. Rosen, and Bruce A. Weinberg. New linked data on research investments: Scientific workforce, productivity, and public value. *Research Policy*, 44:1659–1671, 2015.

[229] Douglas Laney. 3D data management: Controlling data volume, velocity, and variety. Technical report, META Group, February 2001.

[230] David Lazer, Ryan Kennedy, Gary King, and Alessandro Vespignani. The parable of Google Flu: Traps in big data analysis. *Science*, 343(14 March), 2014.

[231] Sinead C. Leahy, William J. Kelly, Eric Altermann, Ron S. Ronimus, Carl J Yeoman, Diana M Pacheco, Dong Li, Zhanhao Kong, Sharla McTavish, Carrie Sang, C. Lambie, Peter H. Janssen, Debjit Dey, and Graeme T. Attwood. The genome sequence of the rumen methanogen *Methanobrevibacter ruminantium* reveals new possibilities for controlling ruminant methane emissions. *PLoS One*, 2010. DOI: 10.1371/journal.pone.0008926.

[232] Whay C. Lee and Edward A. Fox. Experimental comparison of schemes for interpreting Boolean queries. Technical Report TR-88-27, Computer Science, Virginia Polytechnic Institute and State University, 1988.

[233] Yang Lee, WooYoung Chung, Stuart Madnick, Richard Wang, and Hongyun Zhang. On the rise of the Chief Data Officers in a world of big data. In *Pre-ICIS 2012 SIM Academic Workshop, Orlando, Florida*, 2012.

[234] Steven D. Levitt and Thomas J. Miles. Economic contributions to the understanding of crime. *Annual Review of Law Social Science*, 2:147–164, 2006.

[235] David D. Lewis. Naive (Bayes) at forty: The independence assumption in information retrieval. In *Proceedings of European Conference of Machine Learning*, pages 4–15, 1998.

[236] D. Lifka, I. Foster, S. Mehringer, M. Parashar, P. Redfern, C. Stewart, and S. Tuecke. XSEDE cloud survey report. Technical report, National Science Foundation, USA, http://hdl.handle.net/2142/45766, 2013.

[237] Jennifer Lin and Martin Fenner. Altmetrics in evolution: Defining and redefining the ontology of article-level metrics. *Information Standards Quarterly*, 25(2):20, 2013.

[238] Jimmy Lin and Chris Dyer. *Data-Intensive Text Processing with MapReduce*. Morgan & Claypool Publishers, 2010.

[239] Lauro Lins, James T Klosowski, and Carlos Scheidegger. Nanocubes for real-time exploration of spatiotemporal datasets. *IEEE Transactions on Visualization and Computer Graphics*, 19(12):2456–2465, 2013.

[240] Roderick J. A. Little and Donald B Rubin. *Statistical Analysis with Missing Data*. John Wiley & Sons, 2014.

[241] Zhicheng Liu and Jeffrey Heer. The effects of interactive latency on exploratory visual analysis. *IEEE Transactions on Visualization and Computer Graphics*, 20(12):2122–2131, 2014.

[242] Glenn K. Lockwood. Conceptual overview of map-reduce and hadoop. http://www.glennklockwood.com/data-intensive/hadoop/overview.html, October 9, 2015.

[243] Sharon Lohr. *Sampling: Design and Analysis*. Cengage Learning, 2009.

[244] Alan M. MacEachren, Stephen Crawford, Mamata Akella, and Gene Lengerich. Design and implementation of a model, web-based, GIS-enabled cancer atlas. *Cartographic Journal*, 45(4):246–260, 2008.

[245] Jock Mackinlay. Automating the design of graphical presentations of relational information. *ACM Transactions on Graphics*, 5(2):110–141, 1986.

[246] Waqas Ahmed Malik, Antony Unwin, and Alexander Gribov. An interactive graphical system for visualizing data quality–tableplot graphics. In *Classification as a Tool for Research*, pages 331–339. Springer, 2010.

[247] K. Malmkjær. *The Linguistics Encyclopedia*. Routledge, 2002.

[248] Christopher D. Manning, Prabhakar Raghavan, and Hinrich Schütze. *Introduction to Information Retrieval*. Cambridge University Press, 2008.

[249] Christopher D. Manning, Mihai Surdeanu, John Bauer, Jenny Finkel, Steven J. Bethard, and David McClosky. The Stanford CoreNLP natural language processing toolkit. In *Proceedings of 52nd Annual Meeting of the Association for Computational Linguistics: System Demonstrations*, pages 55–60, 2014.

[250] John H Marburger. Wanted: Better benchmarks. *Science*, 308(5725):1087, 2005.

[251] Mitchell P. Marcus, Beatrice Santorini, and Mary A. Marcinkiewicz. Building a large annotated corpus of English: The Penn treebank. *Computational Linguistics*, 19(2):313–330, 1993.

[252] Alexandre Mas and Enrico Moretti. Peers at work. *American Economic Review*, 99(1):112–145, 2009.

[253] Girish Maskeri, Santonu Sarkar, and Kenneth Heafield. Mining business topics in source code using latent Dirichlet allocation. In *Proceedings of the 1st India Software Engineering Conference*, pages 113–120. ACM, 2008.

[254] Erika McCallister, Timothy Grance, and Karen A Scarfone. *SP 800-122. Guide to Protecting the Confidentiality of Personally Identifiable Information (PII).* National Institute of Standards and Technology, 2010.

[255] Andrew Kachites McCallum. Mallet: A machine learning for language toolkit. http://mallet.cs.umass.edu, 2002.

[256] Edgar Meij, Marc Bron, Laura Hollink, Bouke Huurnink, and Maarten Rijke. Learning semantic query suggestions. In *Proceedings of the 8th International Semantic Web Conference*, ISWC '09, pages 424–440. Springer, 2009.

[257] Bruce D. Meyer, Wallace K. C. Mok, and James X. Sullivan. Household surveys in crisis. *Journal of Economic Perspectives*, 29(4):199–226, 2015.

[258] Tom M. Mitchell. *Machine Learning.* McGraw-Hill, 1997.

[259] C. L. Moffatt. Visual representation of SQL joins. http://www.codeproject.com/Articles/33052/Visual-Representation-of-SQL-Joins, February 3, 1999.

[260] Anthony Molinaro. *SQL Cookbook: Query Solutions and Techniques for Database Developers.* O'Reilly Media, 2005.

[261] Stephen L. Morgan and Christopher Winship. *Counterfactuals and Causal Inference.* Cambridge University Press, 2014.

[262] Peter Stendahl Mortensen, Carter Walter Bloch, et al. *Oslo Manual: Guidelines for Collecting and Interpreting Innovation Data.* Organisation for Economic Co-operation and Development, 2005.

[263] Sougata Mukherjea. Information retrieval and knowledge discovery utilising a biomedical semantic web. *Briefings in Bioinformatics*, 6(3):252–262, 2005.

[264] Tamara Munzner. *Visualization Analysis and Design.* CRC Press, 2014.

[265] Joe Murphy, Michael W Link, Jennifer Hunter Childs, Casey Langer Tesfaye, Elizabeth Dean, Michael Stern, Josh Pasek, Jon Cohen, Mario Callegaro, and Paul Harwood. Social media in public opinion research: Report of the AAPOR Task Force on emerging technologies in public opinion research. *Public Opinion Quarterly*, 78(4):788–794, 2014.

[266] Arvind Narayanan and Vitaly Shmatikov. Robust de-anonymization of large sparse datasets. In *IEEE Symposium on Security and Privacy*, pages 111–125. IEEE, 2008.

[267] Kalaivany Natarajan, Jiuyong Li, and Andy Koronios. *Data Mining Techniques for Data Cleaning.* Springer, 2010.

[268] National Science Foundation. Download awards by year. http://nsf.gov/awardsearch/download.jsp. Accessed February 1, 2016.

[269] Roberto Navigli, Stefano Faralli, Aitor Soroa, Oier de Lacalle, and Eneko Agirre. Two birds with one stone: Learning semantic models for text categorization and word sense disambiguation. In *Proceedings of the 20th ACM International Conference on Information and Knowledge Management.* ACM, 2011.

[270] Robert K. Nelson. Mining the dispatch. http://dsl.richmond.edu/dispatch/, 2010.

[271] Mark Newman. A measure of betweenness centrality based on random walks. *Social Networks*, 27(1):39–54, 2005.

[272] Mark Newman. *Networks: An Introduction*. Oxford University Press, 2010.

[273] Cameron Neylon. Altmetrics: What are they good for? http://blogs.plos.org/opens/2014/10/03/altmetrics-what-are-they-good-for/, 2014.

[274] Cameron Neylon. The road less travelled. In Sarita Albagli, Maria Lucia Maciel, and Alexandre Hannud Abdo, editors, *Open Science, Open Issues*. IBICT, UNIRIO, 2015.

[275] Cameron Neylon, Michelle Willmers, and Thomas King. Impact beyond citation: An introduction to Altmetrics. http://hdl.handle.net/11427/2314, 2014.

[276] Viet-An Nguyen, Jordan Boyd-Graber, and Philip Resnik. SITS: A hierarchical nonparametric model using speaker identity for topic segmentation in multiparty conversations. In *Proceedings of the Association for Computational Linguistics*, 2012.

[277] Viet-An Nguyen, Jordan Boyd-Graber, and Philip Resnik. Lexical and hierarchical topic regression. In *Advances in Neural Information Processing Systems*, 2013.

[278] Viet-An Nguyen, Jordan Boyd-Graber, Philip Resnik, and Jonathan Chang. Learning a concept hierarchy from multi-labeled documents. In *Proceedings of the Annual Conference on Neural Information Processing Systems*. Morgan Kaufmann, 2014.

[279] Viet-An Nguyen, Jordan Boyd-Graber, Philip Resnik, and Kristina Miler. Tea Party in the House: A hierarchical ideal point topic model and its application to Republican legislators in the 112th Congress. In *Association for Computational Linguistics*, 2015.

[280] Vlad Niculae, Srijan Kumar, Jordan Boyd-Graber, and Cristian Danescu-Niculescu-Mizil. Linguistic harbingers of betrayal: A case study on an online strategy game. In *Association for Computational Linguistics*, 2015.

[281] Michael Nielsen. *Reinventing Discovery: The New Era of Networked Science*. Princeton University Press, 2012.

[282] NIPS. Annual conference on neural information processing systems (NIPS). https://nips.cc/. Accessed February 1, 2016.

[283] Helen Nissenbaum. A contextual approach to privacy online. *Daedalus*, 140(4):32–48, 2011.

[284] NLTK Project. NLTK: The natural language toolkit. http://www.nltk.org. Accessed February 1, 2016.

[285] Regina O. Obe and Leo S. Hsu. *PostGIS in Action, 2nd Edition*. Manning Publications, 2015.

[286] David Obstfeld. Social networks, the tertius iungens orientation, and involvement in innovation. *Administrative Science Quarterly*, 50(1):100–130, 2005.

[287] President's Council of Advisors on Science and Technology. Big data and privacy: A technological perspective. Technical report, Executive Office of the President, 2014.

[288] Organisation of Economic Co-operation and Development. A summary of the Frascati manual. *Main definitions and conventions for the measurement of research and experimental development*, 84, 2004.

[289] Paul Ohm. Broken promises of privacy: Responding to the surprising failure of anonymization. *UCLA Law Review*, 57:1701, 2010.

[290] Paul Ohm. The legal and regulatory framework: what do the rules say about data analysis? In Julia Lane, Victoria Stodden, Helen Nissenbaum, and Stefan Bender, editors, *Privacy, Big Data, and the Public Good: Frameworks for Engagement*. Cambridge University Press, 2014.

[291] Judy M. Olson and Cynthia A. Brewer. An evaluation of color selections to accommodate map users with color-vision impairments. *Annals of the Association of American Geographers*, 87(1):103–134, 1997.

[292] Myle Ott, Yejin Choi, Claire Cardie, and Jeffrey T. Hancock. Finding deceptive opinion spam by any stretch of the imagination. In *Proceedings of the 49th Annual Meeting of the Association for Computational Linguistics: Human Language Technologies—Volume 1*, HLT '11, pages 309–319, Stroudsburg, PA, 2011. Association for Computational Linguistics.

[293] Jason Owen-Smith and Walter W. Powell. The expanding role of university patenting in the life sciences: Assessing the importance of experience and connectivity. *Research Policy*, 32(9):1695–1711, 2003.

[294] Jason Owen-Smith and Walter W. Powell. Knowledge networks as channels and conduits: The effects of spillovers in the Boston biotechnology community. *Organization Science*, 15(1):5–21, 2004.

[295] Bo Pang and Lillian Lee. *Opinion Mining and Sentiment Analysis*. Now Publishers, 2008.

[296] Hae-Sang Park and Chi-Hyuck Jun. A simple and fast algorithm for k-medoids clustering. *Expert Systems with Applications*, 36(2):3336–3341, 2009.

[297] Norman Paskin. Digital object identifier (doi) system. *Encyclopedia of Library and Information Sciences*, 3:1586–1592, 2008.

[298] Michael Paul and Roxana Girju. A two-dimensional topic-aspect model for discovering multi-faceted topics. In *Association for the Advancement of Artificial Intelligence*, 2010.

[299] James W. Pennebaker and Martha E. Francis. *Linguistic Inquiry and Word Count*. Lawrence Erlbaum, 1999.

[300] Alex Pentland, Daniel Greenwood, Brian Sweatt, Arek Stopczynski, and Yves-Alexandre de Montjoye. Institutional controls: The new deal on data. In Julia Lane, Victoria Stodden, Stefan Bender, and Helen Nissenbaum, editors, *Privacy, Big Data, and the Public Good: Frameworks for Engagement*, pages 98–112. Cambridge University Press, 2014.

[301] PERISCOPIC. A world of terror. http://terror.periscopic.com/. Accessed February 1, 2016.

[302] George Petrakos, Claudio Conversano, Gregory Farmakis, Francesco Mola, Roberta Siciliano, and Photis Stavropoulos. New ways of specifying data edits. *Journal of the Royal Statistical Society, Series A*, 167(2):249–274, 2004.

[303] Catherine Plaisant, Jesse Grosjean, and Benjamin B. Bederson. SpaceTree: Supporting exploration in large node link tree, design evolution and empirical evaluation. In *IEEE Symposium on Information Visualization*, pages 57–64. IEEE, 2002.

[304] Ruth Pordes, Don Petravick, Bill Kramer, Doug Olson, Miron Livny, Alain Roy, Paul Avery, Kent Blackburn, Torre Wenaus, Frank Würthwein, et al. The Open Science Grid. *Journal of Physics: Conference Series*, 78(1):012057, 2007.

[305] Alan L. Porter, Jan Youtie, Philip Shapira, and David J. Schoeneck. Refining search terms for nanotechnology. *Journal of Nanoparticle Research*, 10(5):715–728, 2008.

[306] PostGIS Project Steering Committee. PostGIS documentation. http://postgis.net/documentation/. Accessed December 1, 2015.

[307] Eric Potash, Joe Brew, Alexander Loewi, Subhabrata Majumdar, Andrew Reece, Joe Walsh, Eric Rozier, Emile Jorgenson, Raed Mansour, and Rayid Ghani. Predictive modeling for public health: Preventing childhood lead poisoning. In *Proceedings of the 21th ACM SIGKDD International Conference on Knowledge Discovery and Data Mining*, KDD '15, pages 2039–2047. ACM, 2015.

[308] W. Powell. Neither market nor hierarchy. *Sociology of Organizations: Classic, Contemporary, and Critical Readings*, 315:104–117, 2003.

[309] Walter W. Powell, Douglas R. White, Kenneth W. Koput, and Jason Owen-Smith. Network dynamics and field evolution: The growth of interorganizational collaboration in the life sciences. *American Journal of Sociology*, 110(4):1132–1205, 2005.

[310] Jason Priem, Heather A Piwowar, and Bradley M Hemminger. Altmetrics in the wild: Using social media to explore scholarly impact. Preprint, arXiv 1203.4745, 2012.

[311] Foster Provost and Tom Fawcett. *Data Science for Business: What You Need to Know About Data Mining and Data-analytic Thinking*. O'Reilly Media, 2013.

[312] Marco Puts, Piet Daas, and Ton de Waal. Finding errors in Big Data. *Significance*, 12(3):26–29, 2015.

[313] Lawrence R. Rabiner. A tutorial on hidden Markov models and selected applications in speech recognition. *Proceedings of the IEEE*, 77(2):257–286, 1989.

[314] Karthik Ram. Git can facilitate greater reproducibility and increased transparency in science. *Source Code for Biology and Medicine*, 8(1):7, 2013.

[315] Daniel Ramage, David Hall, Ramesh Nallapati, and Christopher Manning. Labeled LDA: A supervised topic model for credit attribution in multi-labeled corpora. In *Proceedings of Empirical Methods in Natural Language Processing*, 2009.

[316] Raghu Ramakrishnan and Johannes Gehrke. *Database Management Systems, 3rd Edition*. McGraw-Hill, 2002.

[317] Jerome P. Reiter. Statistical approaches to protecting confidentiality for microdata and their effects on the quality of statistical inferences. *Public Opinion Quarterly*, 76(1):163–181, 2012.

[318] Philip Resnik and Jimmy Lin. Evaluation of NLP systems. In Alex Clark, Chris Fox, and Shalom Lappin, editors, *Handbook of Computational Linguistics and Natural Language Processing*. Wiley Blackwell, 2010.

[319] Leonard Richardson. Beautiful Soup. http://www.crummy.com/software/BeautifulSoup/. Accessed February 1, 2016.

[320] Donald B. Rubin. Inference and missing data. *Biometrika*, 63:581–592, 1976.

[321] Bahador Saket, Paolo Simonetto, Stephen Kobourov, and Katy Börner. Node, node-link, and node-link-group diagrams: An evaluation. *IEEE Transactions on Visualization and Computer Graphics*, 20(12):2231–2240, 2014.

[322] Gerard Salton. *Automatic Information Organization and Retrieval*. McGraw-Hill, 1968.

[323] Arthur L. Samuel. Some studies in machine learning using the game of Checkers. *IBM Journal of Research and Development*, 3(3):210–229, 1959.

[324] Evan Sandhaus. The New York Times annotated corpus. *Linguistic Data Consortium*, http://www.ldc.upenn.edu/Catalog/CatalogEntry.jsp?catalogId=LDC2008T19, 2008.

[325] Purvi Saraiya, Chris North, and Karen Duca. An insight-based methodology for evaluating bioinformatics visualizations. *IEEE Transactions on Visualization and Computer Graphics*, 11(4):443–456, 2005.

[326] Joseph L. Schafer. *Analysis of Incomplete Multivariate Data*. CRC Press, 1997.

[327] Joseph L Schafer and John W Graham. Missing data: Our view of the state of the art. *Psychological Methods*, 7(2):147, 2002.

[328] Michael Schermann, Holmer Hemsen, Christoph Buchmüller, Till Bitter, Helmut Krcmar, Volker Markl, and Thomas Hoeren. Big data. *Business & Information Systems Engineering*, 6(5):261–266, 2014.

[329] Fritz Scheuren and William E. Winkler. Regression analysis of data files that are computer matched. *Survey Methodology*, 19(1):39–58, 1993.

[330] Rainer Schnell. An efficient privacy-preserving record linkage technique for administrative data and censuses. *Statistical Journal of the IAOS*, 30:263–270, 2014.

[331] Rainer Schnell. German Record Linkage Center, 2016.

[332] Rainer Schnell, Tobias Bachteler, and Jörg Reiher. Privacy-preserving record linkage using Bloom filters. *BMC Medical Informatics and Decision Making*, 9(1):41, 2009.

[333] Julie A. Schoenman. The concentration of health care spending. NIHCM foundation data brief, National Institute for Health Care Management, 2012. http://www.nihcm.org/pdf/DataBrief3%20Final.pdf.

[334] Bernhard Scholkopf and Alexander J. Smola. *Learning with Kernels: Support Vector Machines, Regularization, Optimization, and Beyond*. MIT Press, 2001.

[335] Steven L. Scott, Alexander W. Blocker, Fernando V. Bonassi, H. Chipman, E. George, and R. McCulloch. Bayes and big data: The consensus Monte Carlo algorithm. In *EFaBBayes 250 conference*, volume 16, 2013. http://bit.ly/1wBqh4w, Accessed January 1, 2016.

[336] Sesame. Sesame RDF triple store. http://rdf4j.org. Accessed February 1, 2016.

[337] James A. Sethian, Jean-Philippe Brunet, Adam Greenberg, and Jill P. Mesirov. Computing turbulent flow in complex geometries on a massively parallel processor. In *Proceedings of the 1991 ACM/IEEE Conference on Supercomputing*, pages 230–241. ACM, 1991.

[338] Charles Severance. Python for informatics: Exploring information. http://www.pythonlearn.com/book.php, 2013.

[339] John Shawe-Taylor and Nello Cristianini. *Kernel Methods for Pattern Analysis*. Cambridge University Press, 2004.

[340] Taylor Shelton, Ate Poorthuis, Mark Graham, and Matthew Zook. Mapping the data shadows of Hurricane Sandy: Uncovering the sociospatial dimensions of 'big data'. *Geoforum*, 52:167–179, 2014.

[341] Aimee Shen, Patrick J. Lupardus, Montse Morell, Elizabeth L. Ponder, A. Masoud Sadaghiani, K. Christopher Garcia, Matthew Bogyo, et al. Simplified, enhanced protein purification using an inducible, autoprocessing enzyme tag. *PLoS One*, 4(12):e8119, 2009.

[342] Ben Shneiderman. Tree visualization with tree-maps: 2-D space-filling approach. *ACM Transactions on Graphics*, 11(1):92–99, 1992.

[343] Ben Shneiderman. Extreme visualization: Squeezing a billion records into a million pixels. In *Proceedings of the 2008 ACM SIGMOD International Conference on Management of Data*, pages 3–12. ACM, 2008.

[344] Ben Shneiderman and Catherine Plaisant. Sharpening analytic focus to cope with big data volume and variety. *Computer Graphics and Applications, IEEE*, 35(3):10–14, 2015. See also http://www.cs.umd.edu/hcil/eventflow/Sharpening-Strategies-Help.pdf.

[345] Konstantin Shvachko, Hairong Kuang, Sanjay Radia, and Robert Chansler. The Hadoop distributed file system. In *IEEE 26th Symposium on Mass Storage Systems and Technologies (MSST)*, pages 1–10. IEEE, 2010.

[346] Helmut Sies. A new parameter for sex education. *Nature*, 332(495), 1988.

[347] Abraham Silberschatz, Henry F. Korth, and S. Sudarshan. *Database System Concepts, 6th Edition*. McGraw-Hill, 2010.

[348] Alex J. Smola and Bernhard Schölkopf. A tutorial on support vector regression. *Statistics and Computing*, 14(3):199–222, August 2004.

[349] John Snow. *On the Mode of Communication of Cholera*. John Churchill, 1855.

[350] Rion Snow, Brendan O'Connor, Daniel Jurafsky, and Andrew Ng. Cheap and fast—but is it good? Evaluating non-expert annotations for natural language tasks. In *Proceedings of Empirical Methods in Natural Language Processing*, 2008.

[351] Solid IT. DB Engines. http://db-engines.com/en/. Accessed February 1, 2016.

[352] SOSP. Science of science policy. http://www.scienceofsciencepolicy.net/. Accessed February 1, 2016.

[353] Peverill Squire. Why the 1936 Literary Digest poll failed. *Public Opinion Quarterly*, 52(1):125–133, 1988.

[354] Stanford. Stanford CoreNLP—a suite of core NLP tools. http://nlp.stanford.edu/software/corenlp.shtml. Accessed February 1, 2016.

[355] Stanford Visualization Group. Dorling cartograms in ProtoVis. http://mbostock.github.io/protovis/ex/cartogram.html. Accessed January 10, 2015.

[356] Mark W Stanton and MK Rutherford. *The High Concentration of US Health Care Expenditures*. Agency for Healthcare Research and Quality, 2006.

[357] John Stasko, Carsten Görg, and Zhicheng Liu. Jigsaw: Supporting investigative analysis through interactive visualization. *Information Visualization*, 7(2):118–132, 2008.

[358] Steve Stemler. An overview of content analysis. *Practical Assessment, Research & Evaluation*, 7(17), 2001.

[359] Rebecca C Steorts, Rob Hall, and Stephen E Fienberg. SMERED: a Bayesian approach to graphical record linkage and de-duplication. Preprint, arXiv 1403.0211, 2014.

[360] S. Stephens-Davidowitz and H. Varian. A hands-on guide to Google data. http://people.ischool.berkeley.edu/~hal/Papers/2015/primer.pdf. Accessed October 12, 2015.

[361] James H. Stock and Mark W. Watson. Forecasting using principal components from a large number of predictors. *Journal of the American Statistical Association*, 97(460):1167–1179, 2002.

[362] Katherine J. Strandburg. Monitoring, datafication and consent: Legal approaches to privacy in the big data context. In Julia Lane, Victoria Stodden, Stefan Bender, and Helen Nissenbaum, editors, *Privacy, Big Data, and the Public Good: Frameworks for Engagement*. Cambridge University Press, 2014.

[363] Carly Strasser. Git/GitHub: A primer for researchers. http://datapub.cdlib.org/2014/05/05/github-a-primer-for-researchers/, May 5, 2014.

[364] Christof Strauch. Nosql databases. http://www.christof-strauch.de/nosqldbs.pdf, 2009.

[365] Elizabeth A. Stuart. Matching methods for causal inference: A review and a look forward. *Statistical Science*, 25(1):1, 2010.

[366] Latanya Sweeney. Computational disclosure control: A primer on data privacy protection. Technical report, MIT, 2001. http://groups.csail.mit.edu/mac/classes/6.805/articles/privacy/sweeney-thesis-draft.pdf.

[367] Alexander S. Szalay, Jim Gray, Ani R. Thakar, Peter Z. Kunszt, Tanu Malik, Jordan Raddick, Christopher Stoughton, and Jan vandenBerg. The SDSS skyserver: Public access to the Sloan digital sky server data. In *Proceedings of the 2002 ACM SIGMOD International Conference on Management of Data*, pages 570–581. ACM, 2002.

[368] Edmund M. Talley, David Newman, David Mimno, Bruce W. Herr II, Hanna M. Wallach, Gully A. P. C. Burns, A. G. Miriam Leenders, and Andrew McCallum. Database of NIH grants using machine-learned categories and graphical clustering. *Nature Methods*, 8(6):443–444, 2011.

[369] Adam Tanner. Harvard professor re-identifies anonymous volunteers in DNA study. *Forbes*, http://www.forbes.com/sites/adamtanner/2013/04/25/harvard-professor-re-identifies-anonymous-volunteers-in-dna-study/#6cc7f6b43e39, April 25, 2013.

[370] TDP. Transactions on Data Privacy. http://www.tdp.cat/. Accessed April 16, 2016.

[371] M. Tennekes, E. de Jonge, and P. Daas. Innovative visual tools for data editing. Presented at the United Nations Economic Commission for Europe Work Session on Statistical Data. Available online at http://www.pietdaas.nl/beta/pubs/pubs/30_Netherlands.pdf, 2012.

[372] Martijn Tennekes and Edwin de Jonge. Top-down data analysis with treemaps. In *Proceedings of the International Conference on Imaging Theory and Applications and International Conference on Information Visualization Theory and Applications*, pages 236–241. SciTePress, 2011.

[373] Martijn Tennekes, Edwin de Jonge, and Piet J. H. Daas. Visualizing and inspecting large datasets with tableplots. *Journal of Data Science*, 11(1):43–58, 2013.

[374] Alexander I. Terekhov. Evaluating the performance of Russia in the research in nanotechnology. *Journal of Nanoparticle Research*, 14(11), 2012.

[375] William W. Thompson, Lorraine Comanor, and David K. Shay. Epidemiology of seasonal influenza: Use of surveillance data and statistical models to estimate the burden of disease. *Journal of Infectious Diseases*, 194(Supplement 2):S82–S91, 2006.

[376] Robert Tibshirani. Regression shrinkage and selection via the lasso. *Journal of the Royal Statistical Society, Series B*, pages 267–288, 1996.

[377] D. Trewin, A. Andersen, T. Beridze, L. Biggeri, I. Fellegi, and T. Toczynski. Managing statistical confidentiality and microdata access: Principles and guidelines of good practice. Technical report, Conference of European Statisticians, United Nations Economic Commision for Europe, 2007.

[378] TSE15. 2015 international total survey error conference website. https://www.tse15.org. Accessed February 1, 2016.

[379] Suppawong Tuarob, Line C. Pouchard, and C. Lee Giles. Automatic tag recommendation for metadata annotation using probabilistic topic modeling. In *Proceedings of the 13th ACM/IEEE-CS Joint Conference on Digital Libraries*, JCDL '13, pages 239–248. ACM, 2013.

[380] Edward Tufte. *The Visual Display of Quantitative information, 2nd Edition*. Graphics Press, 2001.

[381] Edward Tufte. *Beautiful Evidence, 2nd Edition*. Graphics Press, 2006.

[382] United Nations Economic Commission for Europe. Statistical confidentiality and disclosure protection. http://www.unece.org/stats/mos/meth/confidentiality.html. Accessed April 16, 2016.

[383] University of Oxford. British National Corpus. http://www.natcorp.ox.ac.uk/, 2006.

[384] University of Waikato. Weka 3: Data mining software in java. http://www.cs.waikato.ac.nz/ml/weka/. Accessed February 1, 2016.

[385] Richard Valliant, Jill A Dever, and Frauke Kreuter. *Practical Tools for Designing and Weighting Survey Samples*. Springer, 2013.

[386] Hal R. Varian. Big data: New tricks for econometrics. *Journal of Economic Perspectives*, 28(2):3–28, 2014.

[387] Samuel L. Ventura, Rebecca Nugent, and Erica R. H. Fuchs. Seeing the non-stars:(some) sources of bias in past disambiguation approaches and a new public tool leveraging labeled records. *Research Policy*, 2015.

[388] Tyler Vigen. Spurious correlations. http://www.tylervigen.com/spurious-correlations. Accessed February 1, 2016.

[389] Tyler Vigen. *Spurious Correlations*. Hachette Books, 2015.

[390] Hanna Wallach, David Mimno, and Andrew McCallum. Rethinking LDA: Why priors matter. In *Advances in Neural Information Processing Systems*, 2009.

[391] Anders Wallgren and Britt Wallgren. *Register-Based Statistics: Administrative Data for Statistical Purposes*. John Wiley & Sons, 2007.

[392] Chong Wang, David Blei, and Li Fei-Fei. Simultaneous image classification and annotation. In *Computer Vision and Pattern Recognition*, 2009.

[393] Yi Wang, Hongjie Bai, Matt Stanton, Wen-Yen Chen, and Edward Y. Chang. PLDA: parallel latent Dirichlet allocation for large-scale applications. In *International Conference on Algorithmic Aspects in Information and Management*, 2009.

[394] Karl J. Ward. Crossref REST API. http://api.crossref.org. Accessed February 1, 2016.

[395] Matthew O. Ward, Georges Grinstein, and Daniel Keim. *Interactive Data Visualization: Foundations, Techniques, and Applications*. CRC Press, 2010.

[396] Bruce A. Weinberg, Jason Owen-Smith, Rebecca F Rosen, Lou Schwarz, Barbara McFadden Allen, Roy E. Weiss, and Julia Lane. Science funding and short-term economic activity. *Science*, 344(6179):41, 2014.

[397] Steven Euijong Whang, David Menestrina, Georgia Koutrika, Martin Theobald, and Hector Garcia-Molina. Entity resolution with iterative blocking. In *Proceedings of the 2009 ACM SIGMOD International Conference on Management of data*, pages 219–232. ACM, 2009.

[398] Harrison C. White, Scott A. Boorman, and Ronald L. Breiger. Social structure from multiple networks. I. Block models of roles and positions. *American Journal of Sociology*, pages 730–780, 1976.

[399] Tom White. *Hadoop: The Definitive Guide*. O'Reilly, 2012.

[400] Michael Wick, Sameer Singh, Harshal Pandya, and Andrew McCallum. A joint model for discovering and linking entities. In *Proceedings of the 2013 Workshop on Automated Knowledge Base Construction*, pages 67–72. ACM, 2013.

[401] Wikipedia. List of computer science conferences. http://en.wikipedia.org/wiki/List_of_computer_science_conferences. Accessed April 16, 2016.

[402] Wikipedia. Representational state transfer. https://en.wikipedia.org/wiki/Representational_state_transfer. Accessed January 10, 2016.

[403] John Wilbanks. Portable approaches to informed consent and open data. In Julia Lane, Victoria Stodden, Stefan Bender, and Helen Nissenbaum, editors, *Privacy, Big Data, and the Public Good: Frameworks for Engagement*, pages 98–112. Cambridge University Press, 2014.

[404] Dean N. Williams, R. Drach, R. Ananthakrishnan, I. T. Foster, D. Fraser, F. Siebenlist, D. E. Bernholdt, M. Chen, J. Schwidder, S. Bharathi, et al. The earth system grid: Enabling access to multimodel climate simulation data. *Bulletin of the American Meteorological Society*, 90(2):195–205, 2009.

[405] James Wilsdon, Liz Allen, Eleonora Belfiore, Philip Campbell, Stephen Curry, Steven Hill, Richard Jones, Roger Kain, Simon Kerridge, Mike Thelwall, Jane Tinkler, Ian Viney, Paul Wouters, Jude Hill, and Ben Johnson.

The metric tide: Report of the independent review of the role of metrics in research assessment and management. http://www.hefce.ac.uk/pubs/rereports/Year/2015/metrictide/Title,104463,en.html, 2015.

[406] William E. Winkler. Record linkage. In D. Pfeffermann and C. R. Rao, editors, *Handbook of Statistics 29A, Sample Surveys: Design, Methods and Applications*, pages 351–380. Elsevier, 2009.

[407] William E. Winkler. Matching and record linkage. *Wiley Interdisciplinary Reviews: Computational Statistics*, 6(5):313–325, 2014.

[408] Krist Wongsuphasawat and Jimmy Lin. Using visualizations to monitor changes and harvest insights from a global-scale logging infrastructure at twitter. In *Proceedings of the IEEE Conference on Visual Analytics Science and Technology*, pages 113–122. IEEE, 2014.

[409] Xindong Wu, Vipin Kumar, J. Ross Quinlan, Joydeep Ghosh, Qiang Yang, Hiroshi Motoda, Geoffrey J. McLachlan, Angus Ng, Bing Liu, S. Yu Philip, Zhi-Hua Zhou, Michael Steinbach, David J. Hand, and Dan Steinberg. Top 10 algorithms in data mining. *Knowledge and Information Systems*, 14(1):1–37, 2008.

[410] Stefan Wuchty, Benjamin F Jones, and Brian Uzzi. The increasing dominance of teams in production of knowledge. *Science*, 316(5827):1036–1039, 2007.

[411] Beth Yost, Yonca Haciahmetoglu, and Chris North. Beyond visual acuity: The perceptual scalability of information visualizations for large displays. In *Proceedings of the SIGCHI Conference on Human Factors in Computing Systems*, pages 101–110. ACM, 2007.

[412] Zygmunt Z. Machine learning courses online. http://fastml.com/machine-learning-courses-online, January 7, 2013.

[413] Laura Zayatz. Disclosure avoidance practices and research at the US Census Bureau: An update. *Journal of Official Statistics*, 23(2):253, 2007.

[414] Xiaojin Zhu. Semi-supervised learning literature survey. http://pages.cs.wisc.edu/~jerryzhu/pub/ssl_survey.pdf, 2008.

[415] Nikolas Zolas, Nathan Goldschlag, Ron Jarmin, Paula Stephan, Jason Owen-Smith, Rebecca F Rosen, Barbara McFadden Allen, Bruce A Weinberg, and Julia Lane. Wrapping it up in a person: Examining employment and earnings outcomes for Ph.D. recipients. *Science*, 350(6266):1367–1371, 2015.

Index

Note: Page numbers ending in "f" refer to figures. Page numbers ending in "t" refer to tables.

O

P

Q

U

V

W